Managing Pipeline Threats

Managing Pipeline Threats

John Tiratsoo, Editor

CLARION TECHNICAL PUBLISHERS®

TIRATSOO TECHNICAL
A DIVISION OF GREAT SOUTHERN PRESS

Library of Congress Control Number: 2019952926

ISBN 978-0-9906-7006-3

Clarion Technical Publishers
3401 Louisiana Street, Suite 110, Houston, TX 77002, USA
tel: +1 713 521 5929
email: bjlowe@clarion.org

and

Tiratsoo Technical
Level 15, 461 Bourke Street, Melbourne VIC 3000, Australia

Editorial Panel

Contributors

Dr Michael Beller, Rosen Group, Stans, Switzerland

Nigel Curson, Penspen Ltd, Kingston upon Thames, UK

Marcelino Guedes F.M. Gomes, Petrobras, Rio de Janeiro, Brazil

Dr Phil Hopkins, PHL, Whitley Bay, UK

Eric Jas, Atteris, Perth, WA, Australia

Dr Roger King, International Corrosion Services Ltd, Manchester, UK

Dr Gerhard Knauf, Independent Consultant, Duisburg, Germany

Dr Brian Leis, B N Leis, Consultant, Inc., Worthington, OH, USA

Dr Alan Murray, Pipeline Engineering Centre, University of Calgary, Calgary, AB, Canada

Hermann Rosen, Rosen Group, Stans, Switzerland

Dick Williamson, TD Williamson, Tulsa, OK, USA

Andy Young, Rosen (UK) Ltd, Newcastle upon Tyne, UK

Editorial Staff

Editor's note

The QR codes in this book can be scanned using a smartphone or other compatible mobile device to access supplemental reference documents provided by the authors. Codes in the eBook version have clickable links as a further option.

All the supplemental documents linked by the QR codes and the embedded links can be accessed here: http://pipeline-threats.com/plthreats/

Contents

List of Tables

Foreword

Managing pipeline threats: sustaining public trust in energy infrastructure.

It is the most exciting time in history to work in the pipeline industry, and more broadly, to be a part of the energy ecosystem. Pipelines are the safest, most environmentally friendly way to transport oil and gas and the young pipeliner can be passionate that their role helps, in a small way, to fulfill that need.

Due to the long linear nature of pipelines, a successful pipeline project requires extensive stakeholder engagement. Dialogue with all stakeholders starts with a foundation of trust and respect - when pipeline engineers show they can manage all threats to a pipeline, they establish and reinforce the public's trust in the safety of our energy infrastructure.

As the industry and its assets age, there is a need for threat management and other technical knowledge to be transferred to the next generation of emerging pipeliners. Books like this are fundamental to establishing and sustaining pipeline engineering learning. Organizations like the Young Pipeliners Association of Canada (YPAC), for which I am the Director of Event Operations, similarly strive to ensure that industry knowledge is transferred effectively.

The next generation of pipeline professionals needs to be connected to peer networks, to grow technically, and to understand broad industry issues. It is critical that we reach into the minds of our industry experts and learn from the past to improve our environmental and safety performance. I want to note however, that knowledge transfer goes both ways. The young pipeliner can be an effective mentor for more experienced experts as a champion of new ideas, new technology, and new methodologies for today's energy infrastructure.

This dynamic can help protect the public and environment, improve the management of pipeline threats, and help safely deliver the oil and gas that society needs.

Molly Beckel, PEng.

Facilities Integrity and Reliability Engineering,
Canadian Gas Operations, TransCanada Pipelines

Director of Event Operations,
Young Pipeliners Association of Canada (YPAC)

CHAPTER 1

Introduction

Dr Phil Hopkins

We all face 'hazards' or 'threats' (anything with the potential to cause harm); for example, driving, crime, illness, etc. These threats have a probability of occurring (a probability of zero means it will never happen, and a probability of 1 means it will certainly happen).

When we combine this probability of a threat, with its consequences (such as injury, loss of earnings, etc.), we have 'risk.' Risk is a combination of probability and consequence. We need to minimise this risk: we can do this by identifying the threats, and then:

- Reducing our exposure to these threats (e.g., drive less); and also,
- Minimising its consequences (e.g., always wear a seat belt while driving).

This management of personal risks is an important aspect of all our lives. We are all exposed to risk daily, at work and in our private lives, and often manage it subconsciously, but in our businesses we need to ensure that we think about it actively, and act.

This book is an introduction to managing threats in pipelines. It is only a start, as it is not possible to cover all threats and their identification and management in a short book, but all engineers working in the pipeline industry need to be aware of threats and understand how the resulting risks are managed. As Sir Michael Latham (1942 – 2017) said in 1994: 'Risk can be managed, minimized, shared, transferred, or accepted. It cannot be ignored.'

The book starts with this Introduction Chapter, then moves on to 'Pipeline Principles.' This Chapter 2 introduces the reader to the pipeline systems that transport our energy around the world: crude oil and natural gas pipelines. It also gives basic principles in pipeline engineering and explains some pipeline design concepts.

Chapter 3 covers 'Line Pipe Principles.' Pipelines are made using steel tubes called 'line pipe,' and this Chapter covers the manufacture of this line pipe, and the standards we use to ensure high quality.

Chapter 4 is 'Introduction to ILI.' In-line inspection (ILI) means the use of inspection tools within a pipeline. The Chapter provides a comprehensive overview of the in-line inspection tools commercially available on the market today for the inspection of all the types of high-pressure pipelines.

Chapters 5-12 cover some of the main threats to pipelines: corrosion; cracking; mechanical damage; geohazards; material and construction defects; theft; and, specific threats to submarine pipelines.

Chapter 13 is 'Pipeline Defect Assessment Basics.' This Chapter introduces the reader to methods for assessing the significance of pipeline defects such as corrosion and dents.

Chapter 14 presents and discusses 'Pipeline Integrity Management.' 'Integrity management' is part of asset management and includes the many and varied activities pipeline operators must undertake to ensure that releases of products from their pipelines do not occur.

The book ends in Chapter 15, where eminent figures in the pipeline industry are asked the question 'Where Next?'

1.1 Energy in the world today

We live in a world where most of our energy is obtained from fossil fuels such as crude oil, natural gas, and coal, Table 1-1. These non-renewable energy sources (they cannot be replenished) will remain our main source of energy for many more years, although renewable energy sources (they can be easily replenished) will continue their rise.[1]

Table 1-1. World use of primary[1] fuel in 2017[2]

	1990	2017	2040
Liquid Petroleum ('oil')	38%	33%	31%
Natural Gas	21%	22%	25%
Coal and Peat	25%	27%	22%
Renewables	10%	13%	17%
Nuclear	6%	5%	5%

1 'Primary' is all energy consumed by end users, excluding electricity but including the energy consumed at electric utilities to generate electricity.

1.2 Pipelines in the world today

Our crude oil (and products obtained from crude oil such as gasoline and kerosene), and natural gas are mainly transported by pipelines, Figure 1-1. There are approximately 3,500,000 km of high-pressure pipelines operating in the world today, transporting these hydrocarbons, with many entering their 50th or 60th year of service. They will likely have to operate for another 50 years, with the continuing demand for hydrocarbons.

Pipeline economics involve large sums of money. Two examples are:

- A large capacity crude oil pipeline (e.g., transporting 1,000,000 barrels of oil per day, with a barrel of oil costing, say, $67) will generate revenue of $US24,500,000,000/year.
- To replace the vast and ageing oil and gas pipeline system in the USA would cost over $US4,000,000,000,000.

Pipelines are big business and big investments, and not easily or cheaply replaced.

Pipelines have been a relatively safe and reliable mode of transportation of crude oil, crude oil products, and natural gas since they were first used commercially in the 1860s. They continue to be safe, Table 1-2, but they do fail, and will continue to fail, like all engineering structures; therefore, their safety has to be carefully managed, and failures kept to a minimum. Threat identification is a key element in managing pipeline safety, and protecting these critical assets, as threats can lead to failures.

Table 1-2. USA transportation statistics for 2017[3]

	2017	2040
Liquid Petroleum ('oil')	33%	31%
Natural Gas	22%	25%
Coal and Peat	27%	22%
Renewables	13%	17%
Nuclear	5%	5%

Figure 1-1. Most of our oil and gas is transported by pipelines.

1.3 Threats to pipelines

A threat is anything that can cause harm. Most people will know threats to pipelines such as corrosion, and impact damage from excavating machines or ships' anchors, but there are many other threats such as ground movement, organisational change, and incompetent staff, and new threats such as theft and cyber-crime.

Pipelines do not want to fail – they need a reason. Typical threats for onshore pipelines are[4]:

- Corrosion – this is a natural process where materials made from metal deteriorate through an electrochemical reaction known as oxidation.
- Excavation damage – this is accidental damage to a pipeline or its coating that is caused by someone inadvertently digging into a buried pipeline.
- Other outside force[2] – damage to a pipeline, other than excavation damage; for example, vehicle accidents.
- Natural forces – such as floods and wildfires.
- Material/weld failure - defects can exist in the steel used to make the pipe and other components, and their welds.
- Equipment failures - these involve pipeline pumps, compressors, valves, meters, tanks, and other components and devices on pipeline systems.
- Incorrect operation – most pipeline operations are now automated, but humans continue to serve a primary role in many activities, and as a result, human factors can become critical elements in certain types of pipeline and equipment failures.

2 Damage to pipelines caused by impacts is generically called 'external interference.'

1.4 Management of threats

Threat management is contained within the framework of a risk management approach or a safety management system. Safety management systems[7] manage the safety elements of a business, and will include policies, objectives, plans, procedures, organisation, and responsibilities.

1.5 Identifying threats

Threat/hazard identification is the recognition that a hazard exists and knowing its characteristics.

A hazard analysis follows on from the threat identification. A hazard analysis is defined as[9]: '... the identification of material properties, system elements, or events that lead to harm or loss. The term hazard analysis may also include evaluation of consequences from an event or incident.' Obviously, the threat identification and analysis focuses on the most significant threats.

One of the oldest methods of threat identification is the 'what if?' This method is often used in informal design reviews. It involves asking a series of questions beginning with 'what if'; for example, 'what if the compressor stops,' or 'what if the pressure sensor fails?' This method relies heavily on team experience: threats may be missed if the right questions are not asked.

We now place threat identification methods into two groups:

- Reactive threat identification methods – We recognise threats through trend monitoring of incidents, or comparisons.
- Proactive threat identification methods – Threats are identified by analysing a system's performance and functions for threats and potential failures (e.g., by safety surveys or operational safety audits).

In practice we use both, but in successful safety management systems, the proactive approach is used extensively. This allows the threat to be recognised and addressed before it could turn into an occurrence.

Not all threats are 'engineering'

'.. some failures tend to repeat themselves at intervals as staff changes dilute the stock of hard-won corporate experience.' [5].

'It might seem to an outsider that industrial accidents occur because we do not know how to prevent them. In fact, they occur because we do not use the knowledge that is available. Organisations do not learn from the past or, rather, individuals learn but they leave the organisation, taking their knowledge with them, and the organisation as a whole forgets.' [6].

These are various structured methods that can be used to identify threats; for example[9]:

- Failure modes and effects analysis (FMEA); and,
- Hazard and Operability Study ('HAZOP').

1.6 Identifying threats: lagging or leading?

We can use past failures, and failure or 'incident' rates, to identify threats and trends (Figures 1-2 and 1-3), but these trends are now known as 'lagging' indicators of safety.

They are a 'reactive' approach to improve safety, based on a trend. This is because failure statistics – by definition – are historical figures. They are 'out-of-date' as soon as they are published; however, they are essential for 'lessons learnt,' and to help stop repeat incidents. Increasingly, we are looking to 'leading' indicators of safety. 'Leading' indicators of safety are proactive methods, such as: threat identifications; risk assessments; near misses; inspections; training; competency testing; etc. Leading indicators should improve safety, regardless of trends, and help identify new threats.

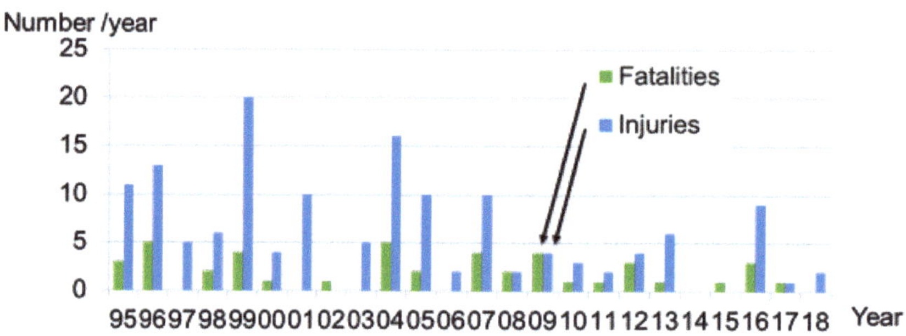

Figure 1-2. Casualty data for USA pipelines transporting liquids (1995 – 2018)[11].

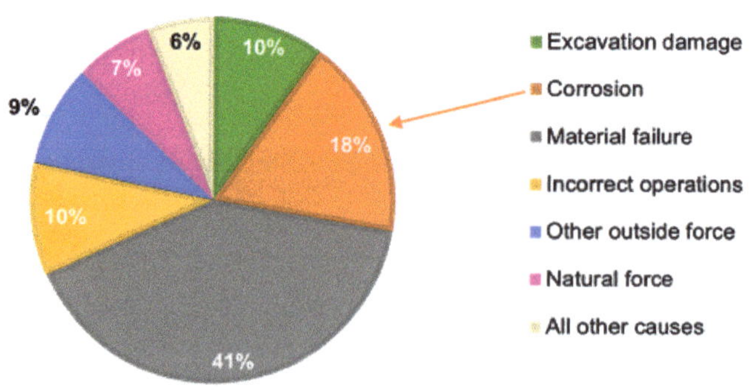

Figure 1-3. Causes of failures in pipelines in the USA (2005 - 2019)[11].

1.7 The aging pipeline – is it a bigger threat?

Many assets we manage today are old: they are 'aging.' 'Aging' is when a component is subjected to some type of material deterioration and damage (usually, but not necessarily, associated with time in service). The significance of this deterioration and damage relates to the potential effect on the equipment's functionality, availability, reliability, and safety.

All types of equipment can be susceptible to ageing mechanisms, but 'aging' is about the plant's condition, not its age[12]. This means that 'old' does not necessarily mean that there is significant deterioration or damage, and 'ageing' is not directly related to chronological age[13].

Indeed, the age of a pipeline is not the major cause of failures[14]:

- 60% of the USA natural gas pipeline system was installed before 1970, and failures do occur with a major leak/rupture rate in this system of approximately 0.00034 per mile per year; but,
- 85% of gas pipeline incidents occurred irrespective of the age of the pipeline, and 15% related in some way to the age of the pipeline.

This leads to the conclusions:

- 'there is no evidence that the ageing of the pipeline system implies a greater risk of spillage'[15];
- 'the safety of a well-maintained and periodically assessed pipeline is ensured regardless of age through diligent and well-planned integrity management practices'[14]; and,
- 'A suitable asset integrity management program should ensure that loss of integrity due to ageing does not occur during the design life of the component or structure'[16].

And remember, managing threats and maintaining safe pipelines is not simply about ensuring the engineering is good[17]: 'Commitment, leadership, and oversight from top management are vital to the overall success of a pipeline safety management system… A safety-oriented culture is essential to enable the effective implementation and continuous improvement of safety management system processes and procedures.'

1.8 Threats change with time

Data from the chemical process industry show how failure causes change with time. Early industry failures were caused by engineering errors (poor materials, poor construction, etc.), Figure 1-4[18]. As the industry matured, and technology improved, the engineering failures reduced. The most recent trends in failures are linked to safety culture[3], and relationships between contractual entities, and the problems that can bring.

Other lessons from other industries on threat identification and management are given in Reference 20.

1.9 The human factor...

Pipeline standards and regulations explicitly require personnel to be both competent and qualified to work on pipelines[21 - 23]. This is both obvious and safety-critical as human error is a contributing factor to pipelines failures[24], and incompetent staff can cause failures[25]. Incompetent staff are a real threat to a pipeline: as we eliminate threats caused by equipment and materials, and poor processes and procedures, we will be left with human error as a leading threat.

Most people accept that making errors is a natural result of being human[26]; hence, the increasing desire to replace humans with automation, where repetitive tasks are performed the same way without deviation or 'error.' Removing the human from the equation and replacing with automated systems is a solution for human error. Robots are a solution... the failures caused by people/management/organisations can be reduced. We are already happy to see robots/artificial intelligence in industry, improving the manufacturing process or maintenance. Robots are always working, always there, always ready. They do not need coffee breaks, have the right (programmed) attitude, and can work in the dark. Artificial intelligence/robots are replacing people in many industries. Most professions will be replaced with robots: accountants, doctors, lawyers.... Any task that is repetitive (e.g., submitting financial accounts), definable (e.g., surgery), or requires little updating (e.g., legal cases) can be performed better by a robot[27]. Robots are reliable, cheap, replaceable, and easily updated as new technology/problems are created.

The threat from human error, and the 'organisation'..

Some of the threats as a pipeline ages can be caused by 'human factors.' Reviews in other industries have concluded[8]:

- 'it is estimated that up to 80% of accidents may be attributed, at least in part, to the actions or omissions of people.'; and,
- 'Organisational factors have the greatest influence on individual and group behaviour....'

3 'Culture... [is] a set of beliefs, values, attitudes and behaviours... a strong safety culture will encompass more than just following regulations...'[19].

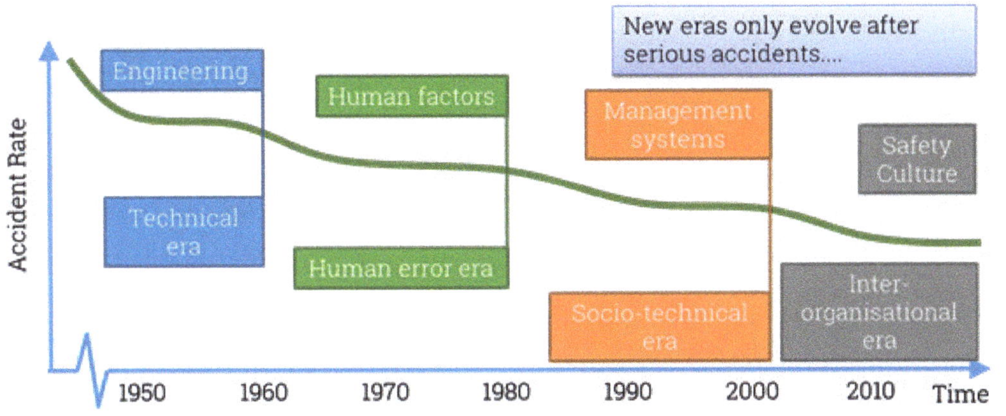

Figure 1-4. Failure threats/causes change with time.

QR Code 1-1. Scan here to learn more from other industries.

1.10 Finally... Do you know your pipeline 'system'?

The first step in identifying and evaluating pipeline threats is to define and gather the necessary data and information that characterize the pipeline and the potential threats[28]. These data will include pipeline operation, maintenance, patrolling, design, operating history, and specific failures. The data to be gathered and integrated should be of sufficient quality and breadth so that it can be used in the threat identification and any associated risk assessment.

Data quality and consistency can be significant issues[29], and collecting the right data is essential[30]. The unavailability of the right data, or poor-quality data, are not a justification for exclusion of a threat[28]. This latter point can be a big issue for older pipelines where data may be missing, or limited.

There is an irony: today we are data 'rich': there are 2.5 exabytes[4] (2.5 x 10^{18} bytes) of data created each day, and 90% of the world's data was generated in the last two years [31]. But most (80%) of today's data are 'unstructured' (e.g., emails, PDFs and other documents) residing in locations that are either inaccessible or cannot be interrogated. We can now have the problem with a pipeline of having too little historical data, and too much current data....

4 A gigabyte is 10^9 bytes.

This data management will be included in a pipeline's safety management system. A management system helps us manage a particular aspect of our business, but having a safety management system in place does not guarantee success in managing safety. Identifying threats is only part of the process. The key to success is how the management system operates in practice, and this depends on[7, 33]:

- Management leadership;
- Ownership at all levels in the business;
- Effective implementation;
- Continuous improvement;
- Assessment of hazard and risks;
- Enhanced controls;
- Communication and consultation processes;
- Monitoring and review.

A failure can occur when any part of this system is faulty: threats exist in all these parts, and not simply the engineering parts of the system.

1.11 References

[1] Anon., 'EIA projects 28% increase in world energy use by 2040,' U.S. Energy Information Administration. 14th September, 2017. https://www.eia. gov/todayinenergy/detail.php?id=32912

[2] Anon., 'Annual Energy Outlook 2018,' U.S. Energy Information Administration. 2018.

[3] http://www.ntsb.gov/investigations/data/Pages/Data_Stats.aspx

[4] https://www.phmsa.dot.gov/incident-reporting/accident-investigation-division/pipeline-failure-causes

[5] A Anderson, 'Making a success out of a museum of failure,' New Scientist. 8th June, 1991.

[6] T Kletz, 'Lessons from Disaster: How Organisations Have No Memory and Accidents Recur,' IChemE. Rugby, UK. 1993.

[7] Anon., 'Managing for health and safety', UK Health and Safety Executive. HSG65. Third edition UK. 2013. http://www.hse.gov.uk/pubns/priced/hsg65.pdf

[8] Anon., 'Reducing error and influencing behaviour', UK Health and Safety Executive. HSG48. Second edition. 1999. Reprinted 2009.

[9] http://www.phmsa.dot.gov/portal/site/PHMSA/menuitem.ebdc7a8a7e39f2e55cf2031050248a0c/?vgnextoid=1448a73ff3d0d110VgnVCM1000009ed07898RCRD

[10] Anon., 'Risk Assessment Techniques', Institution of Gas Engineers, IGEM/G/7. Communication 1655. UK. 1999. https://www.igem.org.uk/media/411610/igem-g-7%20printable.pdf

[11] https://hip.phmsa.dot.gov/analyticsSOAP/saw.dll?Portalpages

[12] Anon., 'Plant ageing. Management of equipment containing hazardous fluids or pressure', HSE Research Report RR509, UK. HSE Books, 2006.

[13] P Horrocks et al., 'Managing Ageing Plant. A Summary Guide', HSE Books. No. RR823. UK. 2010. http://www.hse.gov.uk/research/rrpdf/rr823-summary-guide.pdf

[14 J F Kiefner, M J Rosenfeld, 'The Role Of Pipeline Age In Pipeline Safety', Prepared for the INGAA Foundation, Inc. INGAA. 8th November, 2012.

[15] Anon., 'Performance of European cross-country oil pipelines. Statistical summary of reported spillages in 2011 and since 1971', CONCAWE (conservation of clean air and water in europe). Report 3/13. April 2013. www.concawe.be

[16] Anon., 'Further information on ageing and life extension', UK Health and Safety Executive. http://www.hse.gov.uk/offshore/ageing/information.htm

[17] Anon., 'Pipeline Safety Management Systems', ANSI/API Recommended Practice 1173. 1st Edition. The American Petroleum Institute. USA. July, 2015.

[18] P. Körvers, 'Accident precursors; pro-active identification of safety risks in the chemical process industry', Dissertation. TUE. Eindhoven, Netherlands. 2004.

[19] D Rivera, 'Committing to a Safety Culture', World Pipelines. Vol 17. No. 4. August 2017. p. 45.

[20] P Hopkins, 'Learning from Failures: Moving from 'Failure' Cause to 'Root' Cause", Pipeline Technology Conference. Berlin, Germany, March 2019, and Pipeline Technology Journal. Issue 2. 2019.

[21] M Unger, P Hopkins, 'Competency Standards for the Pipeline Industry', 12th International Pipeline Conference, IPC 2018. Calgary, Alberta, Canada. Paper IPC2018-78477. September, 2018.

[22] M Unger, P Hopkins, 'The Competency Standard Manual for Pipeline Integrity Management'. Journal of Pipeline Engineering. September, 2018.

[23] M Unger, P Hopkins, "Competency' in Engineering'. 28th International Pipeline Pigging and Integrity Management Conference Houston, USA, 8-11th February, 2016.

[24] Anon., 'Pipeline Personnel Qualification', American Society of Mechanical Engineers. New York, USA. ASME B31Q-2016. ASME. 2016.

[25] M Wright, D Turner, C Horbury, 'Competence assessment for the hazardous industries', Health and Safety Executive Research Report 086. UK. 2003.

[26] D L DeMott. 'Human Reliability and the Cost of Doing Business'. https://ntrs.nasa.gov/archive/nasa/casi.ntrs.nasa.gov/20140008715.pdf

[27] R Susskind, D Susskind, "Technology Will Replace Many Doctors, Lawyers, and Other Professionals', Harvard Business Review. 11th October, 2016. ttps://hbr.org/2016/10/robots-will-replace-doctors-lawyers-and-other-professionals

[28] Anon., 'Managing System Integrity of Gas Pipelines', ASME B31.8S – 2018. The American Society of Mechanical Engineers. USA. 2018.

[29] Anon., 'Managing System Integrity for Hazardous Liquid Pipelines', The American Petroleum Institute. API Recommended Practice 1160. Third edition. USA. February, 2019.

[30] J A Drake, 'Integrity Management Continuous Improvement', Proceedings of the 2012 9th International Pipeline Conference. IPC2012. 24-28th September, 2012, Calgary, Alberta, Canada. IPC2012-90406.

[31] B Marr, 'How Much Data Do We Create Every Day? The Mind-Blowing Stats Everyone Should Read', Forbes. 21st May, 2018.

[32] J Rizkallah, 'The Big (Unstructured) Data Problem', Forbes, 5th June, 2017.

[33] Anon., 'PABIAC Strategic Objective 2: Safety Management Systems. A self-assessment tool for SMEs'. http://www.hse.gov.uk/paper/safetymanagement.pdf

CHAPTER 2

Pipeline engineering principles[1]

Dr Phil Hopkins

2.1 Introduction

We are all familiar with pipes in our homes. We have water pipes delivering water around our house and drains and pipes taking the waste from our homes. Many of us have gas pipes in our homes, transporting natural gas to our heating boiler, or cooker.

The water, waste, and gas flow through the pipes because the pipes are pressurised; for example, water pressure is a measure of the force that pushes the water through our pipes. This water pressure is typically over 40 psi (2.8 bar): as a comparison, automobile tire pressures are also around 40 psi (2.8 bar).

The pipes, faucets, bends, meter, and pump that make up our water system are collectively called a pipeline; therefore, we have a complete pipeline in our own homes.

This Chapter introduces you to the much bigger, and far more important, pipelines that transport our energy around the world: crude oil and natural gas pipelines. These pipelines are designed, built, and operated to rigorous standards (e.g., Reference 1) and government regulations.

2.1.1 Energy and its transportation

We use *fossil* fuels (coal, crude oil, or natural gas) for most of our energy needs in the world today:

- Coal, crude oil[2], and natural gas[3] are called fossil fuels because they were formed in the earth's crust from dead plants or animals (*organisms*) from millions of

1 This Chapter is based on the e-learns available at the Competence Club (https://competence.rosen-group.com/) and is courtesy of Rosen Group.

2 Crude oil is a mixture on many different natural (organic) compounds. We extract the crude oil from its underground reservoir and send it to a refinery where we convert the crude oil into refined products such as gasoline or diesel.

3 Natural gas is the gas we extract from underground reservoirs. It is a mixture of gases, but it is mainly methane.

years ago. Most coal, crude oil, and natural gas are located deep underground in rock formations: in *seams* (coal); or, *reservoirs* (crude oil and natural gas).

- They are called *fuels* because they release heat energy when they are burned, and we can change this heat energy; for example, into electricity at a power station, or into motion, in an automobile's combustion engine.

> **'Pressure'**
>
> We see 'pressures' in our everyday lives; for example, the pressures required in our automobile tires or bicycle tires, or pressures in our domestic boilers.
>
> These pressures are quoted in 'psi' or 'bar'. These 'psi' and 'bar' are 'units' of pressure. You will be more familiar with other 'units', such as units of money (e.g., dollars), or units of temperature (e.g., degrees Fahrenheit or Celsius).
>
> Units give us a measure (e.g., $100, or 50°F (10°C)). It does not matter what currency we use: dollars, pesos, Euros are all units of currency, and give a measure of value.
>
> It is the same with pressure. 14.5 psi (or 1 bar) is the pressure (force) that can push water to a height of 33 feet (10 metres).
>
> So… if you want water that enters your house in the basement to reach your attic, you need a high pressure!

The world has huge energy needs and consumes: 93 million barrels of oil/liquid fuels per day (34 billion barrels a year); and, 9.3 billion cubic metres of natural gas per day (3,393 billion cubic metres per year).

The energy we use needs transporting from its supply (e.g., oil and gas from its underground reservoir, or from where it is stored) to its demand (customer markets). Much of our electricity is generated using fossil fuels, which will need transporting to a power station:

- coal is transported to power stations directly from its source (a coal-burning power station is often adjacent to a coal mine), or, by road/rail/sea;
- fossil fuels such as oil and natural gas will be delivered to power stations by road, rail, sea, or pipelines, where the fuels are burned and changed into electricity.

The electricity generated is transported from a power station to our homes by cables on tall pylons.

2.1.2 Crude oil and natural gas

Crude oil and natural gas are called *hydrocarbons* as they consist of a range of compounds made up of hydrogen atoms (H) and carbon atoms (C). Natural gas is the simplest hydrocarbon (it is a mixture of gases, but mainly methane, and methane has the chemical formula of CH_4).

The crude oil we extract from the underground reservoirs is of little use in its *crude* form, as it can contain many different kinds of hydrocarbon, water, gases, sand, etc.

1 barrel (bbl) = 42 US gallons = 159 litres.

1 barrel (bbl) = 35 UK gallons = 159 litres.

1 billion cubic metres = 35.3 billion cubic feet

1 billion cubic metres of natural gas = 6.2 million bbl of crude oil (in terms of energy).

Consequently, the crude oil is sent to a refinery where it is broken down into a variety of products, such as gasoline and aviation fuel, ready for marketing.

Natural gas contains mainly methane but has other gases in it, such as ethane and propane. The natural gas from the reservoir can also contain water, solids such as sand, etc.; therefore, it will usually need some processing (e.g., removal of water) before being sent to market.

2.1.3 Pipes and pipelines

Hydrocarbons, such as crude oil and natural gas, need to be transported between their supply and their demand. Roads, railways, and sea can all be used, but the safest and most efficient mode of transportation is usually a pipeline (see Figure 2-1)[2, 3].

Pipes are cylinders, used to transport fluids. We are all familiar with the metal or plastic water pipes in our homes, and the copper pipes that bring natural gas into our homes, but the pipelines that transport fluids such as natural gas or crude oil are much bigger in size and longer in length.

Ancient history of pipelines. For thousands of years, pipelines have been constructed in various parts of the world to convey water for drinking, and irrigation for agriculture (see Table 2-1, Figures 2-2, 2-3). These pipes include baked clay and hollow bamboo. Indeed, wood (hollowed-out logs) was used as recently as about 100 years ago, to transport brine water in the Developed World.

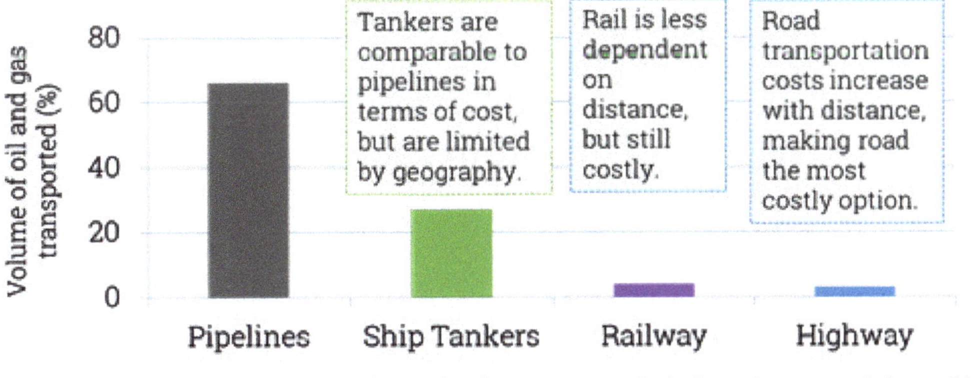

Figure 2-1. Pipelines are the most popular mode of transportation for hydrocarbons around the world.

Table 2-1. Ancient history of pipelines.

Date	Location	Pipeline
3000 BC.	Egypt	Copper pipe to transport water.
2000 BC to 1500 BC.	Crete	Earthenware pipe for water transport.
1600 BC to 300 BC.	Greece	Earthenware, lead, bronze, and stone pipes to transport water. In this era, 'blacksmiths' connected the metal pipes together by simply hammering the red hot pipe ends together.
500 BC.	Persia (now called Iran) and Rome	Metal pipe was used to transport water: the Romans used lead pipes in highly developed towns.
500 BC.	China	The first recorded use of a pipe to transport a hydrocarbon was in China: the Chinese used bamboo pipe to transmit natural gas from shallow wells. They had discovered the gas while they were drilling for salt. They could burn the gas under pans, to boil seawater to separate the salt, and make the water drinkable.
400 BC.	China	Records indicate that the Chinese used bamboo pipe, wrapped in wax, to transport natural gas to light their capital, Peking.

Figure 2-2. Ancient Roman stone water pipeline
(image from www.romanaqueducts.info/picturedictionary/pd_onderwerpen/pipe.htm).

Figure 2-3. Wooden Pipeline circa 1764
(image from www.sewerhistoryorg/photosgraphics/pipes-wood).

Modern history of pipelines. The use of wood, iron, lead, and tin pipes were common into the 1800s to transport water, and in 1821 wood pipe transported natural gas in New York State, USA. In 1843 iron pipe was used, and this reduced the obvious hazards of transporting a flammable, explosive gas in a flammable material.

The first recorded use of metal pipes to transport crude oil was in 1863 in Pennsylvania, USA. This followed the commercial development of petroleum in the USA in 1859, when crude oil was discovered in Pennsylvania, and seen as an alternative to whale oil for domestic and commercial lighting. In the early 1860s, the oil was transported in wooden barrels on rivers by horse-drawn barges. This was dangerous: weather, and labour disputes, often disrupted flow. The railway relieved this, but the oil was now controlled by the rail bosses and their workers... the *teamsters*. Pipelines were an obvious solution to this transport problem, and the early oil workers were familiar with pipes: cast iron and wrought iron pipes of various small diameters were in use around the producing oil wells from the start of the industry. The pipes were used as drive pipe, conductor, casing, tubing, and for conveyance of oil in and around the oil lease; hence, small diameter (2" to 6") pipelines for transporting the oil were constructed.

Later pipelines quickly benefitted from advances in technology: better materials (steel replaced iron); new joining methods (welding replaced screwed joints); introduction of pumps (rather than gravity) to pressurise the fluid; the *telegraph* for communication over long distances; etc.

The 20th century saw long distance oil and gas pipelines built on all continents as oil and gas started to both complement and replace coal as primary energy sources.

Pipelines today. Today, most countries have extensive pipeline systems, delivering huge amounts of energy. These pipelines start close to the oil and gas reservoirs, or their storage facilities, and deliver their contents to refineries, processing plants, storage tanks, or directly to customers. Large diameter, high pressure pipelines that gather the oil and gas from the wellheads above their reservoirs are called *flow lines* and *gathering lines* (with pipe diameters up to about 12" (304.8 mm)). The larger diameter high pressure pipelines that take the oil and gas to refineries, or to customers, are called transmission (*trunk*) lines (see Figure 2-4). These pipelines can be short (a few kilometres) to very long (over 1000 km in length), have a large diameter (e.g., 56" (1442.4 mm)), and operate at high pressures, such as 100 bar (1450 psi). They can cross continents, go under lakes, and go over mountains, they must operate for many decades, and deliver their products safely and securely.

We have over three million kilometres of transmission pipelines around the world transporting crude oil, *products* refined from the crude oil (e.g., diesel), and gases such as natural gas. They can be both onshore (above ground or below ground) and subsea/under water (laid on a seabed or in a trench in the seabed) (see Figure 2-5).

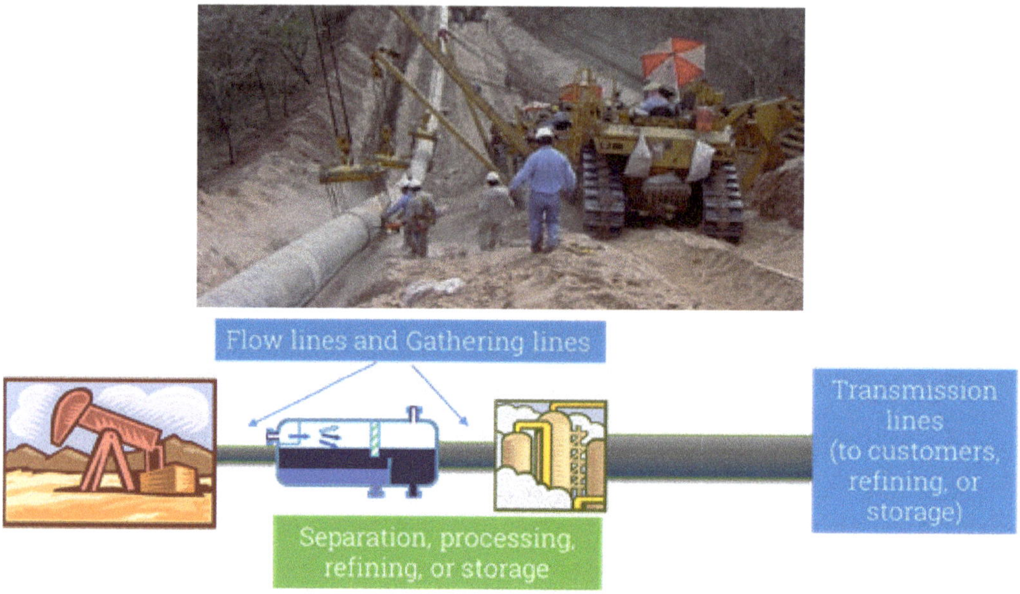

Figure 2-4. Constructing a pipeline (top); types of pipeline (bottom.)

Figure 2-5. Transmission pipelines are usually under the sea or buried underground (top); transmission pipelines (bottom).

QR Code 2-1. Scan here to learn more about pipeline construction.

Most pipelines are onshore, buried below ground (see Figure 2-5). This means that pipelines are rarely seen by the general public, but most are over 40 years of age, and have been silently delivering their contents for many decades.

These pipelines are *systems* which include components (or *fittings*), such as valves, and large installations along the pipeline route; for example, a metering station (to measure the amount of product entering and leaving the pipeline), or a pumping station (to boost the flow of the product in the pipeline), or tanks to store the product[4].

We are not concerned in this chapter with the small diameter, low pressure, gas *distribution* pipelines we see in our streets and along our roads: our focus is the large diameter, long distance transmission pipeline (see Figure 2-5).

2.1.4 Pipelines are safe

Pipelines are a safe and efficient mode of energy transportation, Figure 2-1, compared to other methods such as highways and railways[3]. Pipelines have been the preferred mode of transportation over competing modes such as sea, road, and rail for several reasons:

- they are less damaging to the environment (they are buried, and out of sight, and have a good record in terms of leakages);
- they are less susceptible to theft (but in recent years theft and sabotage have become a major problem in some countries); and,
- they are more economical, safe, convenient, and reliable than other modes.

QR Code 2-2. Scan here to learn more about pipeline safety.

2.1.5 Keeping the pipeline flowing

Fluids and hydraulics. A *fluid* is anything that flows; therefore, liquids and gases are fluids. Fluids are required to flow in a pipeline in an efficient and safe way; therefore, we need to understand this flow. *Hydraulics* is the branch of science concerned with the practical applications of fluids in motion. Hydraulics is a key element in a pipeline's design.

Fluids flow in a pipeline when there is a *pressure differential* between the inlet and outlet of a pipeline (see Figure 2-6). That means the pressure when the fluid enters the pipeline (*upstream*) must be higher than the pressure the fluid leaves the pipeline (*downstream*). This pressure differential can be provided by gravity: a pipeline on a hillside will allow fluid entering the pipeline at the top of the hill, to flow down the hill and exit at the bottom of the hill. We can also provide this pressure differential by using pumps or compressors to increase the pressure in the pipeline and drive the fluids along the pipeline (see Figure 2-6).

What is pressure? Snow is an interesting substance: it is a mixture of ice crystals, and trapped air. Fresh, uncompact snow is typically 90 to 95 percent trapped air. This means that 25 cm (10") of fresh snow can contain only 1.25 cm (0.5") of water. The trapped air gives snow excellent insulating qualities, and many animals take advantage of this by burrowing into the snow to hibernate through the winter.

Snow also illustrates what we mean by pressure. If I walk through deep snow wearing trainers, my feet sink deep into the snow, and I will become trapped. If I wear snowshoes (see Figure 2-7), I am able to walk across the snow, and my snowshoes only sink into the snow by a small amount.

Pipelines and 'High' Pressure

Pipelines transport fluids such as oil and gas at high pressures. These high pressures are needed to allow high flow rates and allow the delivery of huge amounts of the oil and gas required by customers.

But... what is 'high' pressure?

We all know we have to inflate automobile tires. We need to inflate them to a specified pressure; for example, '35 psi' or '2.4 bar'. '35 psi' means we have to put enough air into the tire to give a pressure in the tire of 35 pounds per square inch. We can convert this 35 psi into other units of pressure; for example, 35 psi = 2.4 bar.

We simply watch the pressure gauge on our airline until it reaches this 35 psi or 2.4 bar figure.

Pipelines operate at much higher pressures; for example, 1000 psi, which is 69 bar. This means pipelines operate at pressures over 20 times the pressure in an automobile tire.

Pressure differential

'Pressure differential' is why water streams can only flow down a mountainside and not up: gravity creates the differential pressure between the top of the mountain and the bottom, with the highest pressure at the top of the mountain. The water 'falls' (flows) down the mountain to a lower pressure level.

Figure 2-6. We need a force to drive the fluid through a pipeline.

Figure 2-7. Snowshoes distribute your weight, and reduce the pressure on snow (image from: https://sectionhiker.com/msr-revo-explore-snowshoe-review/).

Snowshoes work by distributing the weight[4] (or *force* or *load*) of a person over a larger area, so that the person's foot does not compress the snow and sink completely into the snow. This means the larger the area of my shoe, the lower the pressure I am exerting on the snow. My trainers have a small area, so, my weight (force) is supported by a small area. This creates a high pressure on the snow, allowing my trainers to penetrate the snow (by compressing it), and I become trapped....

We can represent this pressure-force-area relationship mathematically: when we apply a force (or weight) to an object, we obtain the pressure on the object by:

pressure = force/area.

As I increase force or weight, I increase pressure. As I increase area, I decrease pressure.

Figure 2-8 shows a liquid (green) in a container. As we decrease the amount of liquid in the container (top) we decrease the pressure acting on its base. As we increase the amount of liquid in the container (bottom) we increase the pressure acting on its base.

You can visualise this pressure when you use a bicycle pump. The pressure you create inside the pump is transmitted through the hose of the pump, into the bicycle tire:

- the handle of the pump is acting as a piston;
- you are applying the force with your arm;
- the more air I put into the tire, the greater the pressure in the tire; and,
- the area of the piston is the size of your pump (the cross-sectional area of the piston).

The air is going from the pump into the tire, but all the air is confined in the tire. The pressure in the tire is uniform around the whole tire.

4 Weight is a force caused by the earth's gravity.

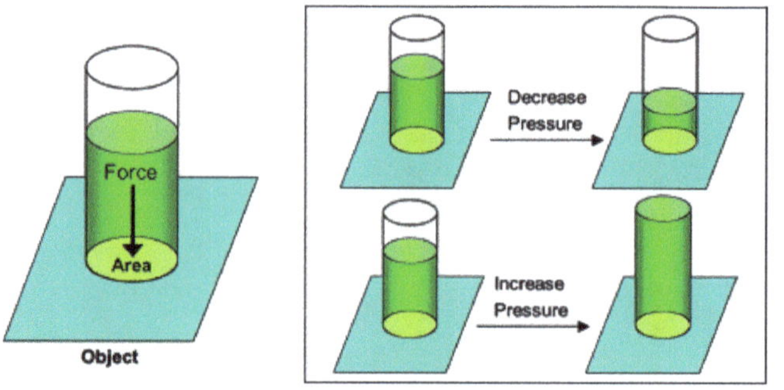

Figure 2-8. Force to pressure.

Pressure and stress. When we apply force to an object, we create a *stress* in the object.

The snowshoe example above can be used to explain stress: as I apply pressure to the snow, the snow becomes compressed. My pressure is causing the snow to deform, and within the snow I have created a high stress that has caused the snow to permanently compress. Mathematically, stress is:

stress = force/area

This is the same as the equation for pressure. The strength of the snow resists this stress. Snow is not very strong, so it is easily compressed. If I replace the snow with a stronger material; for example, concrete, the pressure from my shoes will not compress the concrete. I am exerting the same pressure on the snow, and hence creating the same stress in the snow, but the concrete is very strong and resists the stress and does not compact. You will usually see pressure used with fluids (gases or liquids), whereas stress is more often used with solids.

What is pressure in a pipeline? When we pump a bicycle tire, we are putting air into the tire. The pump allows the air to flow around the tire and fill it completely. The more air we put into the tire, the higher the pressure in the tire. Eventually, we reach the required pressure for the bicycle tire. The air in the tire is under pressure. This pressure creates a force on the tire wall, and this force creates a stress within the tire wall. If we put too much pressure in the tire, it will burst: the stress in the tire wall has exceeded the strength of the tire.

It is the same in a pipeline: a pump raises the pressure in the pipeline, until it reaches the pressure required.

Pumps and compressors. A fluid flows when a *force* is applied to it. This force could be gravity (wastewater flows through your pipes at home because of gravity), or a force is applied to it by mechanical means (e.g., a pump), Figure 2-6. The fluid will flow when this force is applied.

This force is supplied by pumps (for liquids) and compressors (for gases) in pipelines[5]. The force of these pumps and compressors create the pressure in the pipeline. Controlling the speed and load of the pumps and compressor has the effect of changing the pressure on the downstream pipeline, and the amount of flow in the pipeline.

QR Code 2-3. Scan here to learn more about pumps and compressors.

2.2 Line pipe

Pipelines are designed, constructed, and operated using national or international standards and guides (e.g., [1, 5-19]).

A pipeline is made of tubular sections of steel: called *line pipe* (see Figure 2-9). These are metal cylinders specially made for pipeline use, and when they are welded together, they form the pipeline, Figure 2-9. The line pipe is made at specialist *pipe mills* and is usually provided in 12 metre (40 feet) lengths. The line pipe is made from steel. Steel is made by melting iron ore and coke (carbon) in a furnace. Other elements (*alloys*[6], such as manganese) are added to the iron and carbon mix to give the steel good strength and ductility properties. The molten steel is then poured into moulds, where it cools into *ingots* (blocks), or it can be continuously cast. These ingots or castings are then rolled into coils or slabs or billets. These coils/slabs/billets (Figure 2-10) are then formed into cylinders at a pipe mill, to create the line pipe.

We will buy line pipe to a *specification* such as the American Petroleum Institute's API 5L[20]: the specification gives requirements on properties of the steel, tolerances on the pipe dimensions, strength of the pipe, etc. Specifications allow us to produce consistent products from differing producers and countries. Pipeline operators buy their line pipe from these pipe mills. These *orders* can supply many thousands of lengths of line pipe.

5 Both pumps and compressors increase pressure, so they can all be called pumps. Compressors reduce the volume of the gas (they compress it) to move it along a pipeline (just as a bicycle pump moves air into a tire, by compressing it into the tire). A pump does not affect the volume of the liquid it is moving along the pipeline: it forces the flow by pushing the fluid (just as we would have to push a solid piece of concrete along a road, although a fluid is much easier…).

6 Metals are rarely used in structures in their pure form (e.g., pure iron), and are usually mixed with other metals (e.g., manganese) or non-metals (e.g., carbon) to improve their properties: these additions are alloys.

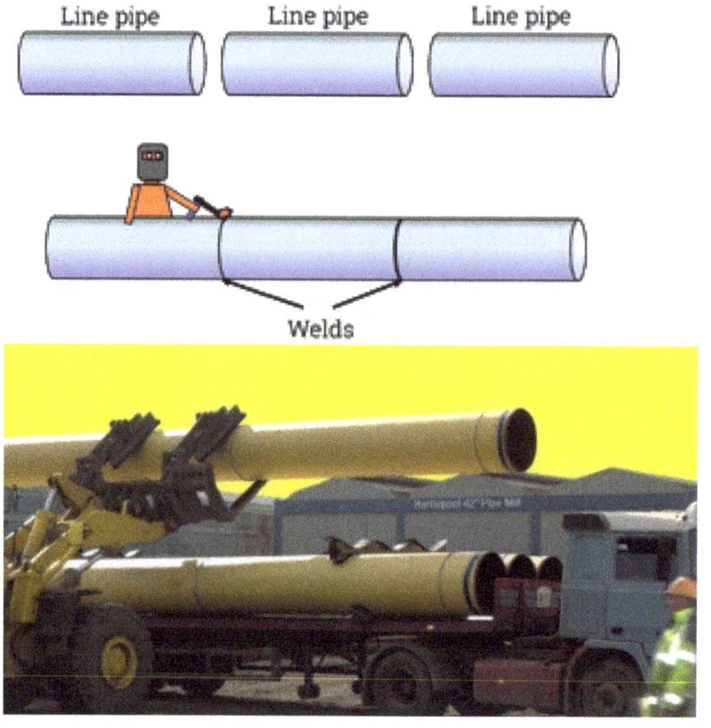

Figure 2-9. Pipe being loaded onto trucks at a pipeline mill (bottom) and being welded together to form a pipeline (top).

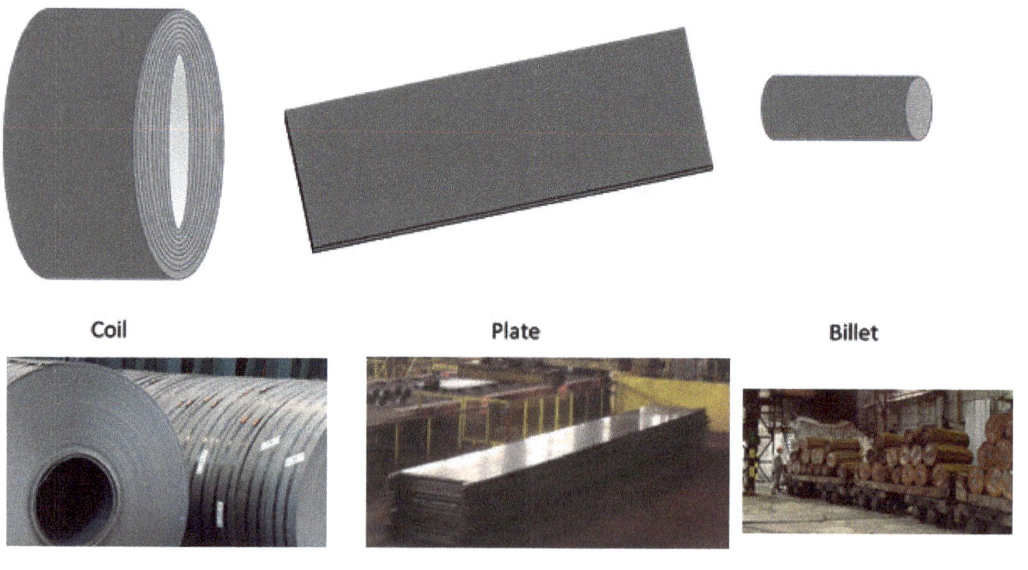

Figure 2-10. Coil, plates, and billets.

2.2.1 Diameter, wall thickness

Line pipe has two key geometry measurements (see Figure 2-11):

- diameter: the diameter dictates how much product we can flow through our pipelines – the larger the diameter, the higher the flow rate; and,
- wall thickness: the thicker the wall, the higher the pressure the pipeline can withstand.

2.2.2 Strength

Our line pipe needs to withstand all the pressures and loads imposed on it during service. The pressure in the pipeline creates stresses in the wall of the pipe. We do not want these stresses to exceed the steel's strength[7], otherwise the pipeline will fail. *Strength* of a material is its ability to withstand an applied load without failure or permanent deformation.

The steel's strength is measured using two parameters (see Figure 2-12):

- the *yield strength* is when the steel starts to deform permanently (and cannot regain its original shape); and,
- the *ultimate tensile strength* is the maximum stress the steel can withstand, and after this strength, the steel fails.

The line pipe manufacturers test the line pipe steel by cutting a small sample from a line pipe length. This small sample is then pulled to failure in a test machine, and the load (which is related to the stress on the sample) applied to the sample, and its elongation (strain), is measured, and plotted on a *stress-strain* graph, (Figure 2-12).

We do not want our line pipe to deform permanently (lose its circular shape), or fail, but we need to operate at high pressures (to give us economic flow rates); therefore, we need strong (high strength) steel, as the high pressures in the pipeline create high stresses in the line pipe.

The line pipe manufacturers cannot measure the strength of every 12 metre section of line pipe they produce; therefore, they conduct sample tests. They measure the yield strength and ultimate tensile strength in the samples. There will be small variations in the strengths measured, as no two samples tested will be the same. This will mean the manufacturer will record a range of yield and ultimate strengths.

7 A material's strength is a measure of its resistance to withstand loads and deformation. A load (force or weight) on a material imposes a stress in the material. The material's strength resists this stress, until it is exceeded. When the stress exceeds the material's strength, the material fails.

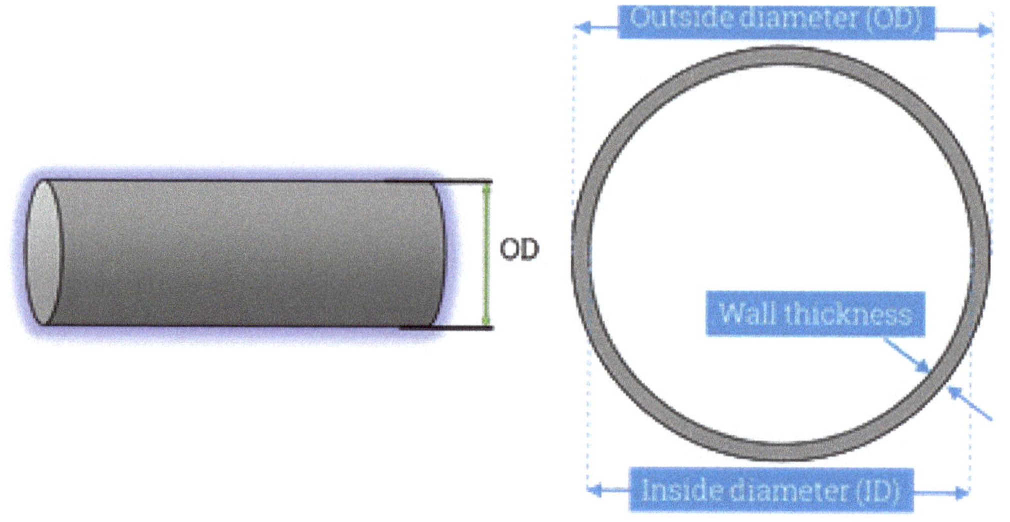

Figure 2-11. Line pipe diameter and wall thickness.

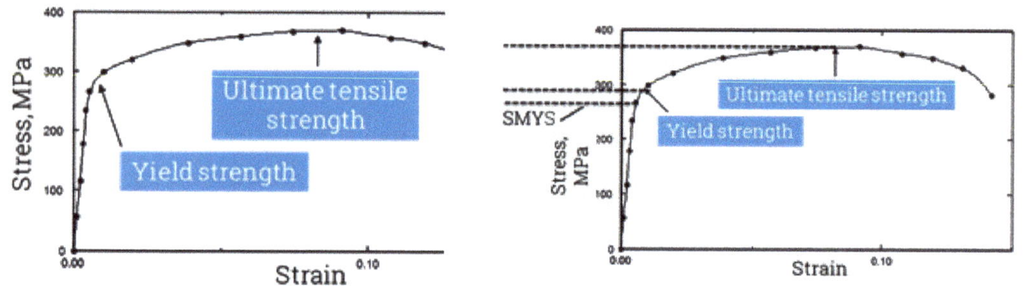

Figure 2-12. Yield strength and 'SMYS' of line pipe.

Line pipe specifications such as API 5l[20] specify a lower bound measure of yield strength for ranges of measured strengths. This is called the *specified minimum yield strength* (*SMYS*), Figure 2-12, and the manufacturer specifies the steel as having this SMYS, based on the measured yield strengths of samples for the line pipe order. The purchaser then buys the order of line pipe knowing its SMYS, but not knowing the actual yield strength of every 12 metre section of line pipe.

Specifications such as API 5L give the yield strengths and SMYSs that are supplied by pipe manufacturers. Strengths are traditionally measured in units of lbf/in^2. We can buy line pipe of various specified minimum yield strengths; for example, 42,000 lbf/in^2, 46,000 lbf/in^2, and 52,000 lbf/in^2.

The pipeline industry usually refers to *grades* rather than strengths, where grade is SMYS/1000; therefore, line pipe with an SMYS of 52,000 lbf/in^2 has a grade of 52. The industry also uses a shorthand form of the strengths (X); for example, a line pipe with an SMYS of 52,000 lbf/in^2 is referred to as grade X52.

2.2.3 Ductility

When we load materials, they behave in different ways:

- when we bend a piece of chalk it quickly breaks and there is no deformation, so it is a *brittle*[8] material;
- when we bend a silver spoon, it does not break and there is extensive deformation, so it is a *ductile* material.

Ductility is the ability to deform. Line pipe needs to be able to bend (we bend pipe during onshore construction (see Figure 2-13)), deform (our pipelines can be dented during operation; for example, by excavating machines accidently hitting them), and stretch (pipelines can be subjected to strains higher than predicted in design; for example, ground movement due to earthquakes around a pipeline can create high loads).

Line pipe specifications require the material to be able to strain (deform) extensively before it fails. The tensile test shows how the line pipe can strain (deform) (Figures 2-12, 2-14). In a tensile test a specimen can deform extensively before failure… this is ductile behaviour, Figure 2-14. It can also show little deformation (brittle behaviour), or, it can show a mixture of brittle and ductile behaviour.

2.2.4 Coatings

Line pipe is made from steel. Steel is mainly made from iron (Fe). All iron or steel structures, such as pipelines, can corrode[9]. Pipelines will be buried in soil or underwater, and steel will corrode if exposed to surrounding soil or sea. Therefore, the line pipe must have an anti-corrosion coating on its outer surface, Figure 2-5.

Most line pipe is now supplied with an anti-corrosion coating on its external surface (see Figure 2-15). The line pipe was not usually coated up to the 1940s. Early coatings (up to the 1950s) were applied to the line pipe in the field, during the pipeline's construction. Gradually, this changed, and now most line pipe coatings are applied in special factories, before delivery to the pipeline's construction site.

8 When we hit (impact) a sheet of glass with a hammer it shatters into many pieces on the first impact. These pieces are not deformed – they have separated along fracture paths. This absence of deformation is called a brittle fracture. If we hit an aluminium can with a hammer, it would dent, but it would not fail. We would need to impact the can many times before it would fail. It would fail after extensive deformation. This is called a ductile fracture.

9 Corrosion is a natural process: metals (such as iron) are not stable in metallic form. They want to corrode and return to more stable forms that are found in their ores, such as oxides (iron is obtained from smelting mined iron oxide). This is why gold does not corrode: it is found in the ground as gold, and not as an oxide.

Figure 2-13. Line Pipe bent during construction, to follow the contour of the land and the route of the pipeline.

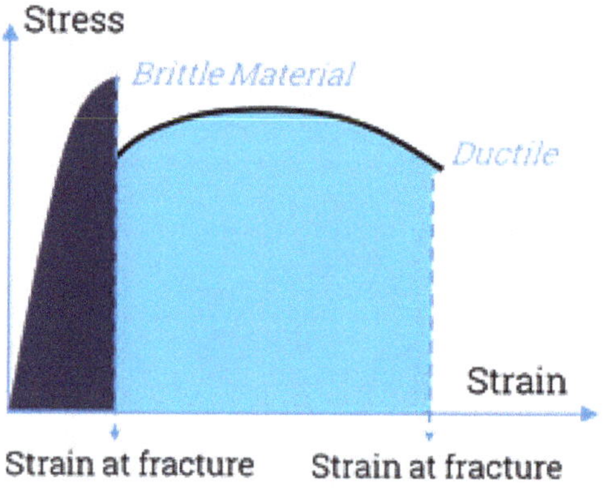

Figure 2-14. Ductility in a tensile test.

Figure 2-15. Pipeline anti-corrosion coating.

2.3 Design of Pipelines

Pipelines need to be designed to operate safely; therefore, published standards are used for a pipeline's design, construction, and operation. There are many pipeline design standards and guides around the world (for example, [1, 5-19]). These standards focus on the safety aspects of a pipeline's design, construction and operation. Other standards focus on the safety management of a pipeline's operation (for example, [21-27]).

2.3.1 Pipeline standards

A *standard* is a set of technical definitions and guidelines that provide consistent and comparable results: it is an agreed, repeatable way of doing something.

Pipeline design standards have the objective of routeing, designing, and constructing a pipeline that will:

- operate safely;
- have minimal impact on the environment; and,
- be cost effective (both in terms of capital expenditure, and operational expenditure).

There are many pipeline standards around the world. Usually, there will be a recognised (by both the pipeline owner and the regulatory authority) standard already in use in the country where the pipeline is to operate.

It should be emphasised that a standard is not a law, it is merely *good practice* and ensures essential and minimum safety requirements. The most popular pipeline design standards are from the American Society of Mechanical Engineers (ASME):

- ASME B31.8[1] for pipelines carrying gases, such as methane;
- ASME B31.4[5] for pipelines carrying liquids, such as crude oil.

These standards focus on the safety aspects of a pipeline's design and operation, and provide generic guidance[16, 17]: they cannot cover unusual conditions or special designs.

2.3.2 Stresses in a pipeline

We need to design our pipeline to meet the demands of its customers. Customers will need a certain amount of crude oil, products, or natural gas delivered on a daily basis. This means the pipeline has to supply its products at a certain flow rate to meet demand. The flow rate will mainly depend on the pipeline diameter and the pipeline pressure.

The internal pressure creates loads in the pipe wall. These loads create stresses in the pipe wall (see Figure 2-16). The pipe wall must be able to withstand these pressures, and pipeline design standards require these stresses to be well below the yield strength of the line pipe, otherwise the pipe will start to distort and fail.

The two main stresses caused by internal pressure are:

- hoop (or circumferential) stress that causes the diameter to increase; and,
- axial (or longitudinal) stress that causes the length of the pipe to change.

You can visualise these stresses in a balloon; a balloon that is not inflated has a short length and small diameter. When we inflate it, it dramatically increases in length, and diameter. These increases are the effect of the hoop and axial stress.

We can calculate these stresses using simple equations:

Hoop stress = $pD/(2t)$
Axial stress = $pD/(4t)$

where: p is the pipeline internal pressure,
D is the pipeline diameter, and
t is the pipe wall thickness.

The wall thickness (t) of the pipeline has to withstand all stresses on it: other stresses such as those from external sources (e.g., if the pipeline is in a mining area, ground subsidence may put extra loads on the pipeline) must also be calculated and considered in the design.

We usually quote the stresses in a pipeline as a percentage of the line pipe's SMYS; for example, a hoop stress of 80% SMYS means the hoop stress in the line pipe is 80% of the line pipe's SMYS.

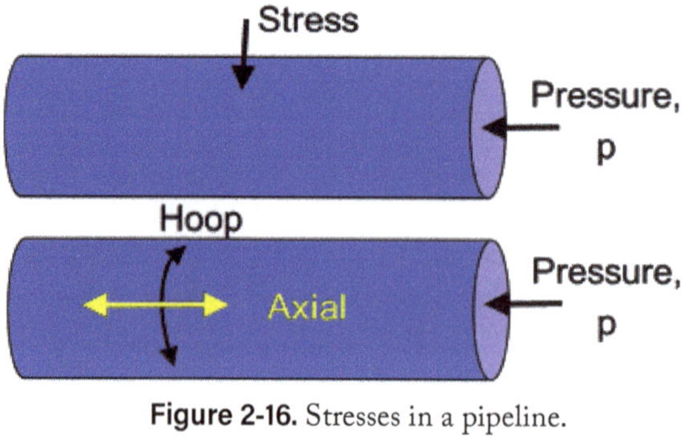

Figure 2-16. Stresses in a pipeline.

2.3.3 Design factor

The pipeline cannot operate at very high stresses, as this would lead to the pipeline yielding and possibly failing. Standards control and limit stresses in pipelines by using *design factors*. These design factors limit the stresses (and hence pressures) in a pipeline. The design factor is a number, but it represents:

- Design factor = Stress in pipeline/SMYS, or
- Stress in pipeline = Design factor x SMYS

The design factors are specified in the standards as a percentage of the specified minimum yield strength, SMYS (e.g., 72% SMYS) or a ratio of pipeline stress to SMYS (e.g., 0.72). We cannot operate our pipeline above these design factors; for example, if a standard specified that the design factor for hoop stress was 0.5, we would only be able to operate our pipeline up to a hoop stress of 50% of the line pipe's SMYS. If the line pipe's SMYS is 60,000 lbf/in^2, we would have to limit our hoop stress to 0.5 x 60,000 lbf/in^2 = 30,000 lbf/in^2.

Our hoop stress is directly related to our pressure, so these design factors limit the pressure in the pipeline.

We do not have very high design factors; for example, a design factor of 1 is very high. A design factor of 1 would mean we could operate our pipeline at a stress equal to the line pipe's yield strength.

We need design factors: they account for uncertainties in the pipeline's design. These uncertainties include:

- variability in materials;
- variability in construction practices;
- uncertainties in loading conditions;
- uncertainties in in-service conditions.

Design and operating pressure.

A pipeline will need to operate at a pressure that will ensure all its customers receive its product at the right time, and correct quantity and quality.

The pipeline cannot operate at a pressure higher than it is allowed in the pipeline standards. These pressures are limited by the standards specifying design factors. We can show how this pressure is limited by mathematics. We have already said:

1. Hoop stress in pipeline = pressure x diameter/(2 x wall thickness), or,
2. Pressure = Stress in pipeline x 2 x wall thickness/diameter.

We have also said:

3. Design factor = Stress in pipeline/SMYS, or
4. Stress in pipeline = Design factor x SMYS.

We can therefore say (combining '2' and '4'):

5. Pressure = Design factor x SMYS x 2 x wall thickness / diameter

This pressure is usually termed the *design pressure*. We should not go above this design pressure during operation of the pipeline.

We can operate at pressures up to this design pressure. We may operate at pressures below the design pressure; for example, if our crude oil or natural gas reservoir is depleting, and the pressure is dropping, or if customer demand does not require us to operate at the design pressure.

Why have 'Design Factors'?

When we design a structure, we have to put limits on its design and operation, otherwise users of the structure may fail it.

We are familiar with speed limits on highways: we impose limits such as 50 miles per hour (80 km per hour), even though our automobiles can travel at much higher speeds.

We impose these speed limits to ensure safety, and to allow for uncertainties in the condition of the highway, weather conditions, drivers' abilities, and the condition of the automobiles.

We would never set a speed limit at the maximum speed attainable by the automobile – the maximum speed is governed by other external factors, and their associated uncertainties.

We would also never set the speed limit based on the highway itself, as this ignores the gradual deterioration of the highway, and uncertain weather/driving conditions.

A speed limit is analogous to a design factor.

2.4 Operation of pipelines

We have to control the flow of product in our pipeline. Pipeline systems now have remote *control centres*. These centres provide the controllers (or *operators*) in the control centres with complete operational information of the pipeline system.

The computerised control centres provide a centralised response-and-control base of day-to-day operations and to facilitate the coordination and response of abnormal or emergency situations. They also provide a centralised base of operations for ongoing maintenance and other field activities occurring on the pipeline system.

In general, a controller's job is to control the entry and exit of natural gas, or liquids into the energy transmission system. A controller monitors the computer for pressure, flow rates, and other parameters at various locations along the pipeline. The computer performs many on-line computations, and sends commands to the field to control the operation of the valves and pumps and ensure a safe and efficient flow of product.

2.5 Integrity management

How do we manage the day-to-day safety of a pipeline system? The past few years has seen an emphasis on *pipeline integrity management*. Its application is through a *pipeline integrity management system*. But... what is a pipeline integrity management system?

- *Integrity* means a pipeline system maintains its structural integrity and does not leak or rupture (i.e., prevent a *release* of product).
- *Integrity management* is the many and varied activities pipeline operators must undertake to ensure that releases do not occur. It is a program of integrity work, within a written plan, within a system.
- A *system* is a group of interacting, interrelated, or interdependent elements (*components*) that carries out a specific activity, performs a duty, or solves a problem.

A management system helps us manage a particular aspect of our business; e.g., pipeline integrity, and is often required by regulators of pipelines [21, 22]. It will need:

- A clear policy;
- Staff organisation;
- A plan, setting targets and objectives, and identifying hazards;
- Assessment of risks, and establishing standards against which you can measure performance (and show continuous improvement); and,
- Periodic audit and review.

An integrity management system will be a comprehensive, systematic, and integrated management *program* providing safety improvements to systems by allowing an operator to effectively allocate resources to prevent/detect/mitigate threats to their pipeline.

We need to *plan* our integrity management within this program (see Figure 2-17). The objective of the plan is to protect people from threats that can cause pipeline failures. Therefore, we need to evaluate all threats, and estimate the risk associated with each threat. This risk assessment will highlight the biggest risks, and we can mitigate these risks by integrity assessments, such as inspection of the pipeline.

The integrity management plan starts with gathering data and identifying *threats* to the pipeline. A hazard, or threat, presents a risk.

Pipelines have many threats such as:

- Activities (e.g., earth moving) around a pipeline;
- Adverse weather conditions;
- Sabotage, theft, vandalism;
- Poor coatings leading to corrosion;
- Poor quality materials or construction;
- Operating faults;
- Equipment faults;
- Internal corrosion caused by a corrosive product.

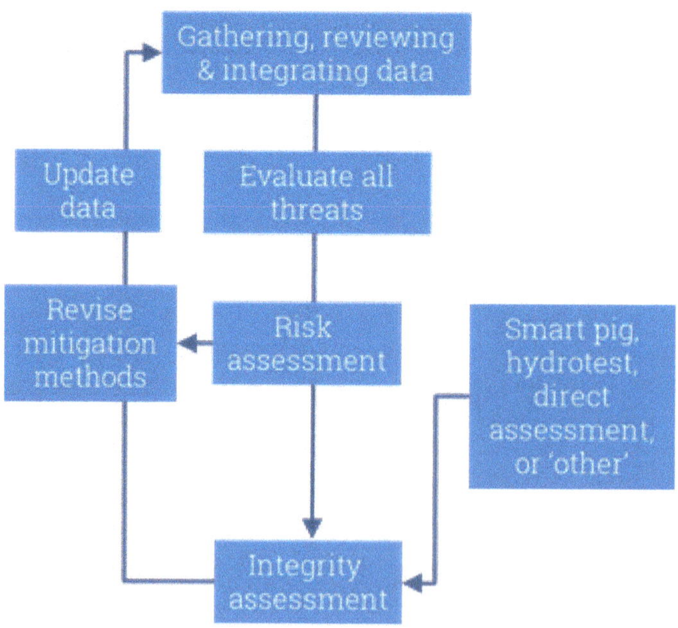

Figure 2-17. Integrity management plan.

We ensure these threats are managed by regularly testing, surveying, or inspecting our pipelines. These are called *integrity assessments*; for example:

- We can *test* our pipelines by removing the product from the pipeline, and then filling it with water (we can use a gas at low pressures). We then pressurise this water (a pressure test or *hydrotest*) to a pressure well above the pipeline's operating pressure. If the pipeline survives this pressure test, we know it can then operate safely at the (lower) operating pressure the pipeline.
- We can survey our pipelines for changes to its condition or surroundings. We can walk along the pipeline looking for changes in ground condition, and we can also do inspections on the pipeline to check the condition of the anti-corrosion coating. We can uncover the pipeline in areas where we have a concern with its condition, and directly inspect it: this is sometimes called *direct assessment*.
- We can send robots (smart *pigs*) down the pipeline. These robots flow with the product and can inspect the pipeline as they move through the pipeline, detecting damage such as dents, or deterioration, such as corrosion.

Standards bodies (such as in America (ASME), the UK (BSI), and Norway (DNV) [21,22,23,25]) have all published standards to assist in the integrity management of pipelines.

QR Code 2-4. Scan here to learn more about pipeline stresses.

2.6 Summary

1. Most of the energy used in the World today is based on fossil fuels. These fossil fuels are oil, natural gas, and coal. These energy forms need transporting from supply to demand.
2. We have various types of pipeline: gathering and flow lines collect the oil and gas from wells; and, transmission lines transport the oil and gas to customers or storage.
3. Our pipelines are made from line pipe: these 12 metre lengths of line pipe are high quality steel tube, made specifically for pipeline use in a pipe mill.
4. The line pipe is welded together to create the pipeline. Onshore pipelines are usually buried below the ground surface, and subsea pipelines reside on the seabed or in a trench; therefore, we rarely see transmission pipelines.

5. Pipelines are designed, constructed and operated using national or international standards.
6. The pipelines have compressors (for gas pipelines), and pumps (for liquid pipelines) to force the product to flow through the pipeline.
7. Their operation is controlled from a central control room.
8. The pipelines are regularly surveyed and inspected to ensure they are operating safely.

QR Code 2-5. Scan here to learn more about pipeline principles.

2.7 References

[1] Anon., 'Gas Transmission and Distribution Piping Systems'. ASME B31.8 – 2018, American Society of Mechanical Engineers. New York USA. 2018.

[2] http://wiki.iploca.com/display/rtswiki/1 and http://wiki.iploca.com/display/rtswiki/1.+Introduction

[3] Anon., 'Pipeline Safety: Collecting Data and Sharing Information on Federally Unregulated Gathering Pipelines Could Help Enhance Safety'. Report to Congressional Requesters. GAO-12-388. United States Government Accountability Office. March 2012.

[4] http://www.phmsa.dot.gov/public/definitions.

[5] Anon., 'Pipeline Transportation Systems for Liquids and Slurries', ASME B31.4 – 2016, American Society of Mechanical Engineers. New York USA. 2016.

[6] Anon.,' 'Petroleum and natural gas industries - Pipeline transportation systems', ISO 13623:2009, International Organization for Standardization, 2009.

[7] Anon., 'Gas infrastructure. Pipelines for maximum operating pressure over 16 bar. Functional requirements', European Harmonised Standard: BS EN 1594:2013. British Standards Institution. 2013.

[8] Anon., 'Petroleum and natural gas industries – Pipeline transportation systems', European Harmonised Standard: BS EN 14161. British Standards Institution. 2015.

[9] Anon., 'Oil and gas pipeline systems', CSA Z662-15, Canadian Standards Association. 2015.

[10] Anon., 'Pipelines - Gas and liquid petroleum - design and construction', AS 2885.1-2012. Standards Australia, 2012.

[11] Anon., 'Code of Practice for Pipelines – Part 1: steel pipelines on land', BSI PD 8010-2. British Standards Institution Published Document, UK. 2015.

[12] Anon., 'Code of Practice for Pipelines – Part 2: subsea pipelines', BSI PD 8010-2. British Standards Institution Published Document, UK. 2015.

[13] Anon., 'Submarine Pipeline Systems', Offshore Standard. Det Norske Veritas. DNV-OS-F101. Norway. October 2013.

[14] Anon., 'Steel pipelines and associated installations for high pressure gas transmission', IGEM/TD/1. Edition 5+A. Communication 1735. Institution of Gas Engineers and Managers. UK. 2019.

[15] Anon., 'Pipeline Route Selection for Rural and Cross-country Pipelines', ASCE Manuals and Reports on Engineering Practice No. 46. American Society of Civil Engineers. USA. 1998.

[16] F A Hough, 'The Gas Industry has Approved its New Safety Code', Gas Magazine, November 1954.

[17] F A Hough, 'The New Gas Transmission and Distribution Piping Code (ASA B31 Section 8)', Gas Magazine, Series in 8 Parts, January through September 1955.

[18] Anon., 'Introduction to ASME Codes and Standards'. American Society of Mechanical Engineers, New York. USA.

[19] http://www.hse.gov.uk/pipelines/resources/designcodes.htm.

[20] Anon., 'Specification for Line Pipe', American Petroleum Institute. API 5L. Edition 46. 2018.

[21] Anon., 'Managing System Integrity of Gas Pipelines', ASME B31.8S-2018. American Society of Mechanical Engineers. New York, USA. 2018.

[22] Anon., 'Managing System Integrity for Hazardous Liquid Pipelines', American Petroleum Institute. API Recommended Practice 1160. Third Edition. 2019.

[23] Anon., 'Pipeline systems Part 4: Steel pipelines on land and subsea pipelines – Code of practice for integrity management' BSI PD 8010-4:2012. British Standards Institution. UK. 2012.

[24] Anon., 'Pipeline systems Part 3: Steel pipelines on land. Guide to the application of pipeline risk assessment to proposed developments in the vicinity of major hazard pipelines containing flammables – Supplement to PD 8010-3:2009', BSI PD 8010-3: 2009. British Standards Institution. UK. 2009.

[25] Anon., 'Integrity Management of Submarine Pipeline Systems'. Recommended Practice. DNV-RP-F116. Det Norske Veritas. Norway. October 2017.

[26] Anon., 'Petroleum and natural gas industries - Pipeline transportation systems - Recommended practice for pipeline life extension', Technical Specification. ISO/TS/12747. First edition. International Organization for Standardization. 2011.

[27] Anon., 'Recommended Practice for Assessment and Management of Cracking in Pipelines', API RP 1176. American Petroleum institute. 2016.

CHAPTER 3

Line pipe principles

Dr Gerhard Knauf

3.1 Introduction

Pipelines consist of a large variety of components and installations such as compressor stations, valves, control and metering stations and – of course – line pipe.

3.1.1 History of line pipes – a brief look back

The technical and cultural achievements of early civilizations in the valleys of the Euphrates and Tigris, the Nile, and the Indus and the Incas, Greeks and Romans are well known. Their preferred building material was stone, and their preferred metal was bronze. They used fabrics that guaranteed good durability but needed a lot of construction material and human and animal work for their processing and laying.[1]

These civilizations could make tubes from clay, lead or wood, but these tubes had a short life and very small cross sections. Since better materials were not available at affordable prices, wooden tubes were used as water pipes, despite their short lifespan, until the 19th century. Also, wooden tubes were used for natural gas transportation.[1]

Cast iron pipes had been used for a long time, but their production costs were high, with high scrap rates. They were only considered for mass production in the 19th century, and then became important components of water and sewage systems.[1]

The rapidly expanding coal gas industry of the 19th century relied on longitudinally welded wrought iron tubes, but these tubes were liable to leaks.[1]

Around 1850/1860 the first pipelines for oil and gas started to be built in the USA and Canada, with lengths of about 30 km.[2,3]

In 1891 a gas pipeline from Indiana to Chicago was built with a length of about 190 km.[4]

Increasing safety requirements, and the desire for higher internal pressure, led to the development of seamless steel tubes. In 1885, the Mannesmann brothers patented the cross-rolling process[1], which for the first time made it possible to economically produce reliable pipes for high pressures.

In 1890, a 25-km-long pipeline (Kedabeg, Caucasus) was completed. It overcame more than 1,000 m of vertical drop and had an operating pressure of approx. 100 bar. One of the first long-distance pipelines using seamless steel pipes with diameters of 8 to 12 inches was built in 1906, connecting Baku (Black Sea) and Batumi (Caspian Sea). At 835 km in length, it was for many years the longest kerosene pipeline in the world.[5]

Steel became by far the most important material for line pipe in the 20th century; however, the construction of large natural gas transmission networks, which began in the 1950s and 1960s, was only possible after major progress had been made in the metallurgy, manufacture and welding of steel pipes.[4]

3.1.2 Line pipes today

To ensure safe and reliable pipeline operation, the pipes must have suitable chemical and mechanical properties, consistent geometry, and corrosion resistance.

Pipes used in gathering and transmission lines for oil and gas are typically made out of carbon steel. The properties of carbon steel pipes can be adjusted within wide limits to the required properties by choosing appropriate chemical compositions and manufacturing processes.

Stresses and strains during construction and operation can pose significant challenges for the pipeline. High internal pressures, pressure fluctuations, ground movement, external mechanical loads and impacts, risk of collapse in deep waters, as well as high or low service temperatures require adequate pipe properties such as optimized geometries, sufficient strength, deformability, and toughness.

Corrosive attack (e.g., by hydrocarbon products with high levels of hydrogen sulphide and water within the pipeline) demand appropriate corrosive resistance of the steel, often in combination with external coating or cladding solutions.

To comply with these demands, line pipes have to be manufactured to high quality standards, with a narrow band of material and geometrical properties within the range of given specifications. The narrow band of properties is of particular importance with regard to the mass production of pipes to allow consistent properties along the entire length of a pipeline.

This chapter on Line Pipe Principles concentrates on line pipe made from carbon steel. It provides an overview of the basic standards and specifications for the production of line pipes and the various steel pipe production routes. Pipe properties, and the fields of application, as well as the tests for determining the pipe properties are presented.

Pipe materials for pipelines

Pipes used in oil and gas production, as well as in pipeline systems for gathering, transporting and distributing of fluids, can be manufactured from various materials such as high alloyed steel, nickel-based alloys, plastics and composites.

The most widely-used material for line pipe, especially in gathering and transmission pipeline networks, is carbon steel.

Common to all pipe making routes is the combination of production and quality control steps in the production flow, required to guarantee appropriate pipe properties. Pipe manufacturing for major pipeline projects is accompanied by third-party auditors to check and certify all production and quality control steps.

All related data and test results are recorded electronically and, together with the pipe delivery, provided to the customer.

3.2 Line pipe specifications

High quality and efficient line pipe manufacturing is based on international standards and specifications. Alternatively, or additionally, compliance with national standards and regulations is often specified. Overviews and comparisons of these main line pipe standards are available (e.g., reference [6,7]). Major pipeline standards that include detailed specifications for pipes are briefly presented below.

International standardization and specification bodies

- ISO - International Organization for standardization [8]
- API - American Petroleum Institute [9]
- CEN - European Committee for Standardization [10]
- DNV GL [11]
- NACE International [12]

networks, is carbon steel.

3.2.1 ISO 3183 and API 5L

ISO 3183[13] and API SPEC 5L[14] are the most important and basic line pipe standards for onshore and offshore service worldwide. Both standards have been harmonized as a result of the joint work of the respective committees of both these organizations. Specific requirements for manufacturing procedure qualification, treatment of surface imperfections, non-destructive testing, ductile fracture propagation, sour and offshore service are given in annexes. Specific European issues, particularly a number of more stringent requirements, are considered in Annex M. The corresponding former separate European standard was withdrawn. The current version, identical with ISO 3183, is therefore now referenced to as EN ISO 3183.[15]

Figure 3-1. Onshore pipe laying[16] (courtesy of Europipe).

Figure 3-2. Offshore pipe laying[16] (courtesy of Europipe).

3.2.2 DNVGL-ST-F101

The DNVGL standard F101[17] is the most important standard for offshore pipeline systems. It covers materials engineering and requirements, design loads and limit states, construction, operation, and abandonment issues, testing, and girth weld requirements.

The material requirements in this standard are generally in line with Annex J for offshore service and Annex H for sour service in ISO3183/API5L. However, additional specific materials and manufacturing procedures for offshore applications are specified in DNVGL- ST-F101.

3.2.3 ISO 15156, NACE MR0175

For pipelines exposed to significant corrosion attacks, the ISO 15156[18] is applicable. This standard deals with corrosion-related material issues and complies with NACE MR0175.[19] It supplements the requirements given in standards for non-corrosive environments, considers the corrosion related annex H in ISO 3183, and provides support to avoid corrosion damage.

3.2.4 General designations and requirements for line pipe materials and production

Line pipe standards specify delivery conditions, ordering information, set requirements for material properties of the starting material and the final product, as well as line pipe related specific material and component tests.

Table 3-1 gives an overview of line pipe grades as specified in ISO 3183[13] and API 5L[14], further referred to as ISO3183/API5L. Line pipe grades are designated with regard to the minimum yield strength level of the pipe material, either in MPa or Ksi.

Designation of line pipe grades

Two designations can be used, which derive from the traditional nomenclature used in ISO 3183 and API 5L prior to the harmonization of both standards.

The ISO designation is referring to SI units in MPa while the API uses the US customary system with rounded Ksi values, (1 Ksi = 6.895 MPa).

Examples for Grade X65 = L450:
L450 >> minimum yield strength 450 MPa
X65 >> minimum yield strength 65.3 Ksi

In delivery specifications, this value is referred to as "specified minimum yield strength" (SMYS).

Additionally, the grades are distinguished with regard to their delivery conditions. These can be 'as rolled' (R) or - after a heat treatment – 'normalized', 'normalizing rolled' or 'normalized and tempered' (N), 'quenched and tempered' (Q) and 'thermomechanically rolled' (M). The letters in brackets are placed at the end of the steel designation to characterize the delivery condition; for example, L450M or X65M would be a line pipe grade with a minimum yield strength of 450 MPa or 65.3 Ksi, manufactured by thermomechanical rolling.

Table 3-1. Line pipe designations and the corresponding strength levels for production specification level PSL 2 in ISO3183/API5L.

Pipe Grade designations		Yield strength, MPa		Tensile strength, MPa	
API 5L / ISO 3185		min	max	min	max
B	L245	245	450	415	655
X42	L290	290	495	415	655
X46	L320	320	525	435	655
X52	L360	360	530	460	760
X56	L390	390	545	490	760
X60	L415	415	565	520	760
X65	L450	450	600	535	760
X70	L485	485	635	570	760
X80	L555	555	705	625	825
X90	L625	625	775	695	915
X100	L690	690	840	760	990
X120	L830	830	1050	915	1145

Figure 3-3. Plate rolling[20] (courtesy of Salzgitter-AG).

Figure 3-4. Coiled strip material[35] (courtesy of Mannesmann Grossrohr).

In the case of seamless pipes, the heat treatment is part of the pipe manufacturing process, in the case of cold formed pipe, the heat treatment and thereby the delivery condition is determined by the plate and strip rolling process in the rolling mill. Plates and strips are rolled from single slabs, produced in the steel mill by cutting individual lengths from steel out of a continuous casting process. From a plate (Figure 3-3) a single pipe is formed. Strip material is coiled after rolling (Figure 3-4) and used for "endless" pipe forming. ISO3183/API5L include two levels of product specifications:

- PSL1 offers a standard quality level;
- PSL2 has additional stringent requirements and is most commonly used for major pipeline projects.

Modern high strength plates and strips for pipe production are generally produced by the thermo-mechanical control process technology, including thermo-mechanical rolling (TMR) and accelerated cooling (ACC).

With a similar chemical composition, the strength of TMCP steels is higher than those of conventional steels, with an improved toughness.[21]

ISO3183/API5L specify a large number of requirements for the product manufacturing and product acceptance criteria in terms of specified properties and their testing.

The most important material requirements for line pipe, their background and their testing, are described later in Section 3.4. Specific small-scale testing and tests on pipes and pipe sections are presented. Details about the various pipe manufacturing routes are presented in Section 3.3.

3.3 Line pipe production

As an overview, Table 3-2 highlights the general differences of production routes, addressed in ISO3183/API5L for the production level PSL2. The type of welding, starting material, pipe forming, heat treatment, final cold forming, and range of grades are listed respectively. According to ISO3183/API5L other production routes, like laser or hybrid welding are possible, but are not discussed below.

Table 3-2. Line pipe types and their production routes (main ISO3183/API5L pipe types, PSL2).

<table>
<tr><th colspan="2">Type of pipe and production route - according to ISO 3183</th><th>Starting material</th><th>Pipe forming</th><th>Pipe heat treatment</th><th>Final cold forming</th><th>Grades (ISO/API)</th></tr>
<tr><td rowspan="2">SAWL</td><td>• UOE
• UOC
• JCOE</td><td rowspan="2">plate
• normalized
• normalizing-rolled
• quenched and tempered
• thermomechanical-rolled</td><td rowspan="4">Cold forming</td><td rowspan="4">heat input by optional coating application may influence mechanical properties, e.g. yield strength [26]</td><td>generally applied</td><td rowspan="2">L245/B to L830/X120</td></tr>
<tr><td>• TRB</td><td>generally, not applied</td></tr>
<tr><td>SAWH</td><td>• continuous spiral forming</td><td rowspan="2">Coil / strip
• as-rolled
• normalized
• normalizing- rolled
• quenched and tempered
• thermomechanical-rolled</td><td>not applied</td><td>L245/B to L830/X120</td></tr>
<tr><td>HFW</td><td>• continuous longitudinal forming and welding without filler material</td><td>sizing by reduction of diameter</td><td>L245/B to L555/X80</td></tr>
<tr><td>SMLS</td><td>• plug mill
• continuous mandrel rolling mill
• multi-stand pipe mill
• pilger mill</td><td>Ingot, bloom, billet
• pierced</td><td>• Hot forming
• Hot forming and cold finishing
• Normalizing forming</td><td>• normalizing
• quenching and tempering</td><td>not applied</td><td>L245/B to L690/X100</td></tr>
<tr><td colspan="7">SAWL: Submerged Arc Welding process for welding of pipe from plate with straight seam
SAWH: Submerged Arc Welding process for welding of pipe from strip with helical seam
HFW: High Frequency induction Welded pipe, formed from strip with straight seam
SMLS: Seamless, without weld
UOE: Pipe formed from plate by U-ing, O-ing and Expanding*
UOC: Pipe formed from plate by U-ing, O-ing and Compression*
JCOE: Pipe formed from plate by J-ing, C-ing and O-ing
TRB: Pipe formed from plate by Three Roll Bending
*For Expansion and Compression see blue box below</td></tr>
</table>

Mechanical expansion, impansion, compression, reduction

Mechanical cold **expansion** (typically 1%) is the last forming step of a **UOE-pipe** and forms the pipe to the final round shape. This is accompanied by a homogenization of strength properties, increase of strength, and reduction of residual stresses.

The reverse process, called mechanical **impansion** or **compression** (UOC) results in similar improvements. An additional improvement of the collapse behavior has been reported [22, 23].

In the case of HFW pipe, the final forming step is a sizing by the **reduction** of the pipe diameter (typically 0.5%) in a rolling stand, resulting in improved roundness and straightness.

The manufacturing processes of the various line pipe types are described in detail in numerous publications and company brochures. In the following sections the major pipe manufacturing processes are described, considering information provided by reference [24], and documents referred to in the respective sections.

3.3.1 UOE pipes

UOE pipes are large diameter pipes manufactured from steel plates, in most cases from thermo-mechanically rolled plates.

Grades. Typical grades for UOE pipe for onshore applications are ISO L450 to X555 (API X65 to X80). The grade X100 is generally available, but after being developed in the late 1990s and early 2000s[26], only a few X100 projects have been launched up to now, similarly for X120. The most commonly used offshore grade is X65.

Welding. The welding procedure is usually 'DSAW', the Double Submerged-Arc Welding Process.

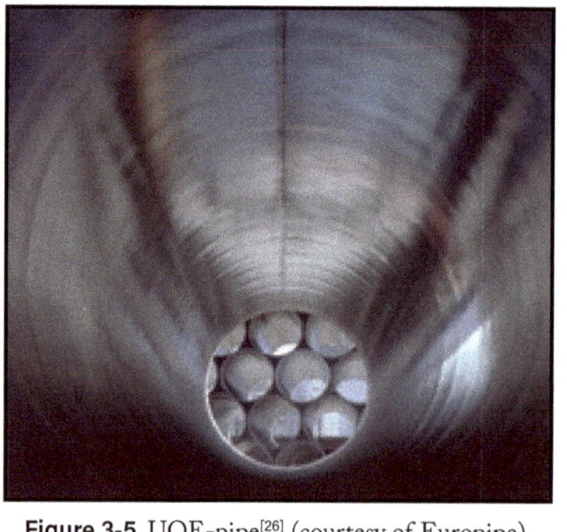

Figure 3-5. UOE-pipe[26] (courtesy of Europipe).

Dimension. The typical length of the pipes is either 12 m or 18 m. While the standard pipe length for offshore pipelines is 12 m to accommodate the spacing of the fixed welding stations on the lay barge, 18 m long pipes are favourable for constructing long-distance onshore pipelines.

The use of these longer pipes will result in a 30% reduction in the number of girth welds and thereby in significant cost savings in terms of construction time and material.[26]

The diameter range of UOE pipe is from 16 in to 60 in (406 – 1,524 mm), and the wall thickness can vary from 7 to 50 mm.[26, 27]

Fields of application. One of the most important applications of UOE pipes is the construction of long-distance trunk lines. The large diameters up to 60 inches, the large range of wall thicknesses, and the high production speed of up to 360 m/h are significant advantages.[28]

By appropriate chemistry and rolling techniques, UOE pipe properties can be adjusted to allow the pipe to be used at low temperature (e.g. -40°C), or sour service. The production process allows the forming of thick walls (up to 40-50 mm) and allows UOE pipe application in deep waters (> 3500 m). Heating up after expansion to around 100°C, e.g. during coating of the pipes, results in optimized collapse resistance.[25] For more details about collapse in deep sea applications see Section 3.4.8.

Manufacturing. UOE pipe manufacturing[29, 30] consists of numerous production and quality control steps, as shown in Figure 3-7. In this overview the typical production steps are indicated in red, the control steps in blue colour.

Submerged - arc welding process

"The submerged-arc welding process or a combination of gas-shielded tack-welding with downstream submerged-arc welding has become widely accepted as the standard method for the welding of large-diameter pipe.

The submerged-arc welding process is an electric fusion welding method performed with a concealed arc.

In contrast to arc welding with welding electrodes, the arc in this case is hidden from view and burns under a blanket of slag and flux.

One of the characteristic features of submerged-arc welding (SAW) is its high deposition rate, which essentially stems from the high current strength which is applied combined with a favorable heat balance.

The filler metal employed takes the form of coiled, bright welding wire which is continuously fed into the molten metal pool by feed rolls at a speed dictated by the deposition rate.

Just above the parent metal (pipe), the welding current is conducted by sliding contacts into the wire electrode and returned via the ground lead connected to the pipe material." [23]

Figure 3-6. Onshore trunk line during laying[31] (courtesy of Europipe).

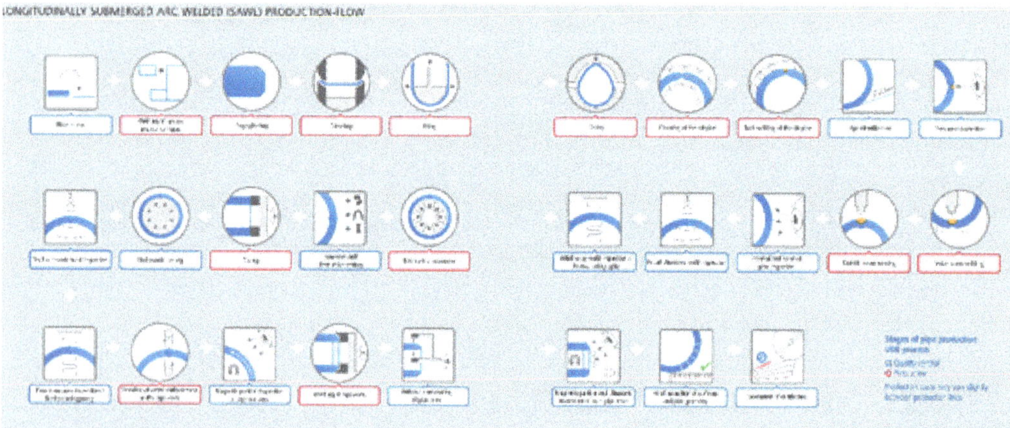

Figure 3-7. Production steps in UOE pipe manufacturing[30] (courtesy of Europipe).

QR Code 3-1. Scan here to view an enlarged version of Figure 3.7

Production steps

- Run-on and run-off tabs – The tabs are welded to the plate, to ensure a stable welding process during start and finish of the seam welding.
- Edge preparation - The edges of the plate are machined so that they are parallel, and a double Y bevel for the longitudinal weld is prepared.
- Crimping – In the area of both edges, the plate is crimped simultaneously to the desired pipe radius. This first forming step leads to an excellent final roundness of UOE pipes, avoiding a roof-shaped weld area, called 'peaking'.
- U-ing – In a second forming step, the plate is bent into a U-shape by pressing the plate between two supports with a tool of circular radius (Figure 3-8). A small amount of overbend is used to counteract the springback effect
- O-ing- The U-shaped plate is formed into an O-shaped slit pipe using two semi-circular dies, each designed and optimized for a tube diameter (Figure 3-9). This forming step requires very high forces of up to 60,000 tons.
- Tack welding - Before submerged-arc welding takes place, the slit pipe is tack welded together under shielding gas. The edges of the open-seam pipe are pressed together to eliminate any offset during tack-welding.
- Submerged arc welding - The submerged arc welding is carried out typically in two steps. First, the inner gap, secondly the outer gap is filled (Figure 3-10). The tack weld is completely dissolved. The welding starts at the attached run-on tabs to stabilize the arc welding process before the welding to the pipe progresses. Respectively, the run-off tabs support to finish properly the long seam welding. Up to five welding wires are simultaneously used and welding parameters are adjusted to optimize welding speed, weld quality and properties.
- Mechanical cold expansion - The as-welded pipes do not generally satisfy the tolerance specifications in relation to diameter and roundness. Therefore, a sizing by cold expansion is applied. This operation is performed by mechanical expanders. The expanding tool is about 1 m in length and is subsequently pushed through the pipe, expanding section by section. (Figure 3-11) The level of expansion is typically adjusted around 1%. This value is considered when determining the initial plate width and circumference of the open seam pipe. This ensures that the final diameter is reached and an optimized roundness is achieved to comply with the demanding requirements from standards and specifications. Cold expansion will increase the yield strength of the pipe[24], which has to be considered when specifying the strength properties of the plate.
- Grinding and bevelling – The production process is completed in the finishing line by grinding the weld reinforcement at the pipe ends and bevelling the pipe ends, which is the final preparation for girth welding in the field.

Figure 3-8. U-ing[30] (courtesy of Europipe).

Figure 3-9. O-ing[30] (courtesy of Europipe).

Figure 3-10. Submerged-arc welding[30] (courtesy of Europipe).

Figure 3-11. Mechanical expansion[30] (courtesy of Europipe).

QR Code 3-2. Scan here for a video showing U-ing

QR Code 3-3. Scan here for a video showing O-ing

QR Code 3-4. Scan here for a video showing inside welding

QR Code 3-5. Scan here for a video showing outside welding

QR Code 3-6. Scan here for a video showing expansion

Quality control. As can be seen in Figure 3-7, various quality control steps are included in the production flow. These tests are of great importance to fulfil the required proof of product properties and to generally guarantee manufacturing of best quality pipes.

- Mechanical testing – After pipe expansion a section of the pipe is cut off from a specified number of pipes out of a production lot. These sections are used to prepare samples for the determination of the mechanical pipe and weld properties. When specified, corrosion tests are also carried out to check the corrosion resistance of the pipe and weld metal.

Within the production flow, a hydrostatic test of the pipes is carried out after expansion. Water is pumped into the pipes so that they are stressed by internal pressure close to their yield point, e.g. 90% of the pipe's specified minimum yield strength. It is assumed that pipes will not see a higher internal pressure during service. More details on hydrotesting and other mechanical and corrosion testing will be given in Section 3.4.

Weld quality

Weld quality plays an important role in the integrity of welded pipes. In order to achieve a corresponding weld quality of a submerged arc weld (Figure 3-10), various aspects must be considered. Beside adequate mechanical and corrosive properties, the weld shape (misalignment, radial offset, weld bead height, weld contact angle, weld bead width, penetration, weld linearity) and avoidance of imperfections (undercuts, slag inclusions, porosities, lack of fusion, penetration, cracks) is important. The mechanical properties and the corrosion resistance is mainly defined by the selection of suitable filler metals, while the weld shape and the occurrence of imperfections or defects are strongly influenced by the welding parameters, and the general quality of the welding process in the mill[32].

- Non-destructive testing – In the course of the production flow, pipes undergo intensive non-destructive inspections, especially, the longitudinal weld which is examined thoroughly by automatic ultrasonic inspection. These tests are carried out before expansion as an initial inspection: pipes with detected weld defects will discarded. After expansion and hydrotesting, a final NDT inspection is carried out.

Where appropriate, or required, manual ultrasonic testing or x-ray inspections are also carried out. After the pipes have successfully passed through all the test and inspection stages, including the dimensions check, they are ready for coating or delivery.

3.3.2 Other pipe production routes with longitudinal submerged-arc welding

Beside UOE forming, other forming processes are used to produce slit pipes to apply submerged-arc welding for pipe manufacturing.

Figure 3-12. JCOE pipe forming[34] (courtesy of Corinth Pipeworks).

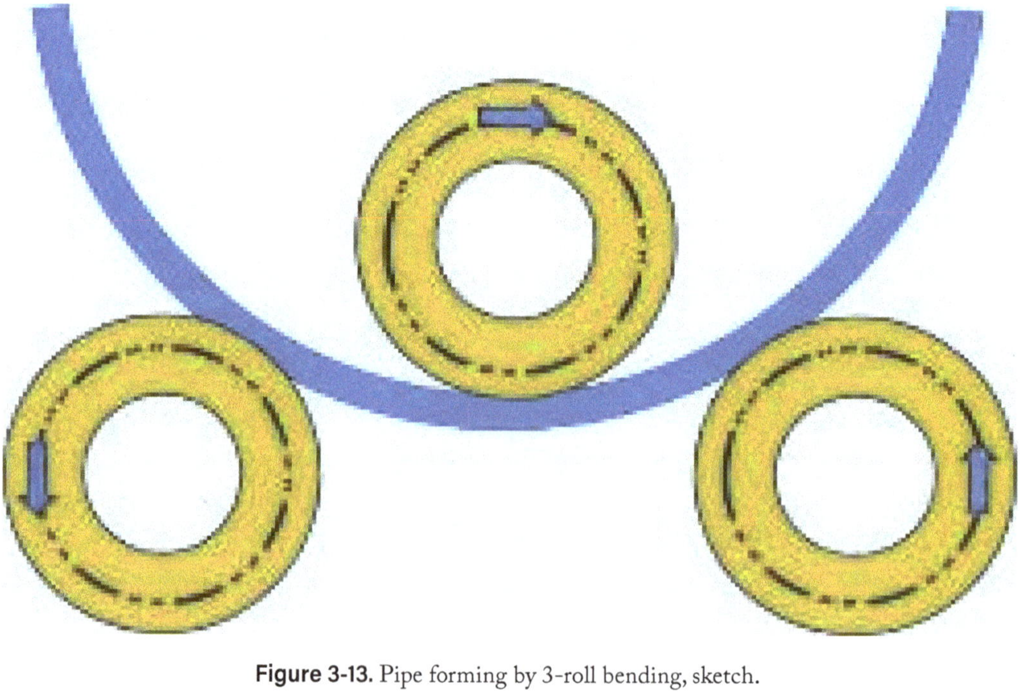

Figure 3-13. Pipe forming by 3-roll bending, sketch.

JCOE forming – The JCOE process is applied to hot rolled plates, and is similar to UOE pipe production. In contrast to the UOE process, plates are shaped by a die with a specially designed diameter and thickness, forming the plate to a 'J' into a 'C' and finally to an 'O' shape (Figure 3-12), before expansion. This type of pipe forming is not as quick as the UOE pipe production, but gives more flexibility with regard to the pipe diameter.[33]

3-roll bending – 3-roll bending facilities are used for pipe forming from plates with thicknesses up to 25 mm. During the forming process, the plates are moved forth and back by the rollers which are continuously moved downwards till the pipe is shaped. A sketch is shown in Figure 3-13. Similar to the JCOE process, this type of forming provides a high degree of flexibility with regard to the pipe diameter and is frequently used for small lots. Typically, no cold expansion is carried out after 3-roll bending, because diameters and roundness can be adjusted sufficiently during the forming process.

3.3.3 Spiral pipe

To manufacture spiral pipes (also known as helical seam pipes) a rolled strip is continuously shaped by spiral forming. An internal view of a spiral pipe is shown in Figure 3-14.

Grades. Spiral pipes are manufactured from Grade L245N / Grade BN to thermo-mechanically rolled high strength steels typically up to L555M / X80M.

Welding. The typical welding procedure is SAW, the Submerged-Arc Welding Process. The same quality aspects apply as mentioned for UOE pipes under "Submerged arc welding."

Dimension. The range of wall thicknesses is from 7.1 to 25.4 mm and the range of diameters for oil and gas pipelines lies between 20 in to 56 in (610-1,422 mm), but for special applications the spiral pipes can be formed up to diameters of 2,500 mm. The limiting dimension for large sizes is the width of the strip material. Because of the continuous production, the pipe length can be easily adjusted (e.g., from 9 to 18 m).

Fields of application. Spiral pipes are used (amongst other applications) as line pipes for oil and gas transportation, mostly for onshore pipelines. As the diameter can be easily adjusted during the production (no additional forming tools are needed), smaller lots can be produced economically.

By appropriate chemistry and strip rolling techniques, spiral pipe properties can be adjusted to sour service, even for higher grades.[36, 37] Offshore applications, particularly for shallow waters, have been reported.[38]

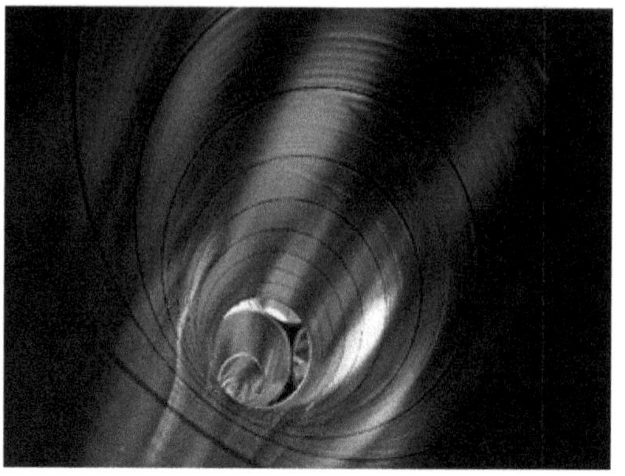

Figure 3-14. Spiral pipe[35] (courtesy of Mannesmann Grossrohr).

Manufacturing. There are two main methods for spiral pipe manufacturing.[35, 39] The integrated forming and SAW welding process are referred to as "One-Step", subsequent forming and SAW welding is referred to as "Two-Step" procedure.

In both production routes, the strip forming is the same, as shown in Figure 3-15. The strip is uncoiled, levelled in a straightening device and welded to the end of the previous coil (skelp end weld) to allow "endless" pipe forming. By machining the edges, the strip is trimmed to a constant width, followed by crimping of the edges. The pipe forming is carried out by 3-roll bending with an inside and/or outside diameter roller cage. The 3-roll bending results in a round shape of the pipe, fulfilling the dimensional requirements without an additional sizing process like expansion.

The pipe diameter depends on the strip width and the angle between the strip and the helically formed pipe. The smaller the inlet angle, the larger the pipe diameter (see small sketch in Figure 3-15).

- One-Step manufacturing – In an integrated production facility the welding is carried out simultaneously with the pipe forming by submerged-arc welding, first inside (around 6 o'clock position) and afterwards outside (around 12o'clock position). Finally, the pipe 'string' (the long continuous section of pipe produced) is cut to the desired pipe length by a 'flying saw'.
- Two-Step manufacturing – in a production line with separate forming and SAW welding, only tack welding is carried out during forming, see Figure 3-16a. The tack weld is applied from the inside of the pipe. After cutting the pipe string to the requested length, pipes are transported to parallel welding stations where inside (6 o'clock) and outside (12 o'clock) SAW welding is carried out (Figure 3-16b). "Two-Step" manufacturing is considered to be more economical because a higher production speed (up to 15 m/min) can be realized.

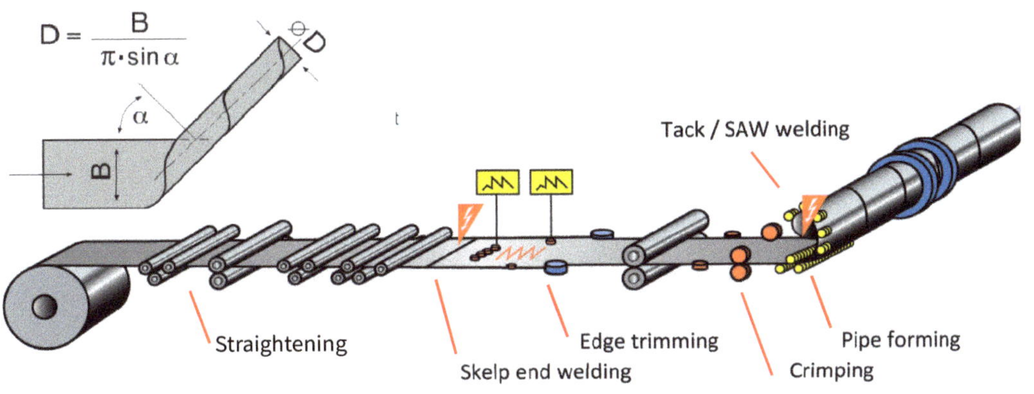

Figure 3-15. Spiral pipe manufacturing[35] (courtesy of Mannesmann Grossrohr).

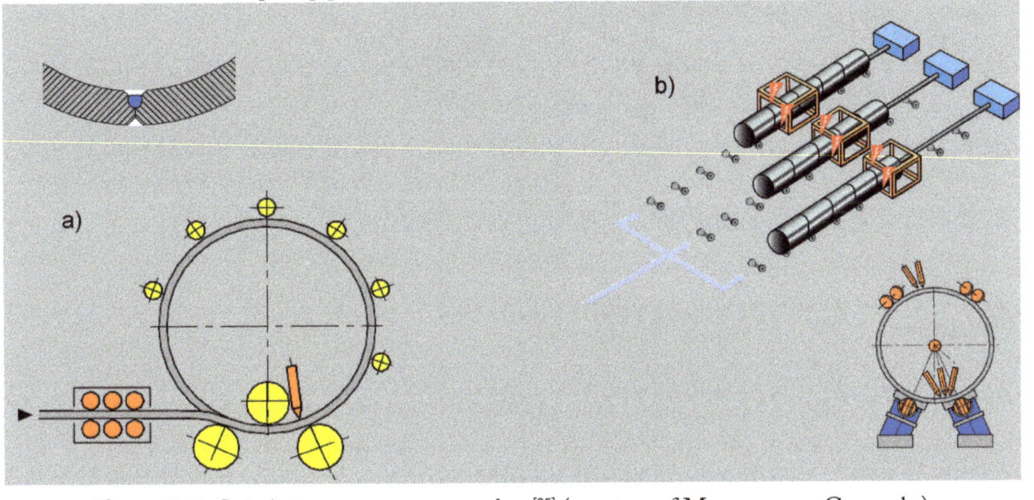

Figure 3-16. Spiral pipe – two step procedure[35] (courtesy of Mannesmann Grossrohr).

Figure 3-17 gives an overview of the entire production flow for spiral pipe production. Similar to the UOE manufacturing process, production steps alternate with quality control steps.

Quality control. As in the case of UOE pipe manufacturing, sampling (Figure 3-17, step 10) for destructive testing takes place before hydrotesting (Figure 3-17, step 12). Ultrasonic inspection is carried out on the strip and skelp welds before forming, and of the weld after the submerged-arc welding (Figure 3-17, steps 3 and 15). Where appropriate, or required, manual ultrasonic testing or x-ray inspection (Figure 3-17, step 16) are also carried out. The skelp welds, connecting the coils need special inspections if they will remain in a pipe section. Otherwise this part has to be cut off. After the pipes have successfully passed through all the test and inspection stages, including the dimensions check, they are ready for coating or delivery.

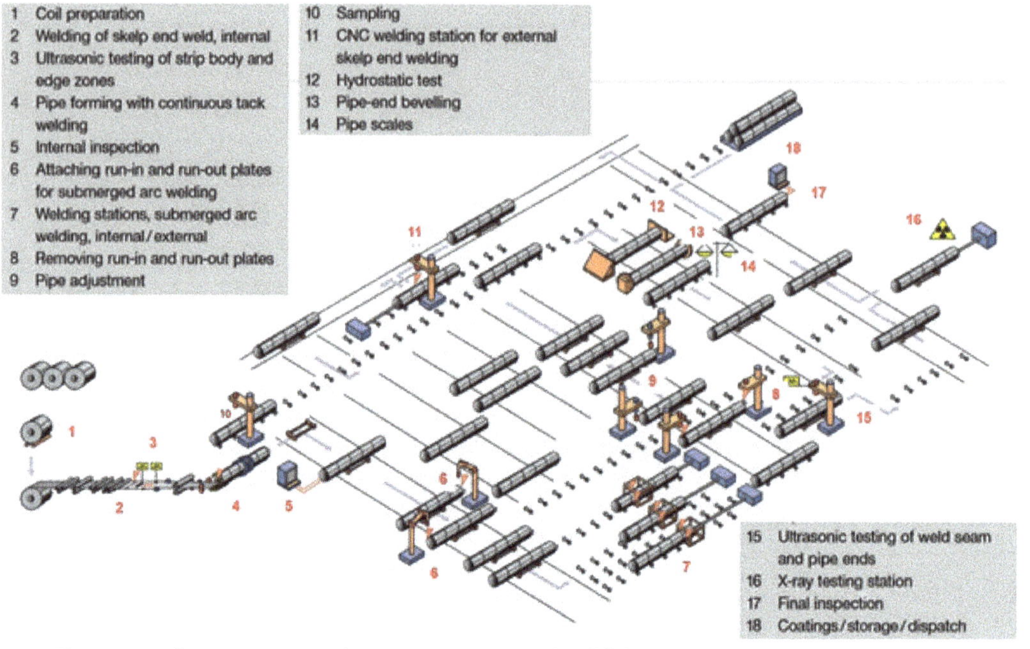

Figure 3-17. Spiral pipe manufacturing, production flow[35] (courtesy Mannesmann Grossrohr).

3.3.4 HFW pipes

High frequency welded (HFW) pipes are manufactured from hot rolled strip material by a continuous forming process from strip to a slit pipe and welded by electric resistance welding, without additional filler material.[40, 41]

Grades. HFW pipes are manufactured from normalizing rolled strip of Grade L245N / Grade BN to thermo-mechanically rolled high strength steels as L555M/ X80M.

Figure 3-18. HFI pipe manufacturing – welding[40].

Welding. According to ISO3183/API5L only high frequency welding is acceptable for the product level PSL2. In the following section the high frequency induction welding process (HFW) is considered.

Electric resistance welding process

Electric resistance welding generally can be carried out by direct current (DCW), low frequency (LFW) and high frequency welding (HFW) processes.

DCW was developed for longitudinal welding of small tubes up to 20 mm in diameter and wall thicknesses up to 2 mm at welding speeds up to 100 m/min (no line pipe application).

LFW is applicable for line pipe product level PLS1 and is performed at alternating current frequencies from 50 to 400 Hz, the welding speed can reach up to 90 m/min.

The alternating current frequency in the case of HFW lies between 200 and 500 kHz. The electric current can be introduced either by conduction welding via copper sliding contacts or by induction welding via an electro-magnetic field. Welding speeds up to 100 m/min can be achieved.

Dimension. The range of wall thicknesses for oil and gas pipelines is from 3.2 to 25.4 mm and the range of diameters from 4.5 in to 24 in (114.3 to 660 mm). Because of the continuous production, the pipe length can be easily adjusted up to pipe lengths of 18 m or longer.

Fields of application. HFW line pipe is used for oil and gas pipelines, both onshore and offshore.[42] The pipe can be used in conventional S-laying of subsea pipelines, and also reeling. By appropriate chemistry and strip rolling techniques, HFW properties can be adjusted to sour service.

Manufacturing. Figure 3-19 gives an overview of the production steps in HFW pipe manufacturing. The strip is uncoiled and straightened and welded to the end of the previous coil (skelp end welding) to allow "endless" pipe forming (Figure 3-19, steps 1-2). Buffering the strip in an accumulator (Figure 3-19, step 3) serves the continuous production flow. The edges are machined to trim the strip to a constant width before the pipe forming process starts. By a series of roll cages, the strip is then continuously formed into the round shape of an open seam pipe (Figure 3-19, step 6).

The so-called CTA (Centralized Tool Adjustment) forming process can significantly reduce the conversion times to produce a different pipe diameter. As a result, small batch production is more economical.

Figure 20 shows a sketch of the welding process. The arrow indicates the direction feed of the open-seam pipe. The high frequency current provided by a welding generator forms an electro-magnetic field around the induction ring which induces an AC voltage in the open-seam tube. The resulting circumferential current concentrates at the open seam edges.

The heated edges are pressed together by rolls and thereby welded. Material is pressed to the inner and outer surface, forming weld flashes. The (still hot) flashes are scraped off to produce a smooth internal and external surface.

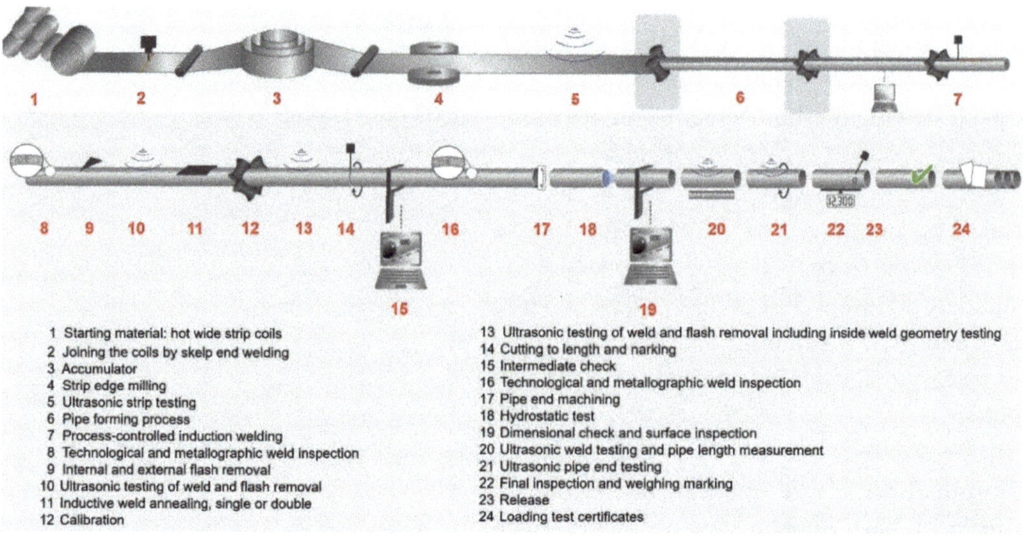

Figure 3-19. HFW pipe manufacturing, production flow[43] (courtesy Mannesmann Line Pipe).

To achieve optimized mechanical properties of the weld area, a subsequent full body or localized heat treatment of the weld and the heat affected zone is carried out in terms of normalizing (e.g. at 950°C), double normalizing, or quenching and tempering.

A final calibration rounds the pipe by compressive stresses around 0.5% and straightens it. Finally, the pipe string is cut to the desired pipe length by a 'flying saw' (Figure 3-19, step 14).

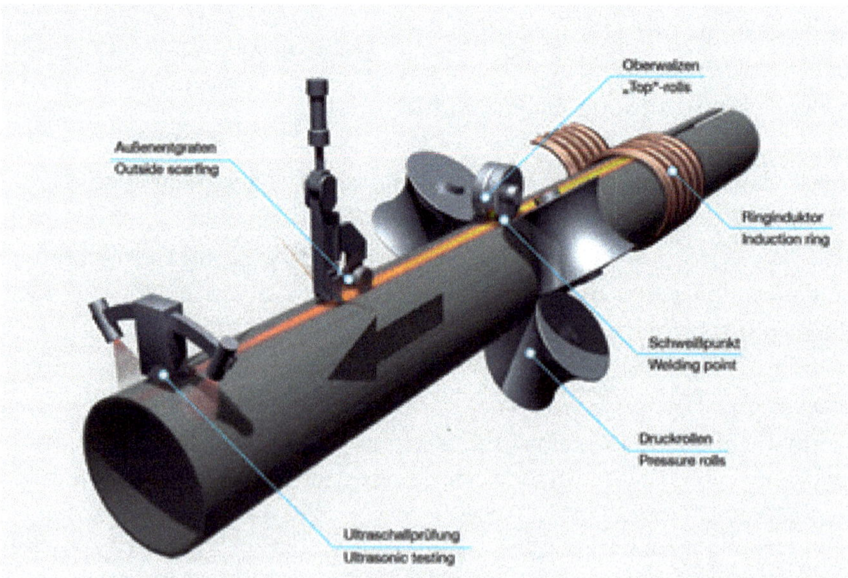

Figure 3-20. HFI welding – sketch[40] (courtesy of Mannesmann Line Pipe).

Quality control. As in the case of the previously described pipe manufacturing routes, sampling for destructive testing (Figure 3-19, step 16) is placed before hydrotesting.

Ultrasonic inspections of the strip and the weld are carried out before forming and after flash removal (Figure 3-19, step 5). Further NDT inspections are placed after the sizing (Figure 3-19, 13) and hydrotesting (Figure 3-19, step 20).

After the pipes have successfully passed through all the test and inspection stages, including the dimensions check, they are ready for coating or delivery.

3.3.5 Seamless pipes

Seamless steel line pipe production is a hot forming process (Figure 3-21), consisting basically of three different forming steps:

1. piercing of the heated billets;
2. elongating/expanding of the pierced hollows by different rolling or forging processes;
3. finish-rolling of the shells by sizing or stretch-reducing.

The pipe rolling processes are named after the respective elongating/expanding process. These processes allow pipe production in diameter ranges from about 20 to 710 mm.[24, 44, 45]

Grades. For production level PSL2, seamless pipes are produced in a wide range of grades from L245/B to L690/X100. Delivery conditions are as-rolled, normalizing formed, normalized, quenched and tempered.

Figure 3-21. Seamless pipe manufacturing[44] (courtesy of Vallourec).

First, second and third-parties...

Why do we usually call mechanical damage 'third-party damage'? It has its origins in contract law/insurance. When you want someone to work for you, you award a contract: you are the 'first party' (the client), and the contract is completed by the 'second party' (the contractor). These are the two parties to the contract. Any other individual or organisation, not bound by the contract, is a 'third-party'.

Fields of application. Seamless pipes are used in onshore and offshore applications; e.g., as flow lines and risers. In most cases, high collapse and corrosion resistance properties are needed and provided by seamless pipes and tubes. By suitable chemical compositions, adapted heat treatments and optimized dimensions, the demanding design requirements can be met. Offshore pipe laying is done using S, J, and reeling laying methods. Connections are made by welding, but also by specially designed threaded connections.

Manufacturing - cross-roll piercing. The cross-roll piercing process is the leading piercing method for seamless tubes. It was patented by the Mannesmann brothers in 1885[46] and is still the basis for major hot rolling processes. Two specially contoured work rolls – driven in the same direction of rotation and placed at an angle of 6° to 12° to the billet axis – thrust the round billet into the process area and form a hollow, by a spiral motion over a piercer plug. Different variants of the cross-roll piercing process were developed. They can be distinguished by the number of the rolls (2 or 3) and the shape of the rolls (cone- type piercer or barrel-type piercer). Furthermore, for the 2-roll piercer, additional guide shoes or rotating guide discs are used. Different dimensions of the pierced hollow can be realised, depending on the size of the rolls and the billets.

Manufacturing - elongating/expanding of the pierced hollows. For the elongation/expanding of the pierced hollows different rolling processes are used. In the following the continuous mandrel rolling, plug mill and the Pilger mill processes are introduced briefly. More detailed information about all the currently applied rolling processes for line pipe and the complex technical background can be found in the literature (e.g., references [24, 45, 49].)

Continuous mandrel rolling. The continuous rolling process using mandrel mills has been successfully used in producing seamless steel tubes since the 1970s. The elongation of the hollow takes place in up to 9 successively arranged roll stands with 2 (conventional layout and 'MPM') or 3 rolls (recent layout) and a mandrel bar as the internal tool. 3 different variants can be distinguished:

- conventional mandrel mills;
- multi-stand pipe mills (MPM),
- state-of-the-art 3-roll Premium Quality Finishing mill, PQF[33], or Fine Quality Mill, FQM.[47]

Most of the seamless pipes are manufactured by the continuous mandrel rolling process.

Plug mill. Figure 3-22 gives an overview of the plug mill production flow. After heating in a preheating and rotary-hearth furnace up to about 1280°C, the billets are pierced in a cross-roll piercing mill, as described above, resulting in thin-walled hollow shells.

The plug rolling mill uses a longitudinal rolling process with two sets of rolls. The first set consist of the working rolls which push the hollow over the plug, causing a reduction of outside diameter and wall thickness. The second set of rolls are stripper rolls which transport the pipe back to the starting position after the first plug rolling. A second plug roll pass starts with a somewhat larger plug. The reeling mill expands the pipe slightly and optimizes its roundness. Final production stages after reheating are sizing and straightening. Required heat treatment will follow.

> **Starting material for seamless pipe production**
>
> Steel works provide either ingots or billets as starting material for seamless pipe forming.
>
> **Ingots** are produced by casting liquid steel in a cast iron mould.
>
> **Billets** are created by continuous casting, preferably into a round cross-section.

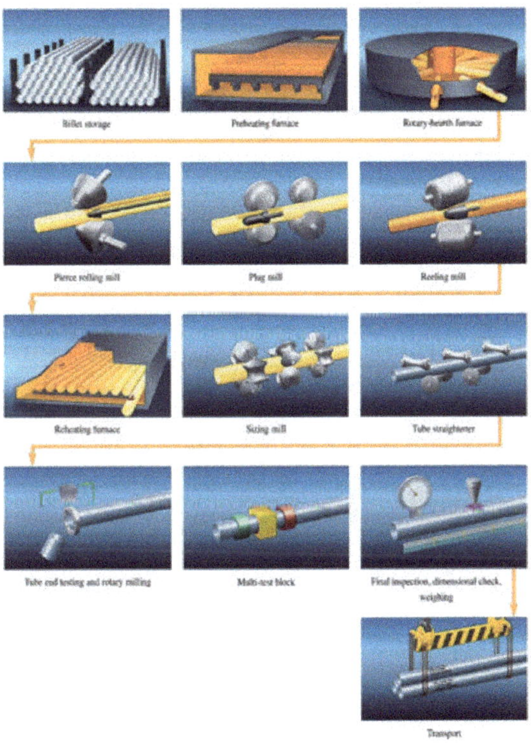

Figure 3-22. Plant layout and process of Vallourec's plug mill in Düsseldorf, Germany[44] (courtesy of Vallourec).

Pilger mill. Figure 3-23 gives an overview of the Pilger mill production flow. After heating in a rotary-hearth furnace up to a maximum of 1300°C, the ingots are pierced in a piercing press, followed by a cross-roll piercing mill as described above, resulting in thick-walled hollow shells.

The pilger stand has two specially formed rolls, rotating against the feed of the thick-wall hollow. The hollow shell is pushed over a mandrel with a diameter close to the desired final internal pipe diameter. The rolls forge the shell stepwise with a recurring backward and forward motion while the hollow is rotated by 90° with each step. This forming results in a uniform wall thickness and excellent concentricity.

This is where the term 'Pilger' originates: it is a sequence of repeated backward and forward steps is a procedure reminiscent of the Echternach (city in Luxemburg) Dancing Procession in which pilgrims (or 'Pilger' in German) proceed in the same rhythm – two steps forward and one back.

Final production stages after reheating are sizing and straightening. The required heat treatment will follow.

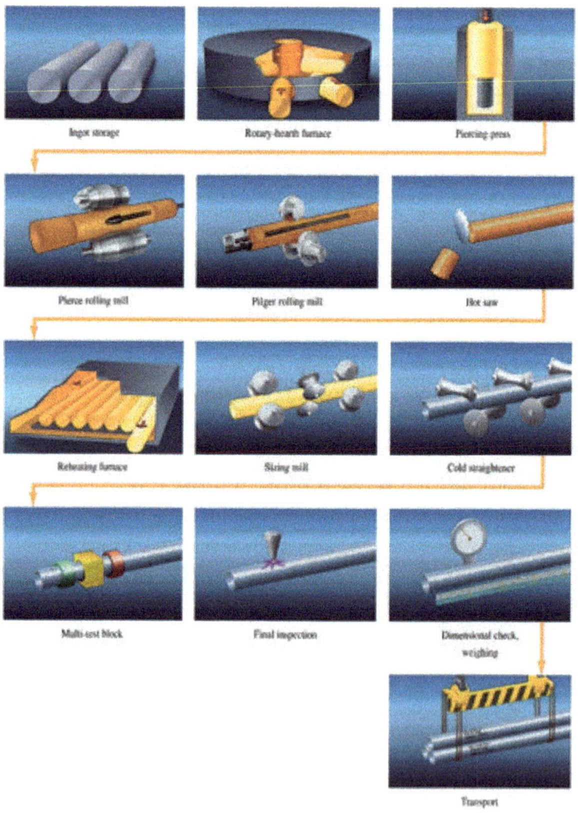

Figure 3-23. Plant layout and process of Vallourec's Pilger mill in Düsseldorf, Germany[44] (courtesy of Vallourec).

Forging. Alternatively, hot forging is applied e.g. by 4-hammer radial forging machines[48], providing diameters between up to 720 mm.

Dimension. For conventional mandrel mills and multi-stand pipe mills the diameter ranges up to 425 mm, whereas up to 508 mm can be realised with the PQF or FQM mills.

The continuous mandrel rolling process provides pipes with an outside diameter up to 180 mm.

The plug mill provides pipes with diameters of about 190 to 410 mm, wall thicknesses of about 6 to 40 mm, and pipe lengths in the order of 16 m.

Pilger mill pipes can be produced with diameters between about 240 and 710 mm, wall thicknesses between 10 and 110 mm, and pipe lengths up to 28 m.

Pipe manufacturing by hot forging provides diameters up to 720 mm.

Heat treatment. Normalizing and quench and tempering is carried out to adjust adequate mechanical properties after the pipe rolling.

Quality control. Extensive quality testing is carried out in the course of seamless pipe production, as in the case of the manufacturing routes for welded line pipe. After completion of the rolling process, the wall thickness is determined by an ultrasonic measurement, and the pipe surface is inspected for surface defects by the flux leakage method (multi-test block in Figures 3-22 and 3-23). Sampling for mechanical and corrosive tests takes place, as well as hydrotesting as specified.

After the pipes have successfully passed through all the test and inspection stages, including the dimensions check, they are ready for any coating or delivery.

3.3.6 Protection of line pipe by coating

To ensure a long service life of a steel pipeline, protection against corrosion and mechanical damage is necessary. This is done by the application of various types of coatings on the outer surface of the pipes. This 'passive' corrosion protection is complemented in the field by the more 'active' cathodic protection of the pipeline.

Also, internal coating or linings are applied, particularly on pipes for gas transportation. The surface roughness of the internal surface of the pipe will resist flow, and a coating will improve product transportation efficiency. Liquid epoxy is sprayed, in thicknesses up to 0.5 mm, onto the inner pipe surface.

The main types of coating are now briefly introduced. More details can be found in company brochures.[50 - 52]

3-layer Polyolefin. A commonly used coating system is the 3-layer coating. Figure 3-24a shows the structure of the three layers. The first layer is a fusion epoxy primer

(FBE) which is applied electrostatically as a powder. The second layer is a bonding layer, while the third layer is either Polyethylene (PE) or Polypropylene (PP). These layers can either be extruded or applied as foil. Usually, and particularly in the case of expected high strains and stresses, PE is used, while in the case of higher service temperatures, PP is applied.

Fusion bonded epoxy coating. Another option is the application of a Mono- or Dual FBE (Fusion Bonded Epoxy) coating, as given in Figure 3-24b.

Cement and rough coating. A final cement mortar may be applied to the coated pipes, especially for offshore/subsea applications. As shown in Figure 3-25a, a specially ribbed layer can be attached to the 3-layer coating and provides an optimal bonding with the cement layer. Cement mortar layers protect offshore pipes and make them heavy to reduce buoyancy. Rough coatings (Figure 3-25b) provide mechanical protection; e.g., during trenchless pipe laying.

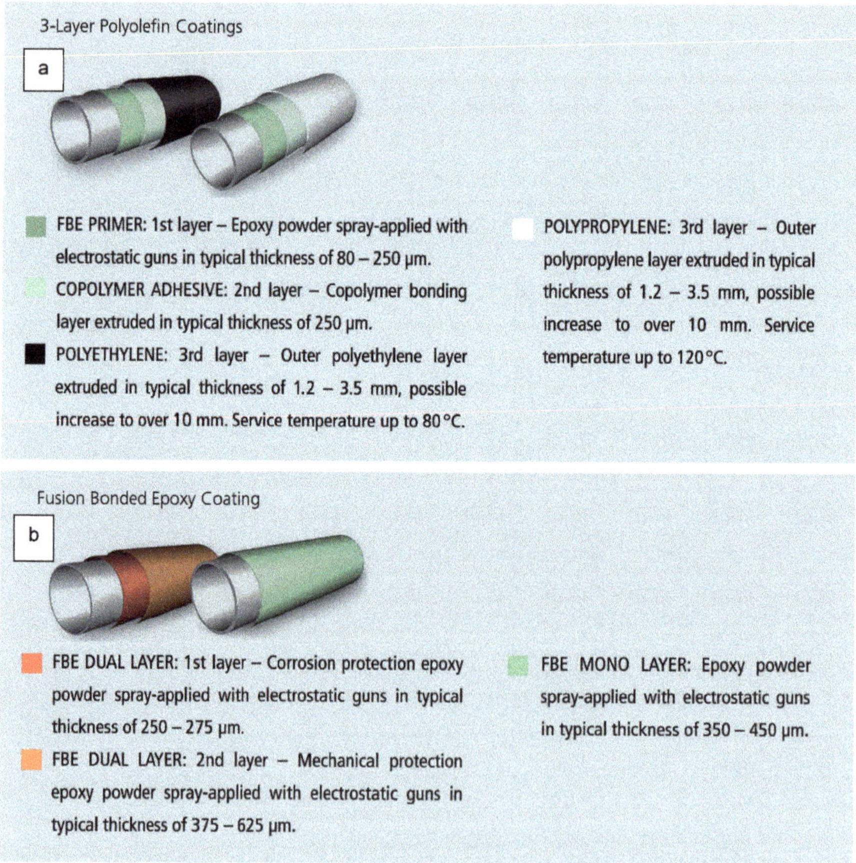

Figure 3-24. Line pipe coating systems[50] (courtesy of Europipe).

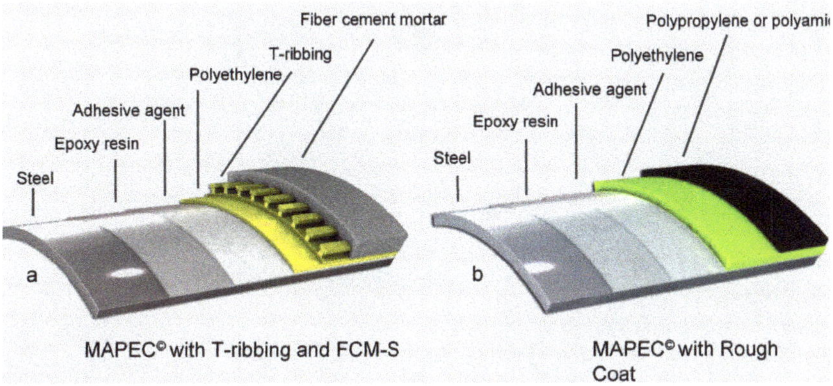

Figure 3-25. Cement mortar and rough coatings[51] (courtesy of Mannesmann Line Pipe).

3.4 Line pipe properties

The demand for natural gas has steadily increased in recent decades. As a result, the gas transmission networks have been greatly expanded worldwide. This was accompanied by continuous development and improvement of pipe properties and pipe quality.

3.4.1 Development and achievements

The focus of the material development of line pipe steels has been on increasing the strength level for a long time. This has been achieved by the development of new steels by improved steelmaking and rolling processes, with new alloying concepts and particularly with thermo-mechanically controlled (TMCP) processing for plate and strip production.

High-strength pipelines from API grade X60 to X70 were built in the 1960s and 1970s. After 2000, higher-strength X80 steels became available, and pipes up to X120 were developed.[53, 57] Figure 3-26 gives an overview of the steps in the development of pipe steels from 1965 onwards, indicating the alloying and rolling concepts used.

The driving force behind the increase in strength was the targeted reduction in project costs for the construction of new pipelines. With increasing strength, a higher operating pressure can be selected with the same pipe dimension. On the other hand, at the same operating pressure, smaller diameters or wall thicknesses are possible. This lowers the project costs by reducing the amount of steel required, and also requires less effort during transport and pipe laying. In the case of diameter and wall thickness reduction, significant cost savings in field welding can also be achieved.

Figure 3-27 shows how increasing the strength levels from X70 to X120 reduces the wall thickness, and thereby the weight, of the pipes for a given pipe diameter.[57]

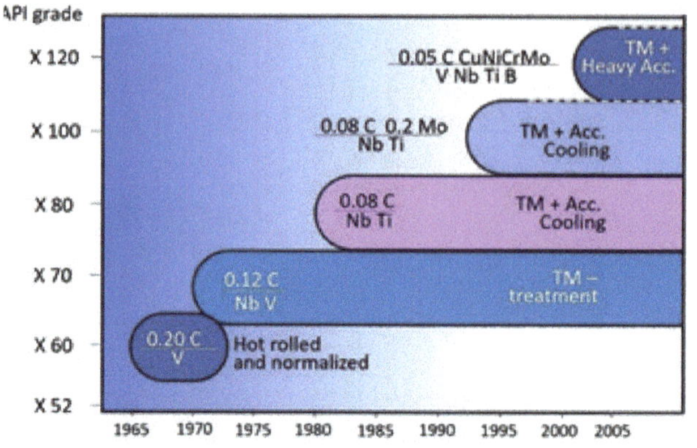

Figure 3-26. Development of HSLA pipeline steels – alloying concepts and rolling technique [after 53] (courtesy of Europipe).

However, with increasing strength, other material properties are affected, such as fracture resistance, ductility, and field weldability. This means additional effort is required in terms of steel composition and processing to provide the best balance of material properties. Up to now, only grades up to X 80 have been shown to be manufactured economically on a large scale, and used for major pipeline projects. In addition to the emphasis on increasing strength, other requirements have influenced and shaped pipe development and production, not only for onshore pipelines but also for offshore applications. Current and future challenges for pipe development and production are:

- pipes and pipelines for areas with ground movement:
- arctic and deep-sea applications;
- sour gas transportation; and,
- transportation of hydrogen and other fluids.[54]

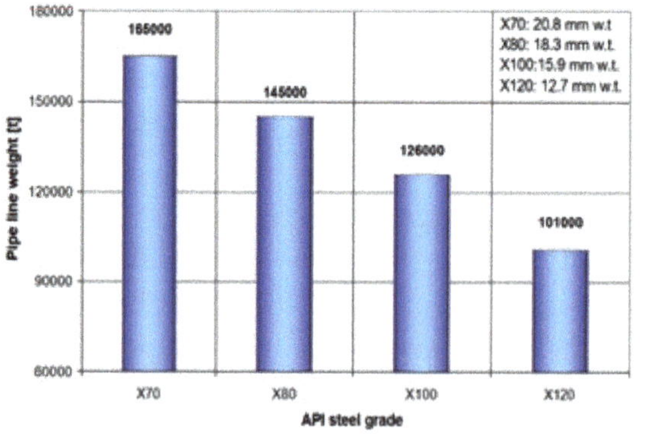

Figure 3-27. Material savings due to the use of high strength steel for a given pipe diameter[57] (courtesy of Europipe).

3.4.2 Determination of line pipe properties

The determination of pipe properties is an essential part of the pipe development and pipe manufacturing. This includes the investigation of the material composition and structure, the determination of mechanical-technological properties, as well as corrosion resistance and non-destructive tests. A number of investigation methods have especially been developed to check and ensure the serviceability of the pipes and pipelines.

Mechanical tests within the frame of pipe manufacturing are typically standard tests on small scale. In addition to testing for strength and toughness (see Sections 3.4.3 and 3.4.6), other tests must also be performed. ISO3183/API5L specifies in detail the applicability and required test results from bend tests[58], flattening tests[59] and guided bend tests.[60]

An exception to the testing of small samples during production is the hydrotest, where the entire pipe is tested in the production flow. More sophisticated laboratory tests on small samples, like fracture mechanics tests, and full-scale tests, are mostly related to research and development, or to pre-production qualification tests.

In the following, tests to determine strength, strain, and toughness, as well as corrosive resistance will be presented.

3.4.3 Determination of strength properties of pipes (tensile test, tube ring hydraulic pressure test, pipe burst test)

The design of a pipeline depends in most cases on the desired operating pressure, resulting in the minimum specified yield strength (SMYS) of the material for a given diameter and wall thickness.

Tensile tests. The yield strength of a material ($R_{p0.2}$, $R_{t0.5}$) as well as tensile strength (R_m) and elongation values are taken from tensile tests, as specified in international standards such as ISO 6892[61] and ASTM A370[62], see Figure 3-28 and the blue box.

- Yield strength values
  An example of a stress-strain curve of a high strength line pipe steel is shown in Figure 3-28. The yield strength, determined on the basis of a 0.2% plastic extension ($R_{p0.2}$) and a 0.5% total extension ($R_{t0.5}$) are often close to each other, but may also differ significantly, depending on the general strength level and shape of the stress-strain curve. In line pipe standards, mostly the total yield strength is used to characterize the yield point of the material.
- Tests in transversal direction
  For the determination of tolerable internal pressures in pipes and pipelines, the tensile properties in the circumferential direction of the pipe are crucial.

$R_{t0.5}$ or $R_{p0.2}$

Yield strength is a material property which indicates the onset of plastic deformation under load.
API materials standards typically use the total yield strength $R_{t0.5}$ (stress value on the stress-strain curve for a total strain/extension of 0.5%). In EN and ISO material standards the plastic yield strength $R_{p0.2}$ (stress value on the stress-strain curve for a plastic strain/extension of 0.2%) is typically used.

$R_{t0.5}$ is not a good measure of yield strength of steel above yield levels around 1000 MPa: in these cases, the yield point would be located on the elastic part of the stress strain curve.

ISO3183/API5L generally require $R_{t0.5}$ values, but above X90 the yield strength at plastic strain ($R_{p0.2}$) has to be applied.

According to ISO 3183/API5L, rectangular samples representing the full wall thickness of the tubes must be taken from the circumferential direction (Figure 3-29a-blue) followed by flattening (Figure 3-29a-yellow). For hot rolled seamless tubes, round bar specimens as shown in green (Figure 3-29a) should be used. For special applications and pipeline designs longitudinal values are required additionally.

The stress-strain curves from unflattened and flattened specimens on the same material are shown in Figure 3-29. The curves from tests on unflattened round bar and flattened strip specimens show a different shape, especially in the yield area. The flattened strip curve exhibits a "round house" shape, while the round bar curve exhibits a pronounced yield point extension, often called 'Lüders band'[63] or 'Lüders plateau'. The "round house" curve of the flattened strip specimen is caused by the cold deformation during flattening.

The flattened strip specimen has nevertheless prevailed as a standard sample, not least because of the simple sampling, but also because the measured strength values reasonably describe the strength of the tube, as investigations have shown e.g. reference [64].

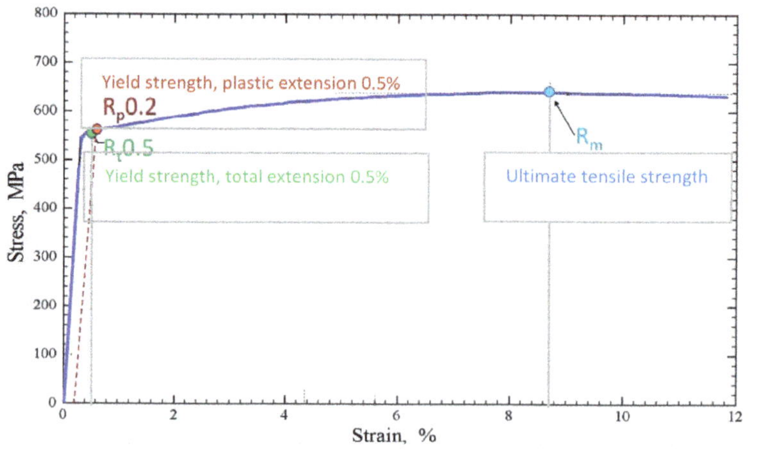

Figure 3-28. Stress-strain curve [after 63] (courtesy Salzgitter Mannesmann Forschung).

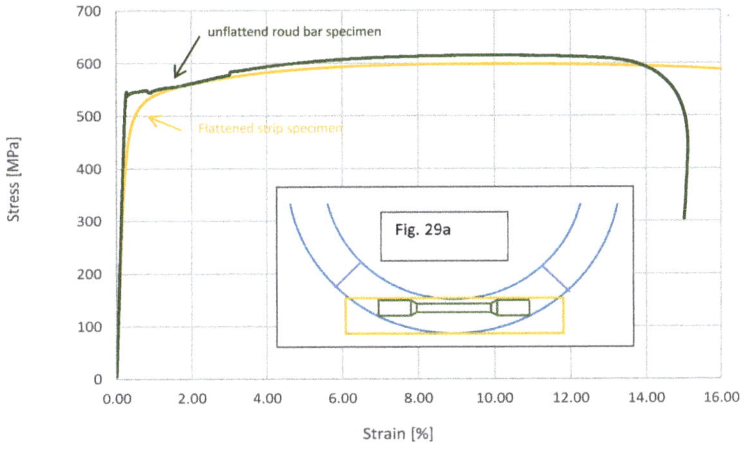

Figure 3-29. Stress-strain curves from tensile test on transverse specimens from X80 (L4854MB) material, OD 1422 mm x WT 22.3 mm [after 65] (courtesy Europipe).

- Tensile test on longitudinal or spiral welds
  Tensile tests on longitudinal or spiral welds are specified in line pipe standards to characterize the strength of the weld metal in relation to the base material. Cross weld tensile tests are required on flattened strips. Only the R_m is specified and the specified minimum tensile strength for the pipe body has to be achieved also for the weld.

Tensile test results of a production lot. In the context of ordering pipes for a pipeline project, agreements are made on pipe properties and test frequencies. The characteristic values to be achieved either correspond to specifications from standards, e.g. ISO3183 or API5L, or may be adapted to the specific requirements of the pipeline project, often based on company specifications. An example for the distribution of strength values in a pipe production lot is given in Figure 3-30. The number of samples are to be tested per production unit, has to be agreed between the purchaser and the pipe mill. The determination of the mechanical properties, in particular the yield strength and tensile strength, is usually carried out with a few or even only one sample per batch or melt.

Thus, the strength values are typically determined per melt on every tenth to fiftieth pipe, mainly depending on pipe sizes, while the wall thickness and pipe diameter are typically determined at the factory on each individual pipe using automatic non-destructive testing techniques.

Although the strength values are determined on only a portion of the pipes, there is a statistically well-conserved database for a complete delivery; e.g., to demonstrate compliance with the required minimum values of yield and tensile strength.

Tube ring hydraulic pressure tests. As described above, only a part of the wall is tested when round bar specimens are used for the tensile test. In the case of flattened strips, the entire wall thickness is tested; however, plastic deformation is introduced into the sample which will influence the resulting tensile properties. (Figure 3-29a). As an alternative, the performance of tube ring hydraulic pressure tests[66] can be considered.

For more information see references [67, 68] and scan QR 3-7.

QR Code 3-7. Scan here for more about tube ring hydraulic pressure tests

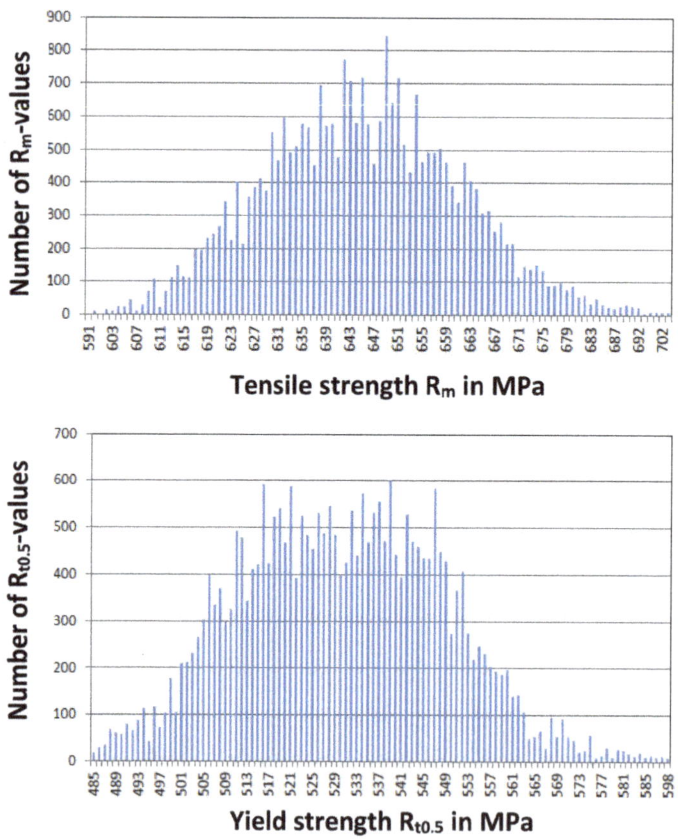

Figure 3-30. Distribution of tensile and yield strength for a pipe production lot, L485MB[65] (courtesy Europipe)

Internal pressure burst tests. Internal pressure burst tests can be part of the investigations or the pre-production testing program for:

- new line pipe developments;
- large pipeline projects; or,
- projects with new or sophisticated pipeline designs.

Deformation properties and failure behaviour of the pipe body, and the integrity of the weld seam can be examined.

In such a burst test a piece of pipe, closed with end caps, is subjected to bursting by a steadily rising internal pressure (Figure 3-31). The result of such a burst test is shown schematically in Figure 3-32, giving the location of the fracture (Figure 3-32a) and the permanent circumferential expansion along the pipe length (Figure 3-32b). The plastic deformations, and a derived mean value of deformation are given.

Burst tests allow comparisons with strength measurements from small-scale tensile tests. The question of whether the strength values determined by tensile tests adequately represent the structural behaviour of a pipe and the influence of the type of tensile test samples has been investigated by the European Pipeline Research Group (EPRG).[64, 69]

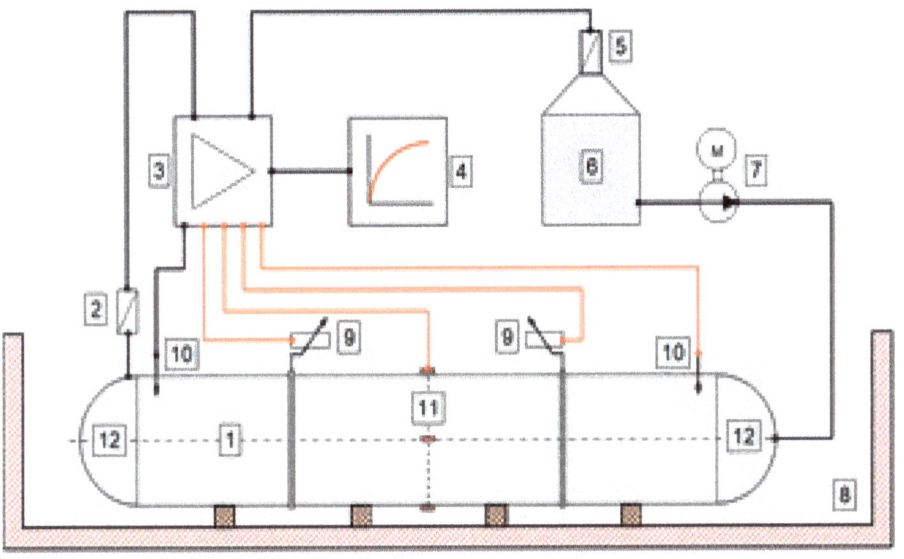

1 Test pipe	4 Signal output	7 Hydraulic pump	10 Thermocouple
2 Pressure transducer	5 Load cell	8 Burst pit	11 Strain gauge
3 Signal conditioning	6 Water reservoir	9 Chain	12 Semi spheircal end cap

Figure 3-31. Burst test set-up[68] (courtesy Salzgitter Mannesmann Forschung).

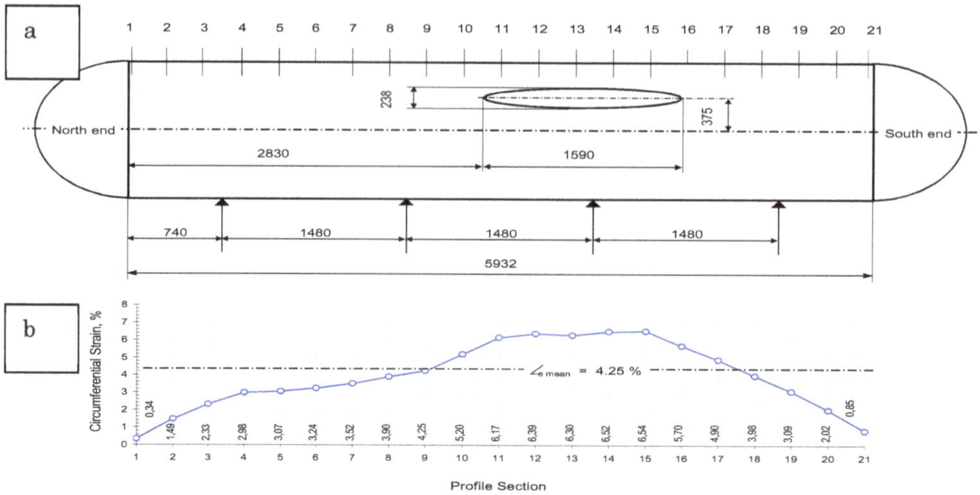

Figure 3-32. Sketch of a line pipe section after burst test with fracture location (a) and measured plastic circumferential strain (b)[68] (courtesy Salzgitter Mannesmann Forschung).

Figure 3-33 provides the comparison between stress-strain curves from a burst test, and a transverse round bar tensile test on a UOE pipe. The $R_{t0.5}$ value measured on the round bar tensile specimen lies typically between the elastic limit ($\sigma_{\varpi}E$) and the stress at 0.5 % total strain ($\sigma_{\varpi}\ \tau 0.5$) measured on the pipe. The ultimate stresses measured in the tensile and burst test lie close to one another, while the uniform elongation value measured on the pipe is generally smaller.

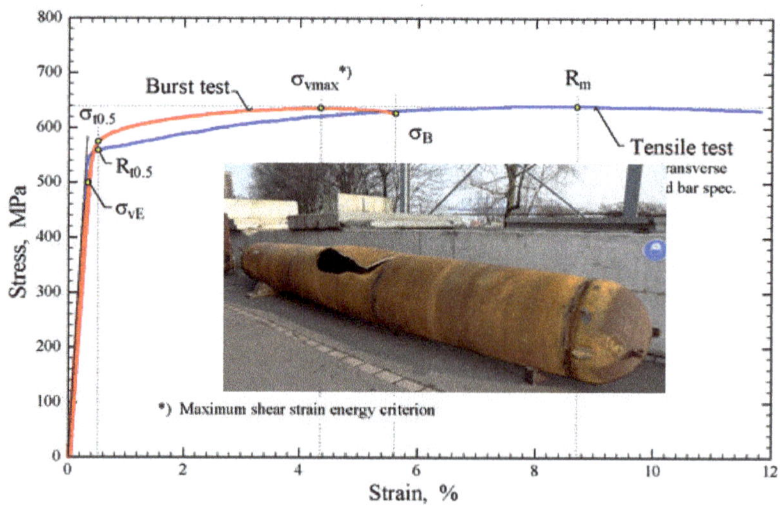

Figure 3-33. Stress-strain curves from a burst and tensile test[63] (courtesy Salzgitter Mannesmann Forschung).

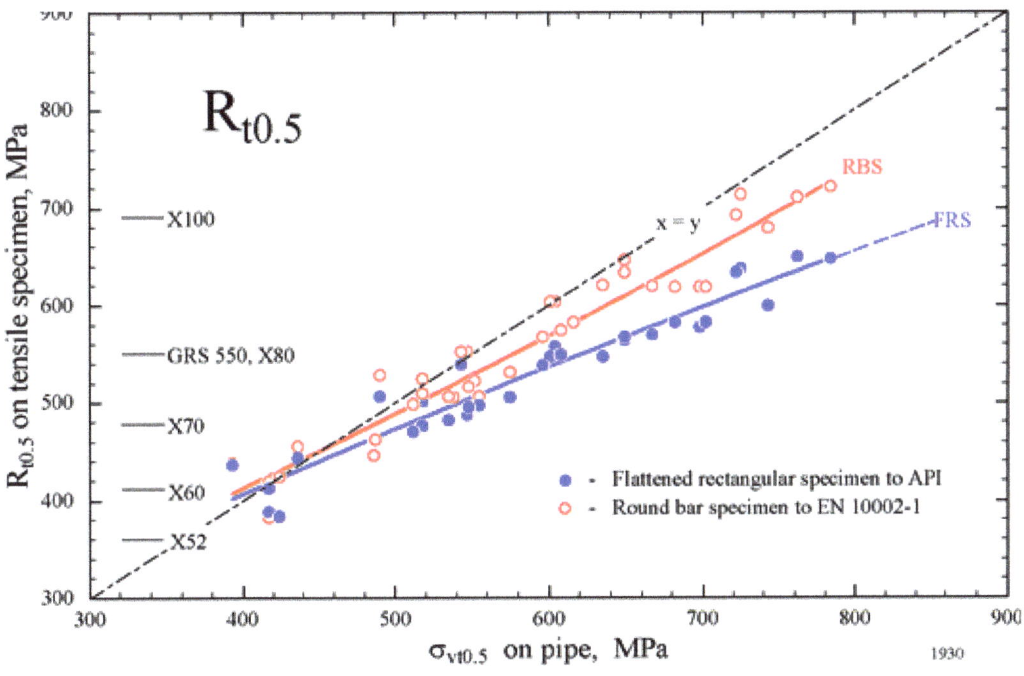

Figure 3-34. Comparison of burst and tensile test results[63] (courtesy Salzgitter Mannesmann Forschung).

The comparison of yield values from burst and tensile tests on UOE and HFW pipes is shown in Figure 3-34. Up to the specified minimum yield strength of grade X 70, the two scatter bands overlap, the divergence increases for higher strength levels. Similar results have been obtained in comparisons of the yield strength values from tensile tests and tube ring hydraulic pressure tests.

'Ihese results are in line with, and support, the current practice of giving general preference to the flattened rectangular specimen in acceptance testing. For high-strength tube grades from X80 (tensile strengths> 490 MPa), testing of round bar samples is proposed.[63]

3.4.4 Mill hydraulic pressure test

According to ISO3183/API5L a hydrostatic pressure test (hydrotest) has to be performed on each pipe during production. As indicated in Figures 3-7, 3-17 and 3-19 these tests are a general part of the production flow in modern pipe manufacturing. Mill hydrostatic pressure tests are the only tests in which, prior to the pipeline construction, the pipes are tested as a whole for their pressure resistance and leak tightness. In the past, this test was also important to eliminate any mix-up with lower grade pipe materials. Nowadays, this purpose is of less importance because of improved quality checks and traceability during production.

Detailed instructions on how to carry out the test are given in ASTM 530.[70] According to this standard, minimum internal pressures have to be used, which produces in the pipe wall a minimum of 60% of the specified minimum yield strength (SMYS) for ferritic steels. The pressure has to be held for a minimum of 5 seconds. In practice, often the holding time is set to 10 seconds and the pressure level is significantly higher at 90% SMYS. In special cases the tests are carried out up to internal pressures of 95% or 100% SMYS, using the nominal wall thickness to calculate the test pressure. There are from time to time considerations to extend the holding time (e.g. to 20 or 30 seconds). There is no evidence available that such longer period results in better detection of defects. On the other side, an extension to 20 sec would make the hydrotest the bottleneck of the production.

3.4.5 Testing to simulate large strains

Pipelines can be routed through areas where there is a risk of ground movements or even earthquakes. For these areas, high demands are placed on the elongation properties of line pipes, in particular in the longitudinal direction. In small-scale tensile testing uniform elongation and elongation at fracture are determined to characterize these properties.

To better understand the failure of pipes and pipelines under these extreme loads, bending tests can be performed on pipe sections, with an attention to the first appearance of bulging in the pipe. Figure 3-35 shows such a test system, as installed at Salzgitter Mannesmann Forschung GmbH.[67] Line pipe of grades up to X100, diameters up to 1422 mm and a wall thickness up to 30 mm can be deformed. Additionally, internal pressure can be applied. Figure 3-35b shows a pipe section with a bulge after such a test. The determined deformation limits are compared with the results of a finite element analysis, and are intended to improve the design of pipelines through challenging areas and the development of pipes with higher elongation limits.

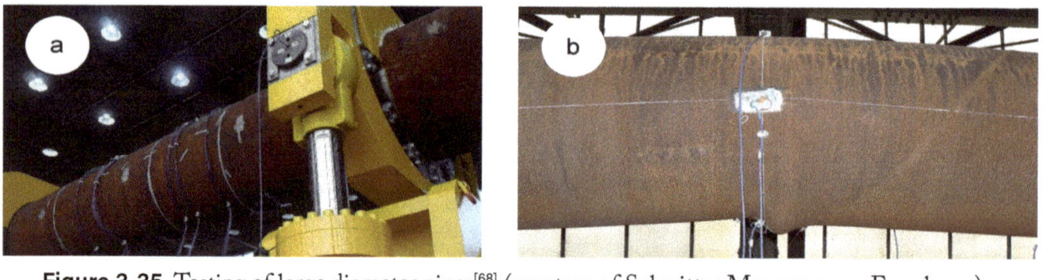

Figure 3-35. Testing of large diameter pipes[68] (courtesy of Salzgitter Mannesmann Forschung): .

a: SZMF test rig for bending large-diameter line pipe (maximum bending force 10 MN, maximum internal pressure 80 bar);

b: pipe with wrinkle after testing

3.4.6 Toughness – determination of properties by small-scale testing

The toughness of a material indicates the ability to resist fracture in terms of crack initiation, crack growth, and the failure of a structure caused by cracks. For steel production, in general, and steel pipe production, in particular, the Charpy-V-notch (CVN) test is the most popular. Line pipe standards request the CVN testing during pipe production.

K_{1c}, CTOD-, J- testing. Fracture mechanics, such as K1c, CTOD, J, and fatigue crack growth tests, are primarily part of pre-production qualification tests or part of development or failure studies. The measured toughness values are needed to calculate allowable defect sizes in structures to prohibit crack extension or failure. Detected or assumed defect sizes in structures can be assessed by applying engineering critical assessment (ECA) methods (e.g. BS 7910[71]). CTOD (crack tip opening displacement) testing is specified in DNVGL-ST F101[17], requiring a minimum of 0.15 mm for the toughness of the weld seam at minimum design temperature. Additionally, ECA procedures are introduced and based on J-R – curves. Detailed information can be taken from[17] and the respective fracture mechanics test standards.[71 - 77]

Different strikers

According to ISO 148, different radii (2mm or 8 mm) for strikers can be used for Charpy testing.

Investigations on the influence of the different strikers, in some cases show no significant difference in the Charpy values [80], but also can show small or larger differences in the values and transition behaviour [81 - 82].

Tests using the 8 mm striker tend to break the specimens up to high toughness values, in contrast to 2 mm strikers.

CVN (Charpy-V-notch) testing. By 1905, George Charpy (1865 – 1945) had proposed a machine design that is remarkably similar to present CVN testing machine designs. Literature containing the first references to "the Charpy test" and "the Charpy method" is given in [78]. Charpy test procedures are described e.g. in ASTM A370[62] and ISO 148.[79]

In the product standards ISO3183/API5L, Charpy values are specified to guarantee sufficient fracture initiation resistance of the pipe material. Depending on the test temperature, the Charpy value will indicate ductile or brittle fracture. Three tests have to be carried out at 0°C on full size test Charpy pieces (10 x 10 x 55 mm) and have to reach minimum average values depending on pipe grade and diameter.

Instrumented DWT testing

The application of strain gages on the striker of a DWT test equipment (force measurement) and measuring the speed of the falling weight, allows the 'instrumented' DWT tests.

Using the resulting force-time diagram it is possible to distinguish between crack initiation and crack propagation and to compare these values with observations from tests on pipes and pipelines.

For welds and the heat affected zones (HAZ), sufficient Charpy toughness must be proven. At 0°C a minimum value of 40 J has to be achieved.

Modern high grade line pipe steels often exhibits Charpy values much higher than 300 J. Friction between specimens and the anvils in the testing machine may cause considerable scatter of the test results and the significance of these test results is called into question.

Background and CVN requirements for the resistance to ductile fracture propagation in gas transmission lines is presented in Section 3.4.7.

In this context, also the significance of high Charpy values will be discussed.

Drop weight tear testing (DWTT). The avoidance of brittle fracture in gas transmission pipelines has to be proven by DWT tests (ISO3183/API5L), and was established in the second half of the sixties based on research work carried out by Battelle.[83] It has been found that the DWT test is an appropriate measure to predict the ductile to brittle transition temperature of a pipe and became a major parameter to ensure crack arrest in gas pipelines.[84]

The DWT specimen is a full wall specimen with a height of 76 mm and width of 305 mm. A comparison of the sizes of a DWT and Charpy specimen can be seen in Figure 3-36. Test procedure and conditions are described in e.g. ISO 10274[85], ASTM E436[86] and API RP 5L3.[87] To simulate a crack, developing in the longitudinal direction of a pipe, samples are taken either from the plate/strip material or from the tube in the transverse direction. Sampling in the transverse/circumferential direction of a pipe section requires flattening of the samples prior to testing (Figure 3-37). It is recommended not to flatten the specimen totally but carry out a so-called 'gull-wing' flattening and avoid plastic deformation in the mid-section of the sample (Figure 3-37b, hatched area). Markings on the centerline of the specimen's cross-section before flattening (Figure 3-37b) should appear in a straight line after flattening (Figure 3-37c) to allow proper testing without tilting the specimen. A set of two specimens is tested per batch and has to show a minimum of 85% shear area to ensure ductile behaviour of the pipe. Figure 3-38 shows fracture surfaces from broken DWT specimens with brittle (a) and ductile (b) behaviour. Specimen (c) shows ductile behaviour, including a separation in the centre line. Controlled rolled steels may have splits (separations) parallel to the specimen surface with cleavage fracture. According to the standards, these areas should not be included in the shear area determination.

With increasing wall thicknesses, high drop weight forces are required to test full wall thickness DWT samples, which in many cases overwhelm existing test equipment. In these cases, wall thickness machining down to 19 mm is permitted.

Instrumented DWT tests have been increasingly used and discussed in recent years in order to better understand the cracking behaviour, especially in the case of high-strength pipe grades e.g. and crack arrest predictions.[84, 88]

Figure 3-36. DWTT and Charpy specimen (courtesy of Salzgitter Mannesmann Forschung).

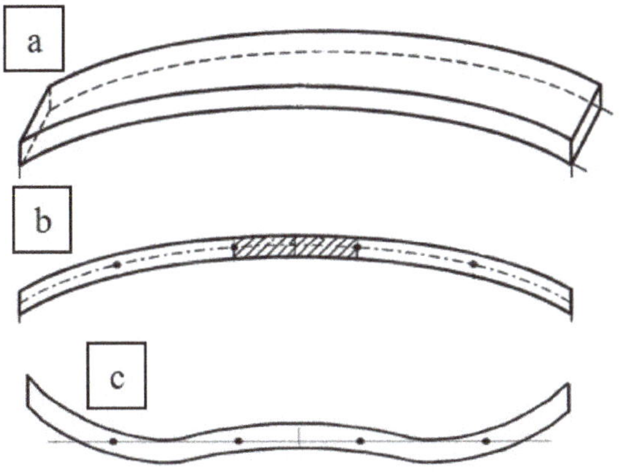

Figure 3-37. Flattening of a DWTT specimen (courtesy of Salzgitter Mannesmann Forschung).

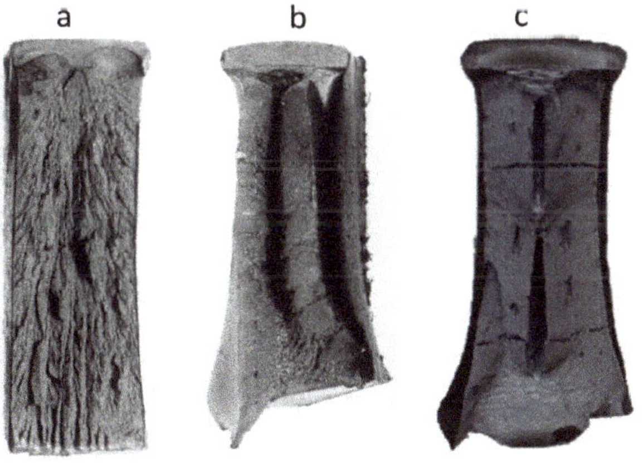

Figure 3-38. Broken DWTT specimens: (a) brittle, (b) ductile (c) including separations (courtesy of Salzgitter).

3.4.7 Propagating ductile fractures in gas pipelines

Propagating ductile fractures in high pressure gas pipelines were first reported in the 1960's in the US[89 - 90], as gas pipelines ruptured over several kilometres in length.

These ruptures can be initiated by defects in the pipe; for example, external damage of pipes (caused by an excavator, ground movement, etc.). When a critical crack length is reached, crack propagation along the pipeline can occur in both directions from the starting, at crack speeds greater than 100 m/s.

For more details about ductile fracture propagation in gas pipelines and how to predict crack arrest see references [69, 81, 91 - 100] and scan the QR code.

QR Code 3-8. Scan here for more about ductile fracture propagation

3.4.8 Offshore applications

Pipes for offshore use have to meet special requirements. Typical pipe types for offshore pipelines are SMLS, HFW, TRB, UOE, JCOE and UOC pipe. The most used standard for offshore pipeline projects is the DNVGL ST F101.[16] A key issue for pipes used for offshore pipe laying are the limitations for the out-of-roundness. With regard to collapse resistance, the pipe body has to show less than 1% out-of-roundness. For the pipe ends even more stringent values are required to support fast girth welding during pipe laying.

For more details about collapse resistance, the calculation of the collapse pressure, compression and collapse testing and the effect of heat treatment on the collapse resistance of pipes see references [16, 101 - 106] and scan the QR code.

QR Code 3-9. Scan here for more about offshore applications

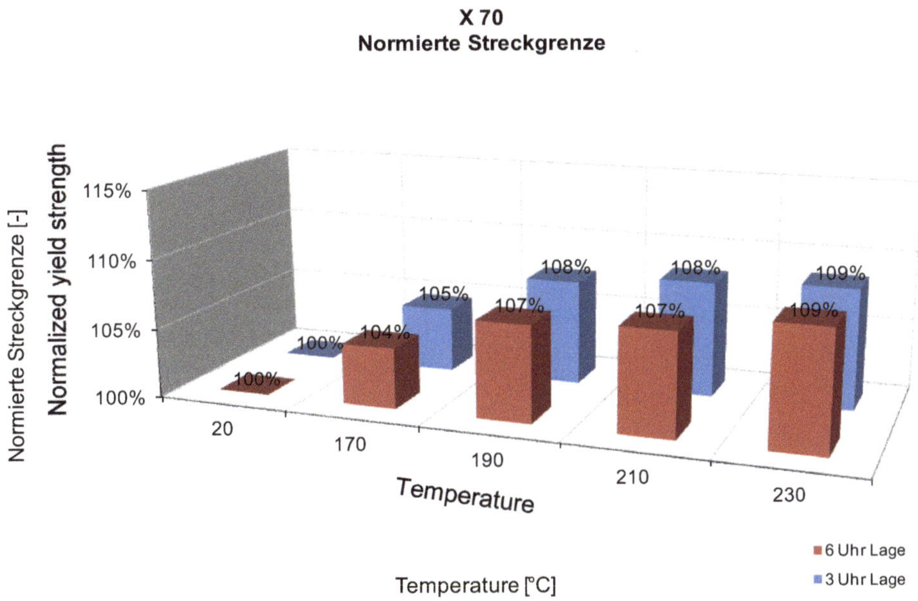

Figure 3-39. Change in yield strength as a function of the coating temperature for Grade X70[63] (courtesy of Europipe).

3.4.9 Property changes due to the coating process

As described in the QR document "Collapse resistance", the heat treatment occurring during the coating can be used to produce improved collapse resistant pipes. The extent of this effect on the compressive yield strength of a pipe was shown to be dependent on grade and the amount of cold deformation as a final production step.

Investigations on UOE pipe[24, 65, 107] have shown that heat treatment also affects the yield strength, but does not significantly affect CVN and DWT test results. The yield strength increases depending on coating temperature, pipe manufacturing conditions and material structure, between approximately 8 and 14% as a result of material aging, Figure 3-39. This means that the yield strength of the delivered component can be up to 14% higher than the value measured according to the specification before the coating process. Because the yield strength increases, the yield to tensile ratio increases, but there is also a higher level of security against plastic deformation.

3.4.10 Corrosion tests

ISO 3183/API5L and other line pipe standards request corrosion testing in the case of pipes ordered for sour service. Tests have to be carried out to evaluate the resistance of line pipe steels to HIC (Hydrogen Induced Cracking) and SSC (Sulphide

Stress Cracking). Background and test conditions are described in references [18, 108 - 110].

For HIC tests, rectangular specimens are taken from the line pipe or from parent material (longitudinal) and from spiral weld areas (perpendicular to the weld) and exposed unstressed in test solutions A, B or C (see blue box). The test solutions are not intended to simulate the operating environment, but to provide comparable and reproducible test conditions. Solution B is rarely used for testing because the conditions are not severe enough for modern sour gas steels.

After an exposure time of 96 hours minimum the specimens are evaluated, the measured crack sensitivity, crack length and crack thickness ratios (CSR, CLR and CTR) have to be reported.

For SSC testing, four test methods are described in reference [107]. In all cases the specimens are subjected to stresses when exposed in a corrosive media. Possible specimen types are Tensile (Method A), Bent-Beam (Method B), C-Ring (Method C) and Double-Cantilever-Beam (Method D) specimens. For SSC testing of line pipe, four-point bend testing, e.g. according to reference [110], is most commonly applied.

Tests can be carried out in acidified aqueous environments containing H_2S. The loading is for a period of a maximum 720 hours and failure/no-failure, time-to-failure, or the appearance of cracks is to be reported. Alternative test procedures can be agreed by the purchaser and pipe mill.

NACE test solutions for HIC tests

Solution A:

Sodium chloride, acetic acid solution, saturated with H_2S at ambient temperature and pressure

Solution B:

Synthetic seawater solution saturated with H_2S at ambient temperature and pressure.

Solution C (Fitness-For-Purpose):

Sodium chloride, acetic acetate solution, saturated with a mixture of H_2S and CO_2, at ambient temperature and pressure.

3.5 Observations

For steel pipes, this chapter introduced the different production methods, materials and test standards, as well as material and component testing methods and their results.

In most cases, small test samples are used, especially in conjunction with other production tests. Of course, it must always be checked whether the selected samples and the experimental procedures are representative of the behaviour of the components.

Large-scale experiments require a lot of effort in terms of experimental work, equipment and - of course - costs. The simulation of these large-scale experiments (for example, by finite element modelling) is becoming more and more popular, and allows load and geometry parameters as well as material and component properties to be varied within wide limits.

Nevertheless, the experimental determination of material and component properties will continue to play a role in ensuring the integrity of structures such as pipelines and pipeline networks. In particular, it will continue to be necessary to adjust and calibrate results on small samples, but also to certify and calibrate the results of simulation and computation models.

Acknowledgements

Thanks for providing information, figures and support to Europipe, Mannesmann Line Pipe, Mannesmann Grossrohr, Salzgitter Mannesmann Forschung, Vallourec, Corinth Pipe Works and the European Pipeline Research Group. Thanks to Dr Phil Hopkins, Dr Roger King, Dr Johannes Groß-Weege, and Dr Christoph Bosch for reviewing.

3.6 References

[1] H A Wessel, 'Kontinuität im Wandel – 100 Jahre Mannesmann 1890 – 1990', Publisher: Mannesmann AG, July 1990, pp. 20-21

[2] http://naturalgas.org/overview/history, accessed 14 December 2018

[3] CEPA – Canadian Energy Pipeline Association; 'How long has the pipeline industry been around?', https://www.aboutpipelines.com/en/pipeline-101/ pipeline-history/history/ accessed 14 December 2018.

[4] G Knauf, J Spiekhout, 'EPRG – 30 years in pipeline research', 3R International, Special Edition, Sep 2002

[5] H. Wessel, 'Mannesmann und Russland, Mannesmannröhren sind seit über 100 Jahren gefragt', Mülheimer Jahrbuch 2005

[6] N A Millwood, R M Andrews, G Knauf, R Howard, P Roovers, 'A comparison of two major linepipe specifications and possibilities for harmonization', Proceedings of the 8th International Pipeline Conference, IPC 2010-31643, Calgary, Alberta, Canada

[7] G Knauf, A Kulgemeyer, 'Major Standards for Line Pipe Manufacturing and Testing', Oil and Gas Pipelines: Integrity and Safety Handbook, First edition, Chapter 16, Edited by R. Winston Revic, © 2015 John Wiley & Sons, Inc. Published 2015 by John Wiley and Sons, Inc.

[8] ISO, International Organization for standardization, https://www.iso.org/standards.html, accessed 14 December 2018

[9] API, American Petroleum Institute, https://www.api.org/Standards/, accessed 14 December 2018

[10] CEN, European Committee for Standardization, https://standards.cen.eu/, accessed 14 December 2018

[11] DNV GL, https://www.dnvgl.com/rules-standards/index.html, accessed 14 December 2018

[12] NACE International, https://nace.org/home.aspx, accessed 14 December 2018

[13] ISO 3183:2012, 'Petroleum and natural gas industries – Steel pipe for pipeline transportation systems', International Organization for Standardization, Geneva, Switzerland (2012)

[14] API SPEC 5L:2018-04, 'Specification for Line Pipe – 46th Edition', American Petroleum Institute, 220 L Street, NW, Washington, DC 20005-4070, USA (2018)

[15] EN ISO 3183:2012/A1:2018, 'Petroleum and natural gas industries - Steel pipe for pipeline transportation systems - Amendment 1 (ISO 3183:2012/ Amd 1:2017), CEN, European Committee for Standardization (08-2018)

[16] Europipe, 'Production range', Company brochure, EP/PRR 12/2013 en 3.000, Download from https://www.europipe.com/fileadmin/europipe-modern/files/brochures/04_EP_Production-Range.pdf, accessed 14 December 2018

[17] DNVGL-ST-F101, "Submarine Pipeline Systems", DNVGL Standard, Edition October 2017, Amended December 2017

[18] ISO 15156-1:2015, 'Petroleum and natural gas industries – Materials for use in H2S-containing environments in oil and gas production', International Organization for Standardization, Geneva, Switzerland (2015)

[19] ANSI/NACE MR0175/ISO 15156-2015, 'Petroleum and natural gas industries – Materials for use in H2S-containing environments in oil and gas production', NACE International, 15835 Park Ten Place, Houston, TX USA (2015)

[20] Salzgitter-AG, 'Stil - The Salzgitter AG Magazine No. 3/2018', Salzgitter AG, Eisenhüttenstr. 99, Salzgitter, Germany

[21] Total Materia, 'Thermo-Mechanical Control Process', https://www.totalmateria.com/page.aspx?ID=CheckArticle&site=kts&LN=EN&NM=366, accessed 14 December 2018

[22] M D Herynk, S Kyriakides, A Onoufriou, H D Yun, 'Effects of the UOE/UOC pipe manufacturing processes on pipe collapse pressure', International Journal of Mechanical Sciences 49 (2007) 533–553

[23] T Reichel, V Pavlyk, J Beissel, S Kyriakides and Wen-Yea Jang, 'Improved Collapse Resistance of Large Diameter Pipe for Deepwater Applications Using a New Impander Technology', 2010 8th International Pipeline Conference, 2010 8th International Pipeline Conference, Volume 2, Calgary, Alberta, Canada, September 27–October 1, 2010

[24] K-H Brensing, B Sommer, ‚Steel Tube and Pipe Manufacturing Processes' Vulkan-Verlag GmbH; Edition 12 (1995)

[25] A Liessem, M Graef, G Knauf, U Marewski, 'Influence of thermal treatment on mechanical properties of UOE linepipe'4th International conference, Pipeline technology; 2004; Ostend, Belgium, PIPELINE TECHNOLOGY PROCEEDINGS; 3; 1263-1282

[26] Europipe, 'Products for the Oil and Gas Industry', Company brochure, EP/POG 02/2012 en 1.500, Download from https://www.europipe.com/ fileadmin/europipe-modern/files/brochures/03_EP_Oil-and-Gas-Industry.pdf, accessed 14 December 2018

[27] Tata, 'Tata Steel first to develop DSAWL pipe through reeling method in the North Sea', https://www.tatasteeleurope.com/es/noticias/noticias/2015/2015-first-to-develop-dsawl-pipe-through-reeling-method, accessed 14 December 2018

[28] H-G Hillenbrand, A Liessem, G Knauf, K Niederhoff, J Bauer, 'Development of large-diameter pipe in grade X 100 - State-of-the-Art Report from the Manufacturer's Point of View', Pipeline Technology Conference, Brugge May 2000

[29] C Kalwa, 'Pipe Manufacture - Longitudinal Submerges Arc Welded Large Diameter Pipe', Oil and Gas Pipelines: Integrity and Safety Handbook, First edition, Chapter 12, Edited by R. Winston Revie, © 2015 John Wiley & Sons, Inc. Published 2015 by John Wiley and Sons, Inc.

[30] Europipe, 'Production and Testing of Large-Diameter Pipes', Company Brochure, https://www.europipe.com/fileadmin/europipe-modern/files/brochures/08_EP_Production-and-Testing.pdf, accessed 14 December 2018

[31] Europipe, 'The world of Europipe', Company brochure, https://www.europipe.com/fileadmin/europipe-modern/files/brochures/11_EP_ GroupWorld.pdf, accessed 14 December 2018

[32] A Liessem, O Reepmeyer, J Schröder, C Kalwa, 'Development and production of large-diameter line pipe for offshore applications', OPT 2006

[33] SMS Group, 'Large diameter Pipe Plants', Company brochure, https://www.sms-group.com/sms-group/downloads/, accessed 14 December 2018

[34] Corinth Pipeworks SA, www.cpw.gr/en/facilities/thisvi-plant/production/lsaw-pipe-mill/, accessed 14 December 2018

[35] Mannesmann Grossrohr, 'Spiral welded large diameter pipes', Company Brochure, https://www.mannesmann-grossrohr.com/fileadmin/footage/MEDIA/gesellschaften/szmgr/documents/MGR_Broschuere_2018_April.pdf

[36] D Mirkovic, V Flaxa, F M Knoop, 'Development and Production of Helical Two-Step (HTS) Pipes: Grades up to API X70 for Sour Service Application', Proceedings of the 9th International Pipeline Conference, September 24-28, 2012, Calgary, Alberta, Canada, IPC2012-90438

[37] T Kurashi, N Yumang, 'Production technology and evaluation of line pipe with helical seam for sour service', Paper presented at 7th Middle East Corrosion Conference, Manama, 26.-28.2.1996, Vol. 1, pp. 429-452

[38] A Eltaher, 'Spiral Welded Pipes for Shallow Offshore Applications ', Oil and Gas Pipelines: Integrity and Safety Handbook, First edition, Chapter 6, Edited by R. Winston Revie, © 2015 John Wiley & Sons, Inc. Published 2015 by John Wiley and Sons, Inc.

[39] F-M Knoop, 'Pipe Manufacture—Spiral Pipe', Oil and Gas Pipelines: Integrity and Safety Handbook, First edition, Chapter 13, Edited by R. Winston Revie, © 2015 John Wiley & Sons, Inc. Published 2015 by John Wiley and Sons, Inc.

[40] Mannesmann Line Pipe GmbH, 'Your specialist for HFI welded steel pipe', Company Brochure, https://www.mannesmann-linepipe.com/fileadmin/footage/MEDIA/gesellschaften/smlp/Documents/MLP_Lieferprogramm.pdf, accessed 14 December 2018

[41] H Brauer, H Löbbe, 'Pipe Manufacture – ERW Pipe', Oil and Gas Pipelines: Integrity and Safety Handbook, First edition, Chapter 14, Edited by R. Winston Revie, © 2015 John Wiley & Sons, Inc. Published 2015 by John Wiley and Sons, Inc.

[42] B Zimmermann, H Brauer, U Marewski, 'Development of HFIW line pipe for offshore applications', 4th Int. Conf. on Pipeline Technology, 9.- 13.05.2004, Oostende, Belgium

[43] Mannesmann Line Pipe, 'Production steps in the welding mill', Company web site, https://www.mannesmann-linepipe.com/en/processes/hfi- welding.html, accessed 14 December 2018

[44] Vallourec, 'The Düsseldorf-Rath works', Company Brochure, http://www.vallourec.com/COUNTRIES/GERMANY/EN/Media/publications/ Lists/Publications/V_A01B0004B-15GB.pdf, accessed 14 December 2018

[45] R Kümmerling, K Kraemer, 'Pipe Manufacture – Seamless Tube and Pipe', Oil and Gas Pipelines: Integrity and Safety Handbook, First edition, Chapter 15, Edited by R. Winston Revie, © 2015 John Wiley & Sons, Inc. Published 2015 by John Wiley and Sons, Inc.
[46] H Koch, '75 Jahre Mannesmann – 1890 – 1965', Mannesmann Aktiengesellschaft, Düsseldorf, 1965, p. 17
[47] Danieli, 'Seamless-pipe mills', Company web-site, https://www.danieli.com/en/products/products-processes-and-technologies/extrusion_26_81. htm, accessed 14 December 2018
[48] R Kümmerling, A Kneissler, 'Radial forging of seamless pipes', Stahl und Eisen 132
[49] Vallourec, 'The Düsseldorf-Rath Works – A comprehensive performance package', Company brochure, http://en.vallourec.com.cn/Uploads/ other/20150920/The%20Dusseldorf%20 Rath_1B0004B_en.pdf, accessed 14 December 2018
[50] Europipe, 'Coating', Company brochure, https://www.europipe.com/fileadmin/europipe-modern/files/brochures/07_EP_Coating.pdf, accessed 14 December 2018
[51] Mannesmann Line Pipe, ‚MAPURE – Field coating system', Company brochure, https://www.mannesmann-linepipe.com/fileadmin/footage/ MEDIA/gesellschaften/smlp/Documents/MLP-Datenblatt-MAPUR_DE_A4_20170926.pdf, accessed 14 December 2018
[52] Mannesmann Grossrohr, 'Coatings & linings ', Company brochure, https://www.mannesmann-grossrohr.com/en/product-range/coatings-linings. html#c60815, accessed 14 December 2018
[53] F Grimpe, S Meimeth, C J Heckmann, A Liessem, A Gehrke, 'Development, production and application of heavy plates in grades up to X120', 1st International Conference Super-High Strength Steels, November 2-4, 2005, Rome, Italy
[54] G Knauf, A Kulgemeyer, 'Pipe Manufacture – Introduction', Oil and Gas Pipelines: Integrity and Safety Handbook, First edition, Chapter 16, Edited by R. Winston Revie, © 2015 John Wiley & Sons, Inc. Published 2015 by John Wiley and Sons, Inc
[55] C W Petersen et al., 'Improving long distance gas transmission economics: X120 development overview', Pipeline Technology Conference 9-13 May, 2004 Ostend, Belgium
[56] H Asahi et al., 'Metallurgical design of high strength steels and development of X120 UOE linepipe', Pipeline Technology Conference 9-13 May, 2004 Ostend, Belgium
[57] H.-G. Hillenbrand et al, 'Development of high strength and pipe production technology for grade X120 line pipe', International Pipeline Conference IPC, 04.-08.10.2004, Calgary (Canada)
[58] ISO 8491:1998, Metallic materials – Tube (in full section) – Bend test', International Organization for Standardization, Geneva, Switzerland (1998)
[59] ISO 8492:2013, 'Metallic materials – Tube, Flattening test', International Organization for Standardization, Geneva, Switzerland (2013)
[60] ISO 7438:2016, 'Guided bend test', International Organization for Standardization, Geneva, Switzerland (2016)
[61] ISO 6892-1:2016, 'Metallic materials -- Tensile testing -- Part 1: Method of test at room temperature', International Organization for Standardization, Geneva, Switzerland (2016)
[62] ASTM A370 - 17a, 'Standard Test Methods and Definitions for Mechanical Testing of Steel Products', ASTM International, West Conshohocken, PA, 2017, www.astm.org, accessed 14 December 2018
[63] Wikipedia, 'Lüders band ', https://en.wikipedia.org/wiki/Lüders_band, accessed 14 December 2018

[64] G Hohl, G Knauf F M Knoop, 'The Effect of Specimen Type on Tensile Test Results and its implications for Linepipe Testing', 3R international, 10-11, 2001, Vulkanverlag GmbH, Essen, Germany

[65] H G Hillenbrand, C Kalwa, O Hilgert, G Knauf, ' Grenzen des Stresstests aus Sicht eines Rohrherstellers', Fachberichte Pipeline, 3R 01-02-2015, pp.48-54

[66] ISO 15363:2017, 'Metallic materials- Tube ring hydraulic pressure test' , International Organization for Standardization, Geneva, Switzerland (2017)

[67] Salzgitter Mannesmann Forschung GmbH, SZMF, https://www.salzgitter-mannesmann-forschung.de/en/index.html, accessed 14 December 2018

[68] G Knauf, 'Rohre und Rohrkomponenten – Bestimmung der Werkstoff- und Bauteileigenschaften', Tagung Werkstoffprüfung 2012

[69] EPRG, European Pipeline Research Group e.V., www.eprg.net, accessed 14 December 2018

[70] ASTM A530 / ASTM 530M - 18, 'Standard Specification for General Requirements for Specialized Carbon and Alloy Steel Pipe', ASTM International, West Conshohocken, PA, 2017, www.astm.org

[71] BS 7910:2013, 'Guide to methods for assessing the acceptability of flaws in metallic structures', British Standards Institution, 2013

[72] BS 7448-1:1991, 'Fracture mechanics toughness tests. Method for determination of KIc, critical CTOD and critical J values of metallic materials', British Standards Institution, 1991

[73] BS 7448-2:1997, 'Fracture Mechanics Toughness tests. Part 1: Methods for determination of KIc, critical CTOD and J-values for welds of metallic materials', British Standards Institution, 1997

[74] BS 7448-4:1997, 'Toughness tests. Method for determination of fracture resistance curves and initiation values for stable crack extension in metallic materials', British Standards Institution, 1997

[75] ISO 12135:2002, 'Metallic materials - Unified method of test for the determination of quasistatic fracture toughness', International Organization for Standardization, Geneva, Switzerland (2002)

[76] ISO 15653:2010, 'Metallic materials - Method of test for the determination of quasistatic fracture toughness of welds', International Organization for Standardization, Geneva, Switzerland (2010)

[77] ASTM E 1820:2009, 'Standard Test Method for Measurement of Fracture Toughness', ASTM International, West Conshohocken, PA, 2009, www. astm.org

[78] T A Siewert, M P Manahan, C N McCowan, J M Holt, F J Marsh, and E A Ruth, 'The History and Importance of Impact Testing,' Pendulum Impact Testing: A Century of Progress, ASTM STP 1380, T. A. Siewert and M. P. Manahan, Sr., Eds., American Society for Testing and Materials, West Conshohocken, PA, 1999

[79] ISO 148-1:2016, 'Metallic materials – Charpy pendulum impact test – Part 1: Test method', International Organization for Standardization, Geneva, Switzerland (2002)

[80] ASTM STP1248, 'The Effect of Charpy V-Notch Striker Radii on the Absorbed Energy', ASTM International, West Conshohocken, PA, 2009, www.astm.org

[81] M Erdelen-Peppler, H-G Hillenbrand, G Knauf, 'Limits of existing crack arrest models', Pipeline Technology Conference, Ostend, 12-14 October 2009, Paper No. 2009-116

[82] E Lucon, 'Influence of striking edge radius (2 vs. 8 mm) on instrumented Charpy data and absorbed energies', International Journal of Fracture, September 2008

[83] Battelle Memorial Institute, Columbus, Ohio, https://www.battelle.org/, accessed 14 December 2018

[84] M Erdelen, R Gehrmann, G Junker, G Knauf, A Liessem, 'Significance of DWT testing for line pipe safety', International Conference on Fracture, ICF11, Italy 2005

[85] EN 10274:1999, 'Metallic materials. Drop weight tear test', CEN, European Committee for Standardization (09-1999)

[86] ASTM E436 - 03(2014), 'Standard Test Method for Drop-Weight Tear Tests of Ferritic Steels', ASTM International, West Conshohocken, PA, 2009, www.astm.org

[87] API RP 5L3: Recommended Practice for Conducting Drop-Weight Tear Tests on Line Pipe. API RP 5L3 3rd Ed. Feb. 1996

[88] I Satoshi, C Guan, B Rothwell and T Hiraide, 'Investigation of Crack Propagation Characteristics Using Instrumented Charpy and DWT Tests for Full-Scale Burst Tested 1219 mm OD Grade 550 (X80) Linepipe', Paper No. IPC2016-64240, IPC2016-64240, 2016 11th International Pipeline Conference, Calgary, Alberta, Canada, September 26–30, 2016, Copyright © 2016 by ASME

[89] A R Duffy, 'Full-Scale Studies', Symposium on Line Pipe Research, American Gas Association, Catalogue No. L30000, 1965

[90] R B Smith R J Eiber, 'Field Failure Survey and Investigations', Fourth Symposium on Line Pipe Research, ppD1-D19. Pipeline Research Committee, Arlington, Virginia. (1969)

[91] Maxey, W.A., Kiefner, J.F. and Eiber, R.J.: Ductile Fracture Arrest in Gas Pipelines. American Gas Association, Catalog No. L32176, December 1975

[92] www.prci.com, accessed 14 December 2018

[93] W A Maxey, 'Fracture initiation, propagation and arrest ', Fifth Symposium on Line Pipe Research. Houston, Texas, USA, 20-22 November 1974. Falls Church, Virginia, USA: Pipeline Research Council International.

[94] G Knauf, 'EPRG – Crack Arrest and Girth Weld Acceptance Criteria for High Pressure Gas Transmission Pipelines' Yokohama Conference 2002

[95] B N Leis, 'Predicting fracture arrest based on a relationship between Charpy V-notch toughness and dynamic crack-propagation resistance', Proc 3rd IPTC, Brugge, May 2000

[96] G Demofonti, G Mannucci and P Roovers, 'Existing methods to evaluate fracture resistance of high-grade steel pipelines', Steel Pipe, 3R international, Special Edition - 1/2008

[97] G Demofonti et al., ECSC Report n. EUR 20330 EN.

[98] G Demofonti et al., ECSC Report n. EUR 21440 EN

[99] R M Andrews et al., Proc IPC 2004, paper 0596, Calgary, Oct. 2004

[100] G Mannucci, 'Fracture Initiation and Propagation', SHSS Conf., Rome, Italy 2005

[101] H Mughrabi, Johann Bauschinger, pioneer of modern materials testing ', Materials Forum, 10 (1): 5–10, 1987

[102] E Tsuru, H Asahi, 'Improved Collapse Pressure of UOE Line Pipe with Thermal Aging for Deep Water Applications'. Proc of 16th International Offshore and Polar Engineering Conference. San Francisco, Vol. 4, pp 187-194, 2006.

[103] A Liessem, J Groß-Weege, G Knauf, S Zimmermann, 'UOE Pipes for Ultra-Deep-Water Application – Analytical and FE Collapse Strength Prediction vs. Full Scale Tests of Thermally Treated Line Pipe', Proceedings of the 17th ISOPE Conference, Lisbon, 2007.

[104] Al-Sharif, A.M. und Preston, R. Improvement in UOE Pipe Collapse Resistance by Thermal Ageing. In: Proceedings of the 28th Annual Offshore Technology Conference, OTC 8211, Houston, pp. 579-588, 1996.

[105] ASTM E9-09(2018), 'Compression Testing of Metals', ASTM International, West Conshohocken, PA, 2009, www.astm.org

[106] A Liessem, U Marewski, J Groß-Weege, G Knauf, 'Methods for collapse pressure prediction of UOE pipe', Proceedings of OMAE, 25th International Conference on Offshore Mechanics and Arctic Engineering, June 4-9, Hamburg, Germany

[107] Y Li, W Zhang, S Gong, L Ji, C Huo, Y Feng, 'Study On Tensile Property And Strain Ageing Behaviour Of X100 Linepipe' Proceedings of the 8th International Pipeline Conference, IPC2010-31180, 2010

[108] NACE TM0284-2016, 'Standard Test Method - Evaluation of Pipeline and Pressure Vessel Steels for Resistance to Hydrogen-Induced Cracking', Revised 2003-01-17, NACE International, 15835 Park Ten Place, Houston, TX USA (2016)

[109] NACE TM0177-2016, 'Standard Test Method - Laboratory Testing of Metals for Resistance to Sulphide Stress Cracking and Stress Corrosion Cracking in H2S Environments ', Revised 1996-Dec-23, NACE International, 15835 Park Ten Place, Houston, TX USA (2016)

[110] NACE TM0316-2016, 'Standard Test Method - Four-Point Bend Testing of Materials for Oil and Gas Applications', NACE International, 15835 Park Ten Place, Houston, TX, USA (2016).

CHAPTER 4

Introduction to ILI

Dr Michael Beller

4.1 Introduction

Pipelines are a valuable asset and need to be protected. In order to achieve this a modern pipeline integrity management program does usually include regular inspections followed by integrity assessment and, if required, repair and rehabilitation measures.

Determining the mechanical integrity of a pipeline is not that dissimilar from going for a medical examination. If you don't feel well and decide to consult a medical doctor, he or she will have to diagnose what is causing the "health" problem before being able to identify the right remedy.

In order to assess if a high-pressure pipeline is fit for purpose the first step is also to diagnose its current status and to identify any defects that may affect the integrity of the line. This implies looking at the entire pipeline, its full circumference and its total length. Inspecting from the inside provides great advantages to achieve these requirements.

Pipeline engineers today have access to a wide range of inspection technologies in order to identify defects in the pipeline wall or the coating or detect pipe movement.

This Chapter will address pipeline inspection and introduce internally applied inspection technologies for onshore and offshore pipelines. It will provide a comprehensive overview of the in-line inspection tools (ILI) commercially available on the market today for the inspection of all the types of high-pressure pipelines found in the up, mid and downstream sectors. Strength and weaknesses of different technologies will be discussed in order to support tool selection.

4.2 Pipeline inspection

The oil, product and gas pipelines considered here are made of steel. A well proven method for the inspection of such pipelines, especially high-pressure transmission pipelines is the use of automated inspection tools which can survey pipelines from within providing full circumferential and axial coverage.

The pipeline industry has inserted equipment, such as cleaning tools, into pipelines for over 150 years. Figure 4-1 shows a typical cleaning tool. These tools travel with the flow of the product in the pipeline and can be retrieved at various exit points along the pipeline.

These free-swimming tools are traditionally called 'pigs' and the process of using a 'pig' in a pipeline is called 'pigging'. Nobody remembers who called them pigs for the first time, but according to a well-known industry anecdote the name was given to them based on their appearance when they were retrieved after a run in a "dirty" pipeline in the late 19th century. Barbed wire was used to hold together the original components of the tools, including bundles of hay and leather clothes. As they entered a line the barbed wire scratched the steel wall and caused a "squeaking" sound.

The inspection tools inserted into pipelines are generally referred to as free swimming in-line inspection tools (ILI) or intelligent pigs, see Figure 4-2. They utilize non-destructive testing techniques such as: Magnetic Flux Leakage (MFL), Ultrasound Technology (UT), Eddy Current (EC), or a combination thereof to detect, size and locate possible anomalies or flaws.

ILI tools have been used for over 50 years and are proven reliable and accurate. Figure 4-3 shows the timeline or history of ILI.

The history of In-Line Inspection (ILI) started in the 1960's with Geometry Inspection. The first Caliper Tools were introduced in the USA. The technology itself had already been used in the inspection of downhole tubing, but now tools were developed to inspect for dents and other changes to the cross-sectional area of high-pressure pipelines.

Figure 4-1. Typical pipeline cleaning pig. (courtesy Rosen-Group, Germany).

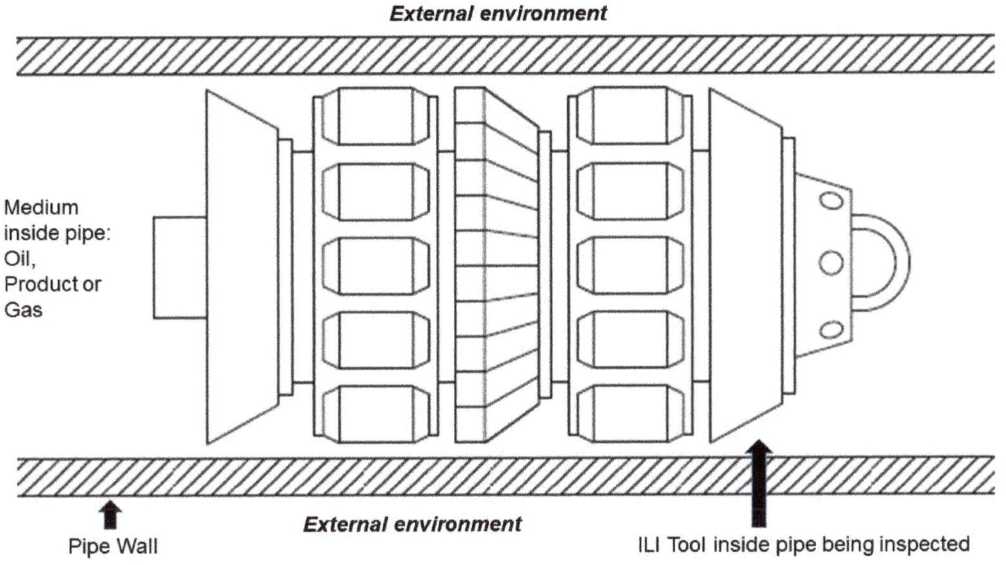

Figure 4-2. Schematic depicting ILI inside pipeline being inspected.

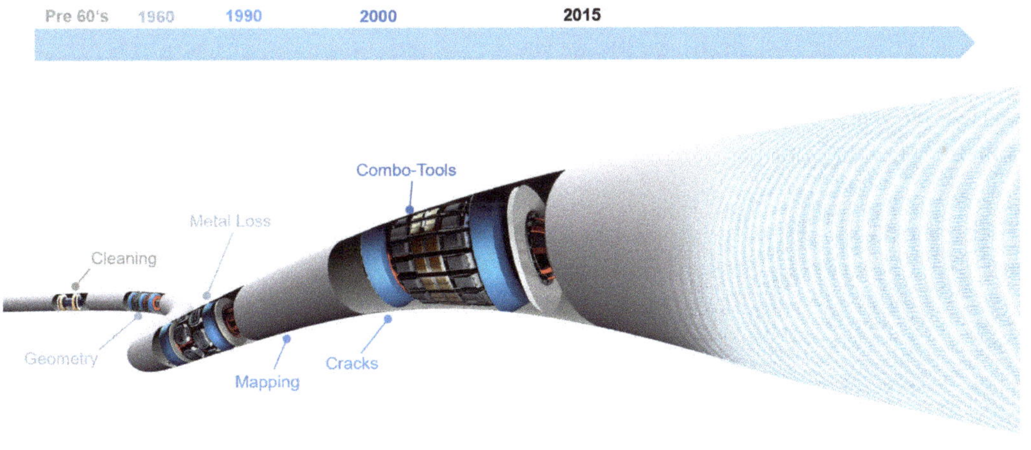

Figure 4-3. History of in-line inspection.

First metal loss inspection tools were introduced in the second half of the 1960`s, also in the USA. They used magnetic flux leakage techniques in order to identify corrosion and gauging features and only had limited circumferential coverage, mainly due to the data storage equipment available at the time. Data was stored on reel-to-reel tape-recorders with a limited number of channels. These implied that signals could only be recorded for a limited number of sensors, and thus full circumferential coverage could not be achieved. The analogue data recorded was printed out on paper strip for analysis purposes. These earlier tools were a major step forward in the inspection of pipelines, although they only offered fairly low resolutions and accuracies compared to modern tools.

In the middle 1970's the first high resolution tools were designed in the UK[1,2]. A mindset change had initiated this development. Earlier tools basically focused on using available non-destructive testing technology for detecting and sizing defects. The new understanding was that the data from ILI tools needed to meet certain minimum, material dependent defect specifications in order to be useful for integrity assessment purposes. The question that needed answering from an integrity perspective was whether the pipeline inspected was fit-for-purpose despite containing a given defect. It was not enough to just know that a defect was present. As a result of this line of thought critical defect specifications were researched, including many costly burst tests of real pipes with a large variety of defects built into them. These critical specifications were then used to define the detection and sizing requirements that an ILI tool had to meet, leading to the development of high-resolution tools. Initially the focus was still on metal loss. Data analysis also progressed and computers started being used for the process. Initially main frame computers were used, because the days of powerful desktop computers had not yet arrived.

The first crack inspection tool was also developed in the 1970s, but is not shown in the timeline above, because it was designed for the detection of a specific type of internal cracking and could therefore not be considered a general crack inspection device[3].

In the mid 1980s the first high resolution wall thickness measurement tool utilizing ultrasound technology was introduced in Germany[4,5], although ultrasound had already been used earlier in a tool developed in the US. Big challenges for all ILI technologies at the time were suitable miniaturized electronics. Digital tape recorders were used, but recording capacity was still limited. The first crack inspection tools utilizing ultrasound were commercially introduced in the mid-1990s, again in Germany[6]. Crack inspection generates a whole order of magnitude larger amount of data requiring use of specialized data reduction and compression algorithms.

The last two decades in the 20th century also saw the introduction of mapping tools, originally developed by a Canadian engineering company, and leak detection tools for pipelines. The latter first introduced in Germany but based on a license from a major oil and gas company.

The first crack inspection tools utilizing electro-magnetic acoustic transducers (EMAT) were introduced round about the millennium change.

Data storage was no longer a real issue as solid state memories became available with larger and larger capacity and prices went down.

All ILI tools up to this point in time were focused on identifying defects in the pipe wall or the coating or measuring center-line.

Only recently a new type of ILI tool has been introduced which focuses on investigating material properties and thus performing a material grade inspection service[7].

New developments in the field of non-destructive testing as well as data analytics are happening all around the globe and the history of ILI will continue and be exciting.

The primary mission of ILI tools is to provide reliable, accurate and robust information on defects that may be present in the pipe wall such as dents, metal loss or cracks. Today there are also tools looking for coating defects. In order to ascertain the condition of a pipeline requires the acquisition of information on both the internal and external surfaces of the pipe wall, plus the body of the pipe wall, as well as all welds over the whole pipeline length.

Non-destructive testing technologies have been developed to look for defects in a material without causing any damage to the sample being inspected.

Different defects that can occur in a pipeline steel will display different geometries or shapes and it is all about detecting and sizing these, and obviously also locating them. This information is then subsequently used for defect assessment. It is therefore very important that any inspection data are precise and the technologies used can reliably identify all the typical types of defects that can potentially appear in a pipeline.

Different defects come with different shapes and there is no single non-destructive testing principle that can detect and size all these different shapes with the same quality. All testing technologies are based on physical principles which in turn come with their respective strengths and weaknesses. This applies to their measurement capabilities as well as the operational conditions prevailing when applying them. They also come with different resolutions, depending on how they are utilized by a specific tool design[8,9,10]. As a matter of fact the choice of non-destructive testing technology is of critical importance because some technologies may be absolutely "blind" to some shapes/types of defects or may not work in certain materials or certain environments. Before introducing the different ILI tools commercially available today, it is therefore important to have a short recap on the potential defects and also look at the most widely used non-destructive testing techniques used in ILI in order to understand their major strength and weaknesses.

As stated, modern ILI tools are designed to inspect pipes made of steel. Depending on the type of non-destructive testing technology applied this would include all typical linepipe steels as well as stainless steels, internally clad pipe and modern corrosion resistant alloys (CRAs).

4.3 Defects

The defects and threats experienced in pipelines are covered in different chapters of this book. From an inspection perspective they can be summarized into the following categories:

- Geometry defects
- Metal loss defects
- Cracks
- Leaks (defects extending through the wall)

The major mission of non-destructive testing (NDT) is to detect such defects and size and locate them. Sizing implies measuring their depth, width and length.

Why is it so important to find and size defects? Pipelines are subjected to loads, the most significant of these usually being the internal pressure of the medium being transported in the pipe. This internal pressure is contained by the wall of the pipe and will cause mechanical stresses inside the pipe wall. We do not have to get into real detail here, just appreciate that out of these, the stresses acting in the circumferential direction are usually the most significant. They are called Hoop Stresses.

The mechanical stresses acting in a wall are carried by the available cross sectional area of that wall. If this area is reduced, the intensity of loading acting on the ligament (i.e. remaining wall thickness) increases. Eventually the stresses acting will exceed the load bearing capacity of the pipe material. This means material properties such as the yield strength or the ultimate tensile strength are exceeded. This will lead to permanent deformation in the first case and fracture in the second. In the case of a cyclic load fatigue may occur. In the presence of cracks things are even more complex. But overall the message is that any change of the locally available thickness of the pipe wall and the extent and shape of that change should be known and measured accurately!

Examples of processes leading to such a change and reduction of the load bearing capacity of the pipe wall are metal loss or cracks.

The above is not being discussed here in order to do any stress analysis, but to highlight that it is very important to know whether the original wall-thickness present in a defect-free pipe is still available to carry the load or not.

The defect types mentioned so far can have a detrimental effect on the mechanical integrity of a pipeline. Ideally they should be discovered and sized by an ILI tool long before they reach a critical size, i.e. a size leading to failure.

Leaks are somewhat special, because they constitute more than a defect. If a pipeline leaks, it has failed, because pipelines are not supposed to spill anything.

Table 4-1 lists the defect categories introduced and provides some further details.

Table 4-1. Overview of types of defect in pipelines that can be detected and sized by ILI tools

Major Categories of Feature	Sub-Categories	Examples
Geometric Anomalies	Diameter Changes	• Dents • Out-of-Roundness • Wrinkles
	Pipeline Position	• Pipeline Movement
Metal Loss	Wall-loss without Corrosion	• Gouging
	Wall-loss due to Corrosion	• General Corrosion • Localized Corrosion • Pitting and Pin-holes
Material Separation	Crack-like Features	• Laminations
	Cracks	• Fatigue Cracks • Hook Cracks • Stress Corrosion Cracking • Hydrogen Induced Cracking

4.4 Major non-destructive testing techniques used in ILI

The most important non-destructive testing technologies being used in ILI are electromagnetic techniques and ultrasound. The two techniques based on electromagnetic principles are magnetic flux leakage and eddy current. Regarding ultrasound the difference is how the ultrasound is generated and introduced into the material being tested. Here we distinguish between piezo-electrically generated ultrasound (usually referred to as UT in the ILI world) and ultrasound generated in the test object by electromagnetic transducers (usually referred to as EMAT).

These technologies are primarily used in ILI tools designed to detect and size defects. The technologies used in specialized ILI tools for leak detection and mapping will be explained later when we look at the tools.

4.4.1 Electromagnetic testing principles

The most widely applied electromagnetic principle in ILI is the magnetic flux leakage technology.

Magnetic flux leakage (MFL) testing is a very reliable and robust method for detecting and sizing metal loss in steel structures and thus is well suited to ILI. The basic principle involves magnetizing a ferrous metal object to saturation level with a powerful magnetic field. Where no flaws are present, the magnetic flux remains undisturbed. In the presence of internal or external metal loss, the magnetic field "leaks" from the object, see Figure 4-4. The magnetic circuit used consists of permanent magnets, a flux link, the wall to be magnetized and a flux return path or back bone. In the schematic shown in Figure 4-4 one magnet would be the north pole and the other the south pole of the magnetic circuit.

In this way metal loss leading to wall thinning, be it due to general or local corrosion, gouging or abrasion can be identified. In an in-line inspection the pipe wall is magnetized to saturation level in the axial or circumferential direction, depending on the type of MFL tool used and its mission. There are also tools magnetizing the wall in a spiral orientation. Modern MFL tools commonly use very powerful permanent magnets.

In the MFL tool, suitable sensors are placed between the poles of the magnetic circuit to record the leakage field. The sensors are oriented within the magnetic field and their number chosen to ensure full circumferential coverage of the pipeline. Changes in the magnetic field due to defects are recorded onboard. In a further step during the analysis process, which happens after a run and after the data has been downloaded from the tool, specialized algorithms are used to analyse and interpret the characteristics of the signals and determine the depth, width and length of the detected metal loss.

Advantages of MFL tools are that they can inspect pipelines transporting liquids, such as crude oil or product as well as those transporting gas. Another strength of MFL is its ability to detect fairly small metal loss features such as pitting corrosion or even pin holes. It has to be noted however that the MFL principle provides an indirect measurement and cannot measure the absolute wall thickness of the pipe wall. Another issue to keep in mind is the need for magnetic saturation of the wall to ensure good data quality. This may limit the wall thickness range MFL tools can inspect. Not really an issue with onshore lines, but possibly with some types of offshore pipelines.

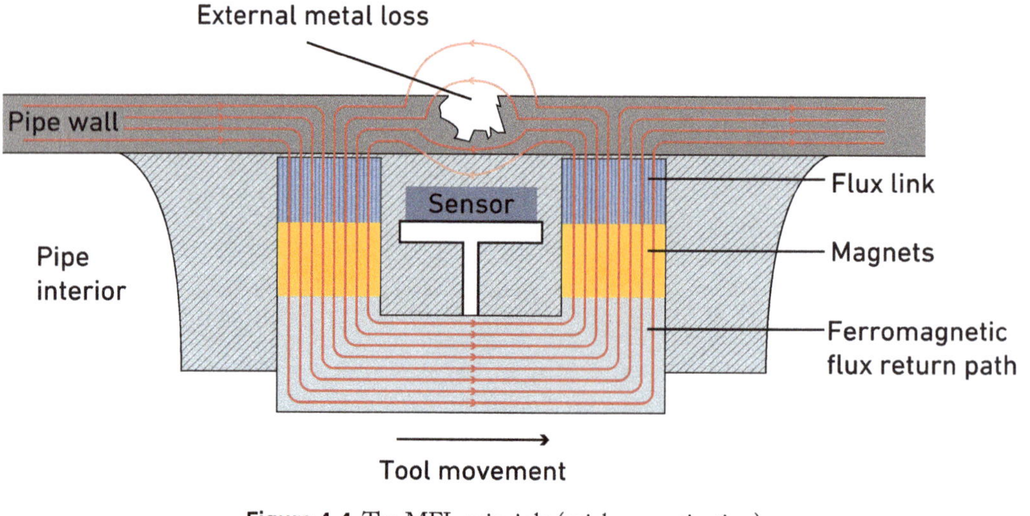

Figure 4-4. The MFL principle (axial magnetization).

Eddy current (EC) testing uses electromagnetic induction to detect flaws in conductive materials. The test piece does not need to be in contact with the test equipment.

In a standard EC testing procedure a cylindrical coil carrying an alternating current is placed close to the test piece, i.e. the pipe wall. The alternating current in the coil generates a changing magnetic field which induces eddy currents in the test piece. Variations in the phase and magnitude of these EC's are monitored using a second coil (search coil), or by measuring changes to the current flowing in the primary coil (excitation coil). Variations in the electrical conductivity or magnetic permeability of the test object, or the presence of any flaws, will cause a change in the flow patterns of the EC and a corresponding change in the phase and amplitude of the measured current, see Figure 4-5.

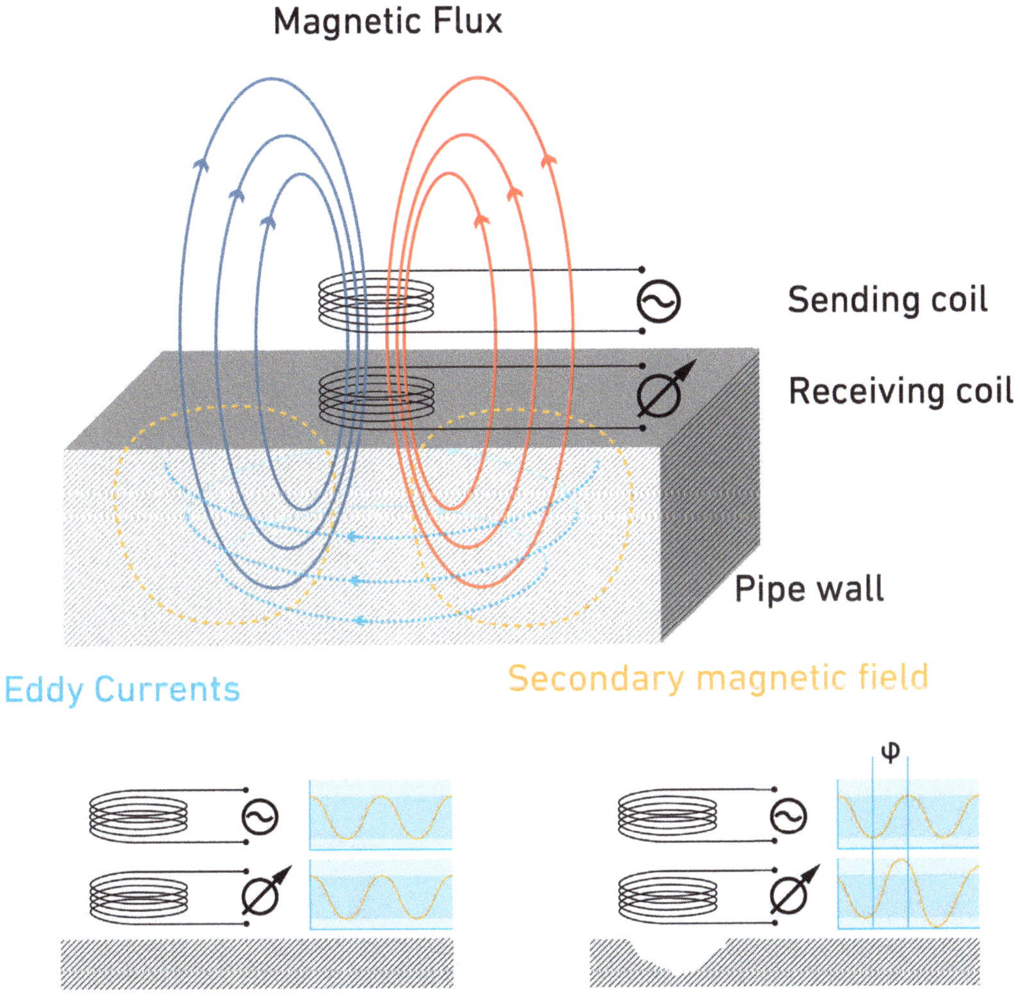

Figure 4-5. Eddy current principle.

EC testing can be used to inspect for metal loss and cracks on the nearside surface of a test piece. The depth up to which features can be detected depends on the permeability[1] of the metal and the frequency of the system used. Advantages of EC are that the principle can be applied in a liquid or gas environment, and that the inspection principle is sensitive to metal loss and cracks. Another advantage is that EC sensors can measure the distance between themselves and the nearside wall (the internal surface of the pipe) and they are therefore suited for geometry inspection, checking the internal diameter of a pipeline or looking for dents. Eddy current sensors are also used on some ILI tools as a complementary technology in order to measure shallow internal metal loss in thick walled pipe.

A drawback is that standard EC utilized in an ILI tool can only inspect the internal surface. There are different types or derivations of EC available today, such as pulsed EC and lately also magnetic bias eddy current, but these technologies are currently not implemented in any commercially available ILI tools. The latter technology is however being investigated for such use.

4.4.2 Ultrasound testing principles

Ultrasound is basically an acoustic energy signal, similar to the acoustic signals we can hear with our ears. The frequencies being used are far above those the human ear can register. Usually the term ultrasound is used for frequencies above 20,000 Hertz (Hz), still within range of the ears of dogs and bats. But the frequencies used in material testing are orders of magnitude higher than that, namely in the Megahertz (MHz) range. Inspecting a pipeline wall implies using ultrasound to investigate a solid, namely the line pipe material. In solids, sound waves can propagate in different principle modes resulting in different wave forms. We will not go into details here, except for mentioning that different wave forms are used depending on the type of defect to be identified.

Two different types of ultrasound technology are used in ILI. Firstly transducers, where the ultrasound is generated in the sensor itself through excitation of a piezo-electric crystal. The second type of technology is based on utilizing electro-magnetic acoustic transducers, where the ultrasound is only generated in the material to be tested itself and not in the sensor. Let us look at both of these now.

Piezo-electric Ultrasound (UT): here an ultrasound transducer incorporating a piezo-electric crystal is used to generate an ultrasonic wave in a sensor probe. The wave is then propagated through a suitable liquid coupling medium and the surface of the coupled test piece. The couplant is required to ensure no air gap between the probe and the material to be tested, given air's relatively poor sound conducting characteristics. Most of us may have experienced this type of technology at the doctors, where UT is often used and the doctor uses a gel between the sensor and our skin. In material

1 Permeability can be seen as a measure of how well a material can be magnetized.

testing in the laboratory the couplant is often water, in the case of ILI it is the liquid medium in the pipeline, such as oil or product. The transducer records the reflections (echoes) caused by both the adjacent and the remote surfaces of the material to be tested – in the case of pipes their internal and external surfaces, see Figure 4-6.

The differences in the arrival times of these reflections are directly related to the distance between the transducer and the surfaces of the test piece. Thus, on the basis of the reflections of the ultrasonic wave, the thickness of the material can be assessed. In ILI, a distinction can be made between metal loss on the internal and external surfaces of the pipe. But UT is also good at detecting mid-wall features such as laminations.

A major strength of ultrasound is its ability to provide true wall thickness measurements. This means that wherever there is a corrosion or wall thinning, the remaining wall thickness can be determined and thus the true contour of the metal loss. Modern assessment codes require this information. A weakness of piezo-electric UT is that a liquid couplant is required. This means that this technology is not really suited for gas pipelines, unless the tool is run within a liquid 'slug' in the gas pipeline. These 'slug' operations are difficult and expensive. However, the additional expense may well be justified if a particular type of defect cannot be reliably detected and sized in any other way. Examples are laminations in gas pipelines or hydrogen induced cracks in a sour gas pipeline. If the UT beam travels through the pipe wall at an angle, the technology can also be used for the identification of cracks. Figure 4-7 shows a schematic of a typical set up.

The sensor carrier of a crack inspection tool is designed in such a way that a suitable incident wave angle is achieved ensuring a 45° travel angle of the refracted wave inside the wall. As the wave meanders from surface to surface of the wall internal and external cracks can be identified. Any cracks (or other reflectors in the material) will reflect the sound wave and return it to the sensor, thus causing registration.

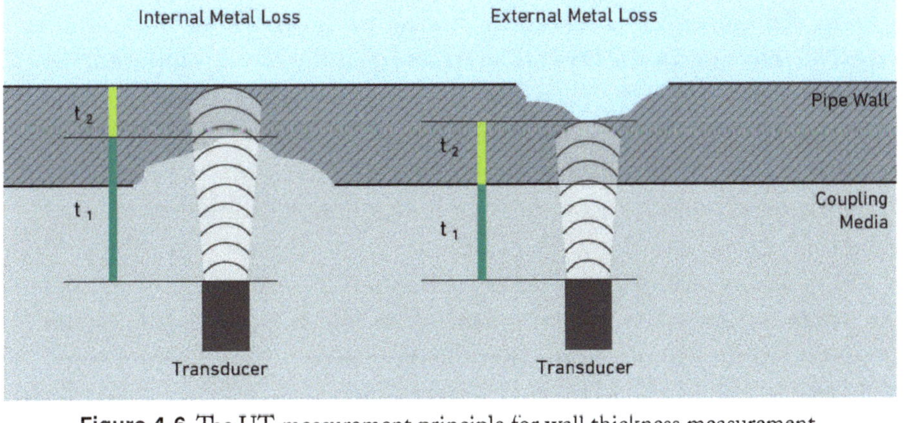

Figure 4-6. The UT measurement principle for wall thickness measurement (t_1 = stand-off, t_2 = wall thickness).

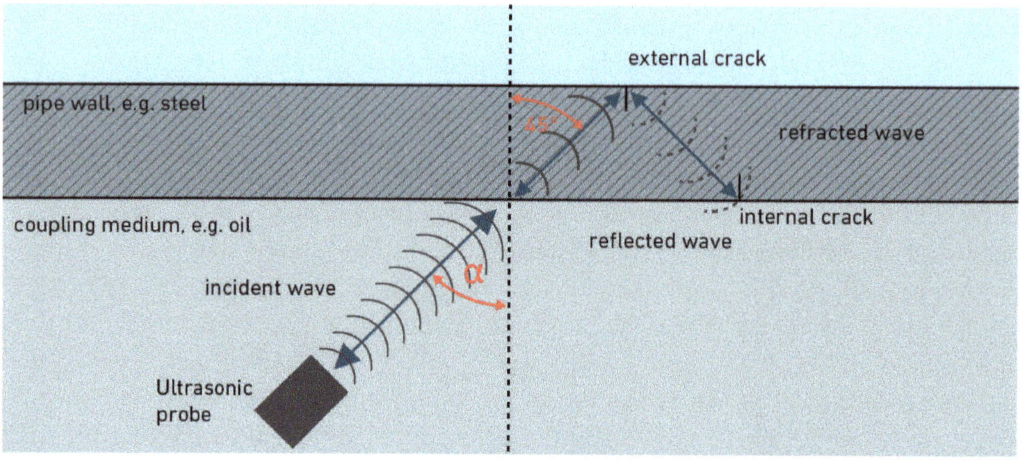

Figure 4-7. UT measurement principle for crack detection in the pipe wall.

Ultrasound utilizing electro-magnetic acoustic transducers (EMAT). EMAT is a non-contacting approach that generates an ultrasonic pulse within the test piece, thus not requiring a couplant. In the case of ILI the ultrasound is generated in the pipe wall itself through electro-magnetic interaction between the sensor on the tool and the pipe wall. The transducer incorporates a coil placed in a magnetic field where an alternating current in the coil generates vibrations in the wall which in turn generate ultrasound. There are two different mechanisms being used, Lorentz Forces and Magnetorestriction, see Figure 4-8. Both phenomena affect the atomic grid of the test sample, thus generating the ultrasound within that material. The Lorentz force is caused by interaction between the electric current inducted through the eddy current coil and the magnetic flux.

In case of magnetorestriction the mechanism is based on change in the size of the metal grid due to the changing orientation of the magnetic domains. Due to the alternating current used a "vibration" of the crystalline structure of the metal is induced, the direction of this vibration controls the propagation direction of the ultrasonic signal. As soon as the ultrasound signal is generated things work just as before. The UT signal is reflected and the reflected waves induce a varying electric current in a receiver. This current signal is interpreted by software to provide indications about the structure of the sample.

Any faults or cracks in a sample constitute a boundary, which results in partial reflection of the ultrasonic pulse. As with standard ultrasonic inspection, the known speed of the ultrasound in the sample means that the dimensions of a crack or a metal loss can be calculated.

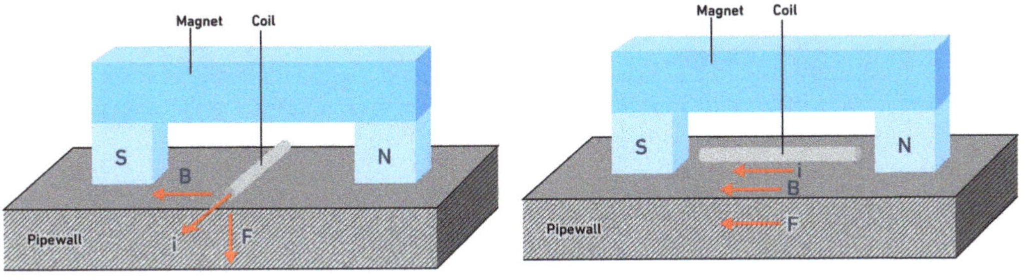

i Current
B Magnetic Flux
F Force

Figure 4-8. The EMAT principle: Inducing an ultrasonic signal without need for a liquid couplant.

The big strength of EMAT is that it does not require a liquid couplant. The advantages of UT can thus be utilized in a gas environment. But there is also a downside. The methodology requires magnetization of the wall in order to generate the ultrasonic waves within the material. Current EMAT tools therefore also have wall thickness limitations.

Ultrasound will be reflected off cracks, but also off the exterior boundaries of pipes, meaning that the technique can be used to determine the pipe thickness in the same way as conventional ultrasound techniques do. However, the sensors have to be aligned close to the wall otherwise not enough energy is available to induce the ultrasonic wave. Furthermore, an EMAT sensor cannot measure the distance between itself and the surface where the UT signal is created (nearside or internal surface of the pipe). These two limitations are the reasons why an EMAT technique is not suitable for identifying internal metal loss in a pipe wall. In order to utilize EMAT for full metal loss inspection and the ability to differentiate internal and external features does therefore require a secondary technology on the tool.

A very prominent feature of EMAT, unmatched by other NDT methods, is detection and location of external pipe coating disbondment so that coating repairs can be made before any structural damage to the body of the pipe can occur. The principle is based on the different attenuation of the signal at the external surface depending on the type and quality of coating present.

Table 4-2 provides a summary of the non-destructive testing technologies mentioned so far.

Table 4-2. Major non-destructive testing technologies applied in ILI tools today.

Method	Application	Main Mission
MFL	• Axial Magnetization • Circumferential Magnetization • Spiral Magnetization	• Metal Loss Inspection
UT	Piezo-electric - vertical beam	• Metal Loss Inspection and Wall Thickness Measurement
	Piezo-electric - angled beam	• Crack Inspection
EMAT	Electro-magnetic acoustic transmission (different wave types used)	Depending on Configurations: • Crack Inspection • Metal Loss Inspection • Coating Inspection
EC	Traditional Eddy Current	• Complimentary technology used for internal metal loss inspection • ID/OD differentiation

4.5 What do ILI tools do?

ILI tools collect data about features present in a pipeline. This process includes detecting a feature, sizing and locating it. The information is basically geometric data on shapes or pipe movement. ILI tools do not identify corrosion or cracks, they identify their geometries. It is then the expertise of a data analyst or also today an intelligent software algorithm which turns these data into useful information. ILI tools therefore provide information which in turn is used during the integrity assessment process.

Looking at this data gathering process implies use of a suitable sensor. This means that we must understand what these sensors can do and what they cannot. We must also understand the quality of the data we will obtain.

Ideally, if we are to investigate the mechanical integrity of a pipeline, we need to know everything (i.e. are there any defects and how big are they), everywhere (i.e. for the entire circumference and length) and at any time about the line in question.

Which of these dimensions do ILI tools fulfil?

Everything addresses data and its quality. This aspect relates directly to sensors used and their capabilities. Resulting technical issues or criteria are threshold, resolution and accuracy.

Everywhere relates to the issue of coverage. Modern ILI tools are very good at coverage, inspecting the entire circumference of the pipe wall and also the entire length of the section being investigated. This can include a total length of over 1,000 km.

At any time would imply a 24/7 monitoring, which ILI tools cannot do. ILI tools are not run continuously through a line, but at intervals determined by either regulatory requirements or a risk-based prioritization.

Using this understanding helps to define what the tools do and what they are: they are automated, mechatronical systems combining the required propulsion for a suitable sensor to get to a location of interest.

4.6 Traditional versus challenging in-line inspection

The oil and gas industry can be divided into three major segments: Upstream, Midstream and Downstream. Without going into details these major segments relate to Exploration and Drilling, Transportation, and finally Distribution of oil, gas and related products, basically the value chain all the way to the end user. Pipeline infrastructure can be found in all segments. The type of pipelines that ILI tools have originally been designed for are high pressure transmission pipelines usually associated with the midstream sector. Today there are approximately two million km (1.55 million miles) of these high pressure transmission pipelines[11]. However, there are also a lot of pipelines associated with the upstream and downstream sector, which are operated at elevated pressures and which need to be inspected. In total it is estimated that this includes roughly another 1.5 million km (937 thousand miles).

The ILI tools most widely used today are the so-called "free-swimming" tools. The word "free-swimming" is used because these tools are pumped through the pipeline to be investigated by the medium (oil, product or gas) transported in the line during normal operations. These tools can also be used for certain pipelines in the upstream and downstream sector. All these lines have been termed "piggable". The term "piggable" implies that a pipeline can be inspected with a free-swimming tool without any need to significantly modify the tool or the pipeline. However, there are pipelines in all three sectors which cannot be inspected by these "traditional" ILI tools, and have therefore been termed "un-piggable" pipelines[12]. The "un-piggable" pipelines are actually often found in the upstream sector, for instance infield (on- and offshore), storage or loading lines, as well as downstream, for instance high pressure distribution lines.

The problem is that the term "un-piggable" is not precisely defined and can be misleading. So what do the terms "piggable" and "un-piggable" really mean? Let's start with "piggable".

The term "piggable" is associated with the ability to use existing free-swimming ILI tools for inspection purposes. Firstly, this means that the inspection device can be inserted into the line at a suitable location and subsequently retrieved downstream. In general, this addresses the accessibility of a line. It implies that a pipeline needs suitable access points for insertion and retrieval. These are called launchers and receivers. Figure 4-9 shows schematically a typical simplified layout for a launcher. The schematic shows the launcher with its trap door. Valves 1 and 2 are used to isolate the launcher. Once the launcher is empty (at ambient pressure) it can be opened in order to load the tool.

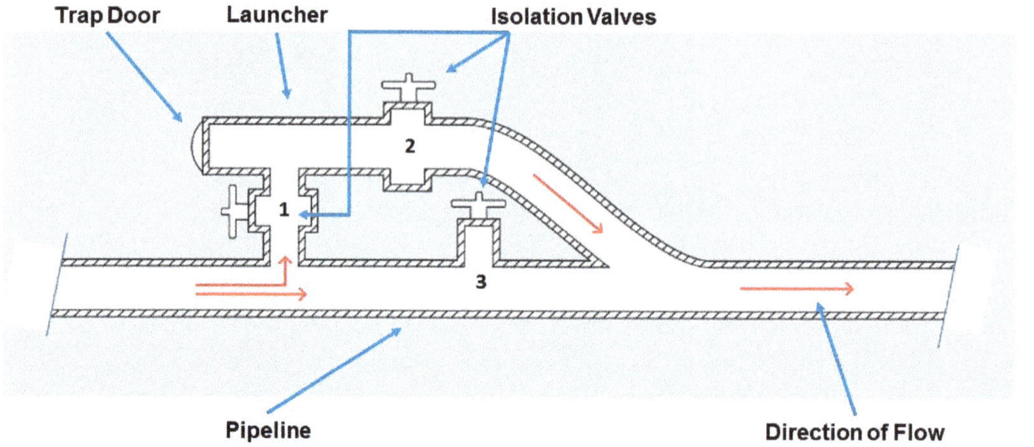

Figure 4-9. Typical launcher set up for a "piggable" pipeline.

The tool is launched by slowly closing valve 3 and opening valves 1 and 2. The flow in the pipeline is then re-directed through the launcher and the tool is pushed into the main line. After launching the trap will be isolated again and de-pressurized. In a real launcher there will be more fittings and pipes needed for example for measuring pressures, emptying the launcher or special devices for checking whether a tool has left the trap or is still present (e.g. pig signalers).

The next consideration is whether a chosen tool can actually travel through the entire line in a controlled manner. This regards the issue of negotiability. Free swimming ILI-tools can negotiate certain pipeline geometries, but not all. If a line to be surveyed has some "unusual" geometric design features, it could be labelled "unpiggable". Unusual geometries, such as very tight bends or mitre bends or significant variations in the internal bore can make existing traditional ILI tools unsuitable.

The third topic to address is the issue of propulsion, i.e. how is the tool moved in the line. Can it be pumped? Does it need to be pulled or pushed by other means? May it even need its own propulsion device? A strength of traditional ILI tools is that they can usually be applied while the pipeline remains in operation, in turn this implies that the operational conditions during a survey must be in-line with its capabilities. This refers to the absolute and differential pressures available during the inspection, as well as pumping speeds. Again, some conditions may prevail during an inspection, e.g. low pressures, low flow or the opposite which are outside the capabilities of the traditional ILI tools.

All in all, many factors come together in determining whether a line is piggable or not. But if a line cannot be inspected with the traditional ILI tools, we still have to ask the question whether the pipeline can be inspected from the inside (internal inspection) or only from the outside (external inspection). As pipelines are in most cases buried or at the bottom of the sea, an internal inspection provides the advantage that the entire pipeline (full circumference and length) can be inspected. In the case of an external inspection only a selected number of points can be inspected, because it is in most cases not feasible to dig up an entire line in order to access it. External inspection therefore usually only provides spot-checks.

The real question is therefore not the distinction between "piggable" and "unpiggable", but whether and how the pipeline can be inspected internally. Table 4-3 illustrates a simple way of defining the classification.

The term "piggable" should therefore be used if a pipeline can be inspected from the inside with a traditional ILI tool and using a standardized process. This implies that neither the tool nor the pipelines need to be modified significantly. If such a modification to the tool, pipeline or the process used is necessary, but the inspection can still be done internally, we will today usually term it a "challenging inspection". The tools used for these challenging inspection are commonly also called Intelligent Pigs or ILI-tools. If the pipeline can really only be inspected from the outside, then we have a truly "un-piggable" situation.

Different categories of internally applied tools are available for challenging pipelines today, see Figure 4-10:

- Speciality free swimming tools
- Tethered tools
- Crawler- and robotic tools

Speciality free swimming tools are modified free swimming ILI tools which can negotiate pipelines with certain geometric restrictions, obstructions or challenges. Examples may be multi-diameter lines, small bend radii, certain pipe installations such as wye- or tee-pieces, or single access pipes that require bi-directional tools in order to be launched and received at the same location. Special designs may also have to be used to accommodate for special operating conditions, such as low flow or low pressure operating conditions.

Table 4-3. Internal vs external inspection.

Piggable vs. Un-Piggable		
"Piggable"	"Challenging"	"Un-Piggable"
Free Swimming ILI - "Traditional ILI"	• Speciality Free Swimming • Tethered Tools • Crawler Tools • Robotic Tools	External Non Destructive Testing (manual or automated)
Internal Access		External Access

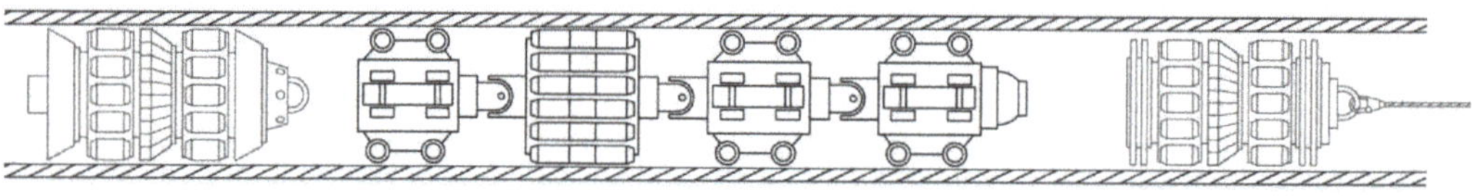

Figure 4-10. Illustrations of free-swimming, crawler and tethered tools.

Another type of ILI tool for challenging pipeline inspection are tethered or cable operated tools. The cable can be used for energy supply, and/or data transfer, or simply as an option to pull the tool back to retrieve it. Tethered tools generally do not have their own drive and need to be pumped in at least one direction, although there are also tools being pulled.

If pumping, provision of a differential pressure or even the use of a tethered tool is not feasible, then an inspection device would need its own propulsion. This introduces the realm of crawlers. Crawlers will move an attached inspection device. The next evolutionary step is a truly autonomous robotic tool, the latter not only in the sense of moving independently of flow or a cable, but also regarding their ability for making decisions. A simple example would be the ability of the robotic tool to independently recognize the end of a pipe section to be investigated and automatically return to its point of entry.

It is important to understand that the inspection of a challenging pipeline requires more than just a specialized tool. It also entails developing and using tailor-made processes for a given inspection task, in other words it requires an optimized solution engineering approach. This not only includes the measurement systems but also the required cleaning and pipeline preparation, as well as tool tracking procedures[13].

QR Code 4-1. Scan here for flowchart on accessibility, negotiability, and propulsion.

4.7 Types of tool commercially available

Table 4-4 provides an overview of the ILI tools commercially available today. The Table is sorted by inspection task. It also indicates which types of propulsion are utilized. The largest variety of ILI tools commercially available are free swimming tools. Tethered and robotic tools are not used as widely and therefore are currently also still limited in the inspection tasks they can be used for.

Table 4-4. ILI tools commercially available today

Type of Tool	Working Principle	Tool Mission	Available Form of Propulsion
Caliper/ Deformation	Mechanical spider, electro-magnetic principles	Geometry inspection	Free-swimming
Mapping/Center Line	Inertial measurement unit	Center line, pipe movement, mapping	Free-swimming
Leak Detection	Pressure difference, acoustic	Detection and location of leaks	Free-swimming
Metal Loss	Ultrasound, Magnetic Flux Leakage, Eddy Current	Detection, sizing and location of metal loss	Free-swimming, Tethered, Crawler
Crack Inspection	Ultrasound	Detection, sizing and location of cracks and crack-like defects	Free-swimming, Tethered
Material Grading	Electro-Magnetic	Identifying steel strength of linepipe	Free-swimming
Coating Inspection	Ultrasound	Identification of coating disbondment and coating type	Free-swimming

4.7.1 Caliper inspection

Caliper tools, in some regions also referred to as Deformation tools, are designed to detect, size and locate geometric features such as dents, out-of-roundness or basically any change to the internal cross section of the pipe. Most tools on the market use a mechanical spider. Figure 4-11 depicts a typical Caliper tool.

The fingers of the spider, often fitted with little wheels to avoid any scratching of the pipe wall, are pressed against the inner surface of the pipe with stiff springs. Any change to the internal bore of the pipe will be registered and measured and recorded electronically.

Today there are two types of such tools. Single channel tools, where the information is recorded into one channel and multi-channel tools, where the information from each finger is recorded individually. The latter allows the determination of the circumferential position of any defect identified. There are also Caliper tools on the market which utilize electro-magnetic sensors, again to detect and measure changes to the internal cross section of the pipe. Modern tools are very sensitive and accurate. They can pick up local defects with a very high resolution. Information from such tools can be fed into Finite Element Analysis software in order to calculate local strains and subsequently mechanical stress distributions.

Caliper tools can also be fitted with bend measurement systems in order to identify or confirm bend radii.

Figure 4-11. Caliper – ILI inspection tool for geometric anomalies (courtesy MEFS, Germany).

4.7.2 Mapping tools

This type of tool, often also referred to as Inertia or Centre-Line Tools, use inertia measurement units (IMU`s) in order to record accelerations in all three coordinate axis directions as the tool travels through the line. By using mathematical procedures this data can be used to determine x-, y-, z- coordinates for every point of a pipeline to a sub-meter accuracy. Earlier tools of this kind actually used gyroscopes.

The accuracy of these tools depends on various factors, as the IMUs tend to drift. The accuracy increases if the system is given more reference points, i.e. benchmarks for which the exact position is known. These benchmarks can be provided by marker positions or other known locations in the line. Markering will be explained in the chapter on ILI operations.

Repeated runs with these tools can be used to identify pipe movement. In turn this information can be used to calculate induced global strains and resulting stresses.

4.7.3 Leak detection tools

Leaks really constitute a failure rather than a defect. If they occur they need to be located. Leak detection tools do exactly that. They detect and locate leaks, provided the leak rate is higher than the detection threshold of the tool. Internal free swimming leak detection tools are currently only available for liquid lines. There are two types of tools. The first type uses a pressure difference approach. The theory here is that there is a local minimum pressure in the region of a leak. This local lowest pressure can be measured and a leak can thus be identified. The second type use an acoustic approach.

Any leak causes a characteristic sound. As the liquid transported in the pipe leaks out a "noise" is created with a characteristic frequency. The tools are fitted with hydrophones, a special type of microphone, and record any noise in the line. Filters are then used to identify a location where the typical "leak" sound is present. Odometer wheels are used for distance measurement and thus the ability to also locate the leak. The tools do not measure the leak rate itself and are therefore screening tools rather than full blown inspection tools.

QR Code 4-2. Scan here for more information on tool selection.

Figure 4-12 depicts a typical acoustic leak detection tool for liquid pipelines.

Modern leak detection tools are usually also fitted with other measurement equipment and can often additionally record temperatures and pressures. Some are also fitted with IMU systems and can also record travel velocities.

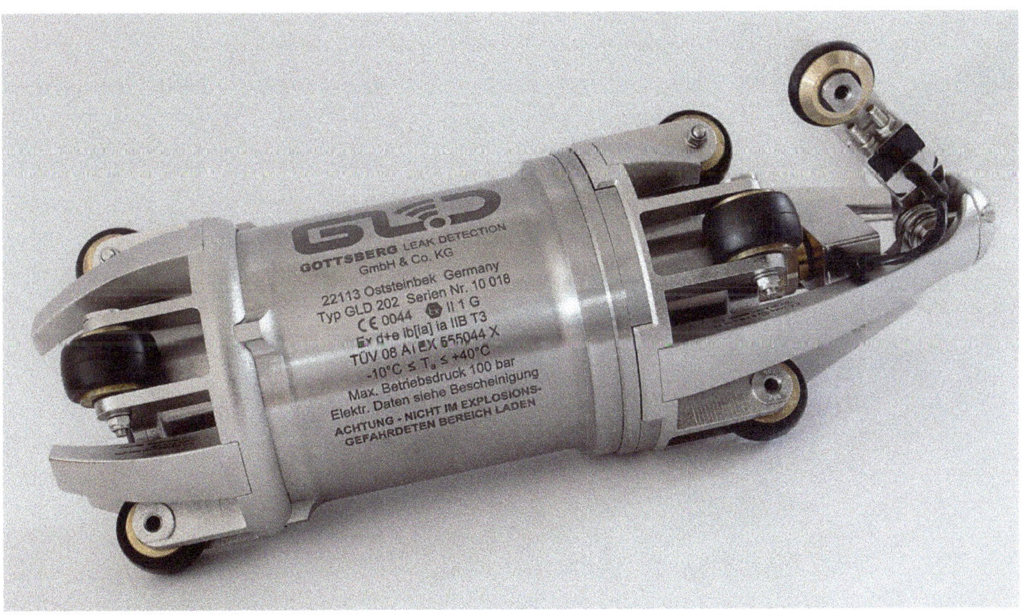

Figure 4-12. Free swimming acoustic leak detection tool (courtesy Gottsberg, Germany).

4.7.4 Metal loss inspection tools

Most of the ILI inspections performed every year are aimed at identifying metal loss. The major technologies utilized are magnetic flux leakage (MFL) and ultrasound (UT). There is a wide range of metal loss inspection tools available on the market today. Choosing the best tool for a given inspection task is not trivial. Depending on what type of metal loss you are looking for and which resolution is required for subsequent integrity assessment there are different MFL tools and UT tools. The differences relate to the direction of magnetization for MFL tools, shape of the ultrasonic beam for UT tools, detection thresholds, accuracies and resolutions. This implies that different types or configurations of these tools are available on the market and defect specifications of the different vendors should be compared carefully when planning an inspection. On the MFL side there are high resolution tools and recently also tools termed as ultra-high resolution tools. There are also MFL tools on the market that utilize different levels of magnetization. A high magnetic saturation level is required for reliable metal loss detection and sizing, whilst a lower magnetization level is useful to detect hard spots. Figure 4-13 depicts an advanced high resolution MFL tool utilizing axial and spiral magnetization to provide multiple data sets.

Regarding wall thickness measurement tools utilizing ultrasound technology there are also two configurations on the market today. High resolution UT tools and tools utilizing an even higher resolution. This latter type of tool was originally designed for the detection and sizing of pitting, and they are therefore often referred to as UT pitting tools.

Figure 4-14 shows a typical UT wall thickness measurement tool.

More recently another type of UT tool has been introduced. It uses the principle of acoustic resonance. Here a UT signal with a very much lower frequency is sent from the transducer to the wall to be inspected. If the signal has a wavelength corresponding to the wall thickness it will resonate. This provides another approach to measuring wall thickness and thus metal loss. This particular approach is not widely used within the ILI industry and has therefore not been included earlier in the chapter on non-destructive testing technologies.

Figure 4-13. Rendered picture of an advanced MFL metal loss ILI tool (courtesy of TDWilliamson).

Figure 4-14. Typical UT wall thickness measurement tool. (courtesy of Baker Hughes, a GE company).

4.7.5 Crack inspection tools

Crack Inspection tools generally use ultrasound techniques, although some specific types of cracks can also be detected by MFL. The two UT techniques or rather sensor types used are piezo-electric transducers and electro-magnetic acoustic transducers. Both techniques come with their individual and particular strengths and weaknesses from a defect as well from an operational perspective. The number of vendors offering crack inspection services is rather limited as compared to metal loss or geometry inspection. Crack inspection tools are technically more complex which can also be said for the data analysis process. This means that crack inspection is also the inspection methodology with the highest costs, i.e. crack inspection costs for a given length and diameter of pipeline are usually higher than for metal loss inspections. Figure 4-15 shows an EMAT crack inspection tool.

Figure 4-15. EMAT crack inspection tool (courtesy of Rosen-Group, Germany).

4.7.6 Material grading inspection

In order to assess a pipeline it is necessary to know its material specifications. Yield strength and Ultimate Tensile Strength are two such parameters. Sometimes, especially for older lines, these values are not known. A material grading inspection will utilize an ILI tool which can compare a signature measured for a given line with a reference value and thus help to identify the material characteristics of that line. It is a qualitative measurement based on an electro-magnetic measurement principle. What the tools cannot do is a quantitative measurement of a material property, say similar to a value determined via destructive testing in the laboratory. Even so this type of service is very valuable to help and understand what type of pipe the operator has. The technology was originally developed especially for vintage pipelines where documentation was not available or needed to be confirmed.[14].

4.7.7 Coating inspection

There is also a class of tools utilizing electro-magnetically induced ultrasound to detect coating disbondment. The theory behind this measurement is that any material touching the external surface of the pipeline attenuates the ultrasound signal. It is therefore possible to identify areas where coating is missing. Different types of coating influence the ultrasonic reflections. The technique can therefore also be used to distinguish different types of coating and even identify a coating type if it can be compared to a reference value.

4.8 ILI operations

The major steps in the process of an in-line inspection of a pipeline are:

- Pipeline preparation and cleaning
- Tool preparation
- Work at site – the inspection
- Data analysis and reporting

Let us now look at these individual steps.

Prior to inspection a pipeline must be checked to determine if it is ready to be inspected. This addresses the issues of cleanliness, accessibility, negotiability and propulsion. ILI vendors will ask the operator to fill in a Pipeline Questionnaire addressing all related questions.

Cleaning. In order to collect good quality data the internal surface of the pipe should be as clean as possible. That means removing anything adhering to the pipe, e.g. debris, scale, wax, etc. If the pipe wall is not clean then sensors can get damaged or are prevented from collecting inspection data correctly. Reasons can be for instance signal dispersion for UT tools or incorrect sensor spacing with respect to the internal wall for MFL. Pipelines should also be clean to avoid the whole tool being damaged or even getting stuck. For instance, crude oil lines with heavy wax build-up or gas lines with excessive amounts of dust from corrosion processes can cause operational issues regarding inspection.

Accessibility. Tools need to be introduced into a line, therefore access points must be available, referred to as launchers and receivers. The geometries of these must be suitable for the tool to be used. Sometimes only single access is possible or tool traps may be missing all together. Access points must be suited regarding the tools or the use of any auxiliary equipment that may be required.

Negotiability. The tools come with their own tool specifications. They need to be operated within these specifications in order to collect data with the highest possible quality. Inspection speed, pressures, temperatures, bend geometries, available internal diameters are important parameters.

Propulsion. Can the tool be pumped? Does it need to be pulled? May it require its own drive? These are important questions regarding the issue of propulsion.

Tool Preparation. Tools will be prepared at the premises of the vendor. Everything must work properly and consumables or wear parts, such as batteries or disks and cups, must be replaced. Preparation also includes calibrations and other procedures necessary prior to an inspection. After this step tools will be transported to site, wherever the pipeline may be. The crew operating the tool will usually travel and arrive separately.

Work at Site. At site the tool will be finally prepared and checked again prior to launching it in the pipeline to be inspected. Crew members will liaise closely with personnel from the pipeline operator, the latter always being responsible for the pipeline operation during the survey. During inspection, especially for onshore lines, markering operations are performed. This implies that one or more teams place above ground marker systems (AGM) along the pipeline route. These AGMs can communicate with the tools, as they travel through the line and can record the exact time at which the ILI tool has passed a given position. The coordinates of this position are known or measured. The information is then used as a benchmark regarding the distance measurement of the tool. The location accuracies achieved today are in the range of approximately ± 50 cm. Markering during offshore inspection is not usually performed, partly due to the complexity involved and partly because pipe books of subsea pipelines are generally accurate and the girth welds positions can be used. After an inspection run, the tool is retrieved. The tool is checked for any damage and data checks are also performed. The data cannot usually be analysed at site, but it is checked for completeness. These checks are necessary in order to determine whether a tool can be demobilized and sent back or a re-run may be needed.

Data Analysis and Reporting. The data collected is then sent back to the offices of the respective ILI vendor and then analysed. Based on the data collected the ILI vendor will prepare a report on the findings. This report can come in any number of formats, for instance printed or in an electronic format or ready for inclusion into a pipeline integrity management software.

4.9 Data analysis and reporting

The final step in the inspection process is data analysis and creation of a report. This report will include the results of the inspection. In the past different vendors have used their individual formats for these reports and sometimes they will still do this today. In order to make it easier to compare reports and findings from different vendors the pipeline industry has set up the Pipeline Operator Forum (POF)[15]. A large number of operators from around the world, onshore and offshore, are members of this group. They have worked out recommendations on how the findings of ILI inspections should be reported. The recommendations include a proposed structure, definitions of defect categories, the required general content as well as specific items a good quality report should contain. These recommendations are regularly reviewed and made available on the web site of the Pipeline Operator Forum.

In the past the final Reports issued used to be printed. This format is still available today but complemented by electronic formats such as digital version of the report, formats which can be fed into spreadsheets, data bases, GIS systems or even full Pipeline Integrity Management Systems (PIMS). Figure 4-16 depicts a screenshot of a typical data visualization of a metal loss defect, in this case from an MFL inspection.

It is important to understand that the data collected is of critical importance for subsequent integrity assessment or fitness-for-purpose investigations.

A question often raised in practice is why it takes so long for reports to be published. There are two major reasons for this and why reporting times also depend on the technologies applied. The more data an inspection tool collects, the longer the analysis process takes. The amount of data collected depends on the length and diameter of line inspected and also the state of the line, i.e. lots of defects mean more data, less defects mean less data. The other factor is the technology used. Crack inspection tools collect more data than metal loss tools which in turn collect more data than geometry inspection tools for a given length and state of line. Thus, final reports on crack inspections will take longer than those for geometry inspection. Typical reporting times for metal loss are in the order of forty to fifty working days, whilst reporting times for crack inspection may go up to 180 days.

The total time needed for reporting not only includes the actual data analysis process, but also the speed at which the inspection data collected at site is made available for analysis. This point addresses checking of data collected, translating and transforming the data collected by the tool to a format that the computers used for data analysis understand.

There is a lot of ongoing work within the industry to speed up these processes and use automated software routines and in future also the principles of machine learning and big data analytics to shorten reporting times.

Today the reported ILI data is often fed into and used in a Pipeline Integrity Management System. The process is not be confused with the various types of asset management software programs available, often incorporating a GIS[2] functionality. Asset integrity management software allows all relevant linear (spatial) pipeline information and documentation to be archived, queried, correlated, compared. Such packages often include risk analysis capabilities, allow integrity assessment based on standardized maximum allowable pressure calculations or provide a functionality to plan and optimize dig programs and can thus help to simplify, formalize and combine components of Integrity Management processes into structured workflows[16].

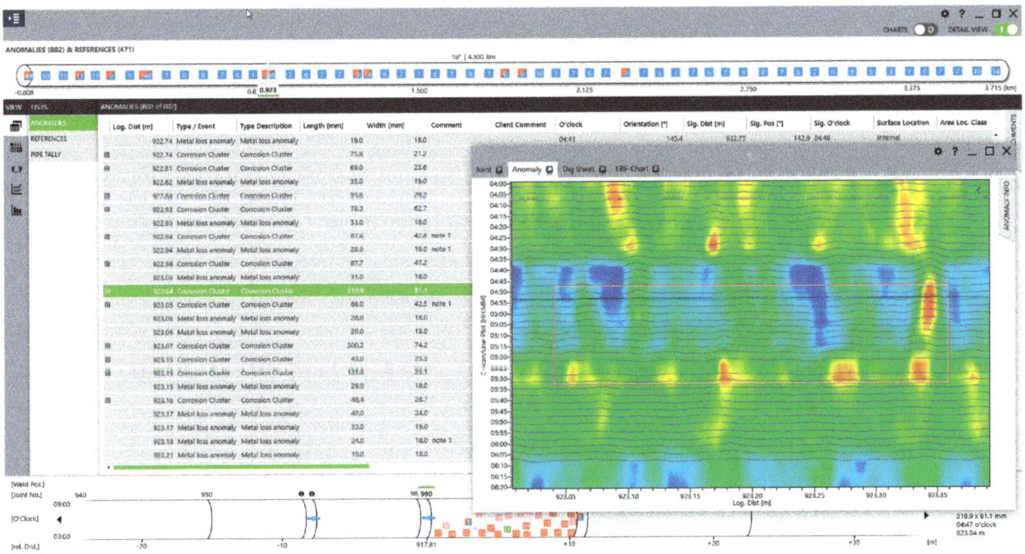

Figure 4-16. Screenshot of data collected with an MFL metal loss inspection tool (courtesy of Rosen-Group, Germany).

2 GIS (Geographical Information System).

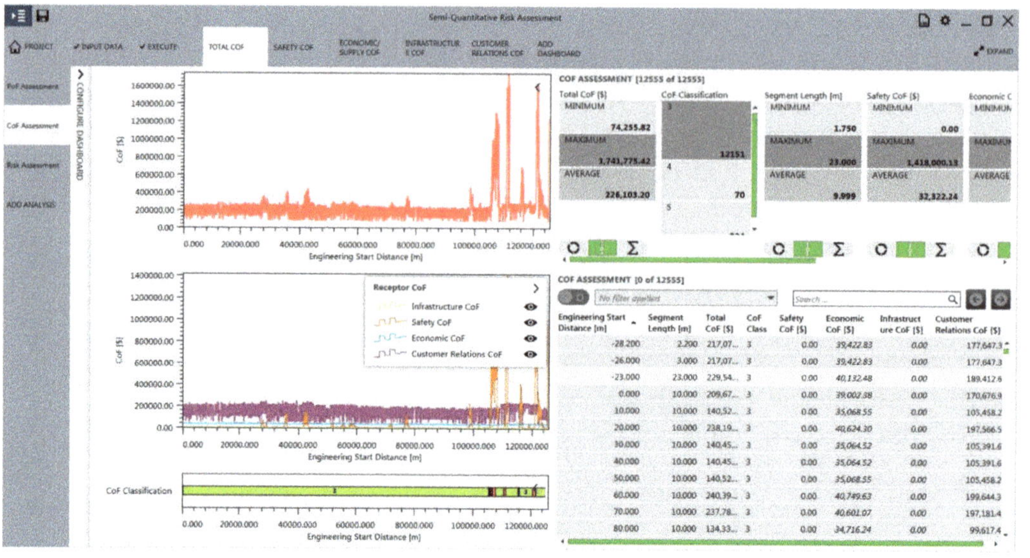

Figure 4-17. Screenshot of pipeline risk data displayed by pipeline integrity management software (courtesy of Rosen-Group, Germany).

4.10 Verification

Subsequent to an ILI inspection defects found are often validated and/or verified. This does not apply to every single defect reported during a survey, but a chosen sample. There can be different reasons for doing this, but usually the idea is to verify whether a chosen ILI tool has performed to the defect specifications contractually agreed.

In case of metal loss for instance this would mean validating the reported depth, length and width as well as the reported location.

Common complaints from operators performing such verifications can be that a defect identified by the ILI survey does not exist or that the sizing is incorrect. On the other hand, it may well be that the verification team, sent to validate a defect, have gone to the wrong location.

Not finding a reported defect may happen for two reasons: an error in the estimation of the location of the defect, either regarding distance along the line or orientation around the circumference. The other reason could be that the inspection equipment used for validation was not sufficiently accurate. The equipment used for verification of ILI inspection data should generally have a higher resolution than the ILI tool itself[17].

Planning an effective verification or validation is not trivial and comprises of the following major steps: identifying the true location of a reported defect, identifying the most suitable non-destructive testing equipment for the verification based on the resolution of the ILI tool. Planning the excavation: for onshore validation work or possible repair, it is necessary to excavate a bell hole for access. Depending on the environment the pipeline is in, this can be a very costly exercise. Offshore validation is not easier, involving diver time and issues around accessing the pipe, for instance if an offshore pipeline is trenched or weight coated with reinforced concrete.

4.11 References

[1] Shannon, R.W.E., Development Operating Experience Prove Value of British Gas ' On Line Inspection, Pipeline & Gas Journal, July 1985.

[2] Shannon, R.W.E, Braithwaite, J.C., Morgan, L.L., Flux-leakage vehicle pass tests for pipeline inspection, OGJ Report, Oil & Gas Journal, Aug.8, 1988.

[3] Kittel, A., Der Rißprüfmolch – ein Gerät zum Auffinden von Fehlern bei erdverlegten Rohrleitungen – Crack Detection Pig for Buried Pipelines, 3R-International, 15(2/3), Essen, 1976.

[4] Goedecke, H., Krieg, G., Ultraschall-Molchsystem zur Korrosionsprüfung in Pipelines, Technische Überwachung (TÜ), Bd.28 Nr.1, pp.9-11, 1987.

[5] Barbian, A.O., Beller, M., Goedecke, H., Advances in Pipeline Integrity Assessment Using In-Line Inspection Tools, Proceedings of the Pipeline Pigging and Inspection Technology Conference, Houston, 1992.

[6] Willems, H., Barbian, A.O., Operational Experience with In-Line Crack Inspection in German Crude Oil Pipelines, Proceedings ECNDT, Copenhagen, 1998.

[7] Eiken, T., Thale, W., A Novel Non Destructive Technology for Pipe Grade Determination and MAOP/Design Pressure Validation of Operating Pipelines, Proceedings, 19th World Conference on Non-Dectructive Testing (WCNDT), Munich, 2016.

[8] Beller, M., Reber, K., Uzelac, N., How Do Defect Assessment Methods Influence The Choice And Construction Of In-Line Inspection Tools, IPC02- 27391, Proceedings, International Pipeline Conference, Calgary, 2002.

[9] Beller, M., Pipeline Inspection Utilizing Ultrasound Technology: On The Issue Of Resolution, PPSA-Seminar, Aberdeen, 2007.

[10] Barbian, A.O., Beller, M., Hugger, A., Jäger, C., Pfanger, A., Thresholds, Accuracies, And Resolution: The Advantages Of Quantitative Measurement For Metal Loss Inspection, Proceedings, Pipeline Pigging and Integrity Management Conference (PPIM), Houston, 2010.

[11] Global Data, Energy, Data Base, © GlobalData 2018. John Carpenter House, 7 Carmelite Street, London EC4 0BS, UK.

[12] Shie, T., Koch, G., Bubenik, T., Unpiggable Pipelines, Proceedings, Pipeline Pigging and Integrity Management Conference, Houston, 2010.

[13] Beller, M., Steinvoorte, T., Vages, S., Mastering the Inspection of Challenging Pipelines, Part 1 and 2, Pipeline & Gas Journal, Oildom Publishing, October and November, 2015.

[14] Slater, S., Dwyer, K., Ginten, M., Eiken, T., Burkinshaw, O., An Integrated Approach For The Verification Of Pipeline Material Using The Latest State-Of-The-Art Technologies And Engineering Processes, IPC2018-78776, Proceedings IPC 2018, Calgary, 2018.

[15] Pipeline Operator Forum, Specifications and requirements for in-line inspection of pipelines, Version 2016, www.pipelineoperators.org (accessed September 19th, 2018).

[16] Huisman, O., Reyes, A., Formalizing Integrity Management Workflows: Towards Integrity Process Modelling, IPC2018-78512, Proceedings IPC 2018, Calgary, 2018.

[17] Beller, M., King, R.A., Reid, G, Location and Validation of Metal Loss Defects Identified by In-Line Inspection, Proc. 31st Pipeline Pigging and Integrity Management Conference (PPIM), Houston, 2019.

4.12 Recommended reading and further information

listed in alphabetical order.

- Encyclopedia of Pipelines Defects, 3rd. edition, Rosen-Group, published by Tiratsoo Technical, 2018.
- In-Line Inspection Systems Qualification, API Standard 1163, API, 2013.
- Managing Systems Integrity for Hazardous Liquid Pipelines, API Recommended Practice 1160, Second Edition, September 2013.
- Managing Systems Integrity of Gas Pipelines, ASME Code for Pressure Piping, B31, Supplement to ASME B31.8, 2014.
- NACE SPO102-2010, Standard Practice In-Line Inspection of Pipelines, NACE International, 2010.
- Oil and Gas Pipelines, Integrity and Safety Handbook, Edited by Winston Revie, John Wiley & Sons, 2015.
- Pigging Products & Services Association, web-site: www.ppsa-online.com (accessed December 4th, 2018).
- Pipeline & Gas Journal (printed Journal), Gulf Energy (USA), www.pgjonline.com (accessed Dec. 5th, 2018).
- Pipeline Integrity Assurance – A Practical Approach, edited by M. Mohitpour, A. Murray, M. McManus, I. Colqhoun, ASME-Press, 2010.
- Pipeline Pigging and Integrity Technology, 4th. edition, ed. John Tiratsoo, Tiratsoo Technical.
- Pipeline Pigging Handbook, 3rd edition, Jim Cordell, Hershel Vanzant, Clarion and Scientific Survey, 2003.
- Pipeline Technology Journal (e-Journal), EITEP (Germany), www.pipeline-journal.net (accessed Dec. 5th, 2018).
- Pipelines International (printed Journal), Great Southern Press (Australia), www.pipelinesinternational.com (accessed Dec. 5th, 2018)
- World Pipelines (printed Journal), Palladian Publications (UK), www.worldpipelines.com (accessed Dec. 5th, 2018).

CHAPTER 5

External corrosion

Dr Roger King

5.1 Introduction

Pipelines are protected from external corrosion by the combination of a coating and cathodic protection. A corrosion allowance is not provided. External corrosion rarely occurs on submarine pipelines, except on riser sections above sea level, because the sacrificial cathodic protection systems used to protect submarine pipelines are very effective in the low resistance, relatively homogenous subsea environment. External corrosion is an onshore pipeline hazard and occurs because there has been a combination of failure of the coating and inadequate cathodic protection. In some cases, corrosion may occur where a coating has disbonded from the pipeline surface but remains sufficiently integral to prevent adequate CP current from reaching the pipe surface, this is termed 'shielding'. An example is given in Figure 5-1. Shielding may also occur if a large stone or rock is tight against the pipeline as the stone will prevent the CP current reaching the pipe.

The external corrosion mechanisms may vary along the pipeline route depending on the soil conditions. Corrosion occurs by the following mechanisms:

- oxygen corrosion is the most common mechanism and occurs in most soils, in particular in sandy and loam soils;
- the carbon dioxide (CO_2) concentration in soils may be high, up to 10%, and the CO_2 dissolving in the soil water will generate carbonic acid;
- organic acid corrosion occurs in areas where there is decaying vegetation, e.g. marshland, peat, mangrove swamps;
- growth of aerobic acid producing bacteria in soils containing sulphides may generate sulphuric acid;
- microbiologically induced corrosion occurs in clays, silts and heavy loams where anaerobic conditions can occur and when sulphate is present;
- stray current corrosion and telluric currents.

Corrosion is found at the areas where the corrosion coating has failed and exposed the pipeline; this often occurs at the field joints because these are applied in the field often under difficult conditions. Because there is a potential difference between weld material and pipeline steel there may be preferential weld corrosion at the longitudinal and girth welds and the heat affected zones. Corrosion may also occur as a result of shielding.

Routine maintenance involves regular monitoring of the CP potentials of the pipeline at the test points. These are typically 1 - 5 km apart along the complete length of the pipeline with additional test points at areas of especial interest, e.g. road crossings, third-party pipeline crossings and isolation flanges. The routine monitoring is used to provide information to fine tune the CP system to be as efficient as possible. In many modern systems the CP potentials at the test points is automatically uploaded to a control centre for evaluation.

At the test points the potential is measured by connecting a high impedance voltmeter to the pipeline through the connection at the test point and placing a reference electrode on the ground above the pipeline. The reference electrode is designed to be stable provided current is not drained from it. The most common electrode for use with onshore pipelines is made from copper in a copper sulphate solution (see Figure 5-2).

In an ideal world the reference electrode would be placed very close to the pipeline, but this is not practical, and the reference electrode will be 1 - 2 m distance from the top of the buried pipeline (see Figure 5-3). This measure of the potential is subject to a systematic error because there is electric current flowing through the soil to the pipeline and this introduces a voltage error in the reading of the potential. The voltage is the result of the current and the resistivity of the soil, i.e. V = I x R and is termed the IR drop or IR error. The IR drop always makes the potential of the pipeline look more negative than it really is, so it is necessary to measure the potential of the pipeline avoiding the IR drop. The CP potential measured when the CP system is operating are termed CP ON potentials.

One way to exclude the IR drop is to switch off the CP system and measure the potential of the pipeline immediately that the current is off. Because current is zero, the IR term is also zero. This method is commonly used and involves voltmeters with synchronized timers that match the switch at the CP station. These potentials, termed CP OFF potentials, are the correct potentials on which to judge the effectiveness of the CP system.

The limitation of routine monitoring is that the potential of the pipeline is only measured at a few locations. To ensure the pipeline is fully protected it is necessary to measure the potential at close intervals along the length of the pipeline to ensure all the pipeline is at the necessary protection potential. This type of survey is described as a close interval potential survey, CIPS.

The CIPS will identify if there are areas of inadequate potential and where there may be external corrosion occurring. The most common reason why the potential is inadequate is that there is coating damage. The location and extent of coating damage can be identified by a second type of survey: Pearson or DCVG survey. In some instances, a DCVG survey can indicate if corrosion is occurring at the coating defect or not. Where shielding occurs the CP potential may be correct and a CIPS, Pearson or DCVG survey will not identify corrosion. Corrosion can only be identified by an inline inspection survey.

To determine which corrosion mechanism has occurred it is necessary to obtain soil from the corroded area and also some corrosion product. The soil should be tested for pH, total acidity, sulphate and chloride. If possible, soil resistivity should be measured at the site. When an inline inspection survey is not possible then an evaluation of the pipeline may be made following recommendations in NACE SP-0502[1] and NACE SP-0204[2].

QR Code 5-1. Scan here for additional information on external corrosion.

QR Code 5-2. Scan here for additional information on CP potential surveying.

Figure 5-1. Corrosion under failed field joint coating showing enhanced attack on both sides of the girth weld.

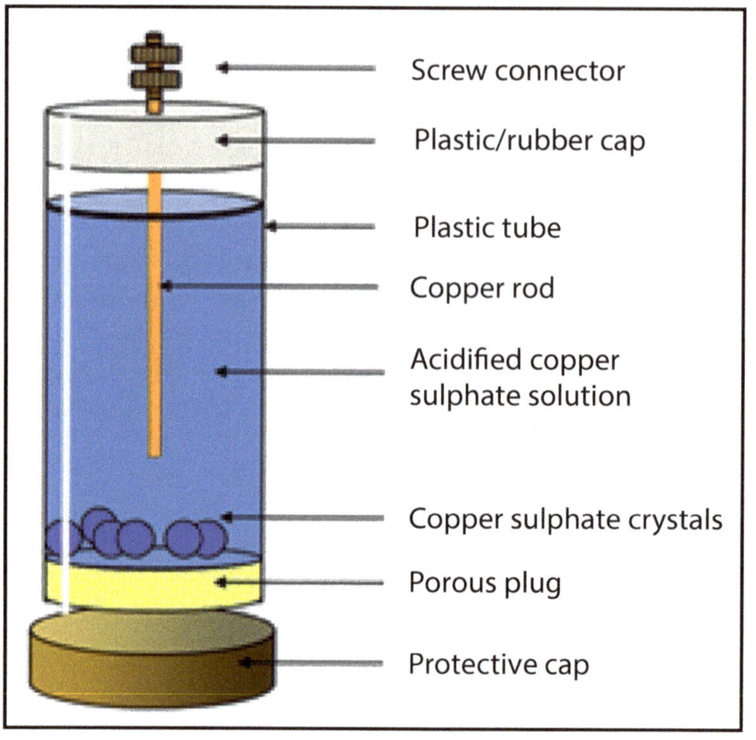

Figure 5-2. Schematic of a copper-copper sulphate reference electrode showing copper rod, acidified saturated solution of copper sulphate within glass cylinder sealed with permeable lower seal.

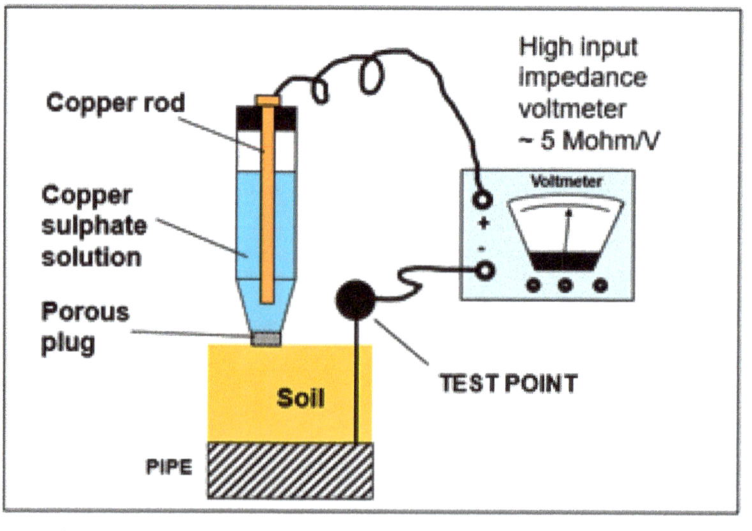

Figure 5-3. Measure of pipeline potential using a reference electrode and voltmeter connected to buried pipeline through a test point. Typically test points are every 1 - 5 km apart.

5.2 General external corrosion

5.2.1 Oxygen corrosion in near neutral soils

Mechanism. If the pH is near neutral (>6) or alkaline then corrosion is most likely caused by dissolved oxygen. Typically, these soils are sandy, sandy loam or light loam. Corrosion will be general with superimposed pitting (see Figure 5-4). The corrosion occurs when there is coating damage to expose the steel and there is water and oxygen which react with the pipeline steel to form ferrous hydroxides and oxides though other salts may be formed depending on what anions are in the soil.

The corrosion rate is not linear but declines over time as corrosion product films build up on the pipeline surface. The depth of corrosion, penetration *P*, follows a power law:

$$P = k\,(\text{time})^n$$

Where *k* is related to the nature of the soil and partly to the steel composition, *n* is the exponent related to the nature of the soil and other conditions, e.g. water content, seasonal cycles. In many soils the exponent *n* is in the range 0.4 - 0.6. With bacterial corrosion the exponent is often closer to 0.9. Figure 5-5 illustrates the effect of the exponent on perforation rate.

The corrosion rate will be determined by soil resistivity, increasing as the resistivity decreases. Corrosion rate will also be related to the chloride and sulphate concentrations; high concentrations increase corrosion rate. Corrosion products will be brown with black magnetite close to the pipeline surface. If the pH is alkaline then there is the risk of carbonate cracking, see relevant section.

Moderating oxygen corrosion

- A CP potential more negative than -850 mV (CCSE) is required but in most cases a potential more negative than -950 mV (CCSE) is specified as it is possible that microbiological influenced corrosion may occur.
- CP current requirement will be related to the corrosion rate.
- Assess pipeline using NACE SP-0502.

Figure 5-4. Regular corrosion perforations as a result of unfurling of a tape wrap coating.

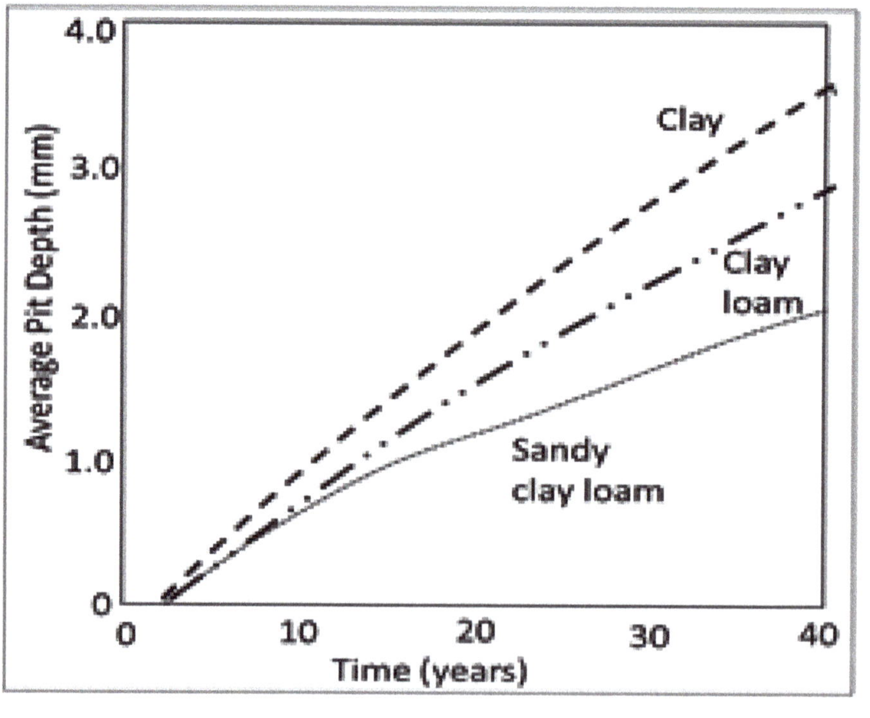

Figure 5-5. Penetration rate of buried pipeline depends on the nature of the soil.

5.3 Acid soils

Acid corrosion occurs when a pipeline is buried in acid sulphate soil and in bogs, marshes, fens where there is decaying vegetation that releases organic (humic) acids. If the pH is below 6 and the total acidity high, though this may not be reflected in the pH value of the soil water, then organic acids are likely to be a major reason for corrosion. Corrosion by oxygen may also be involved. Corrosion will be general with limited pitting and the potential corrosion rate will be high.

The corrosion morphology is general corrosion with less pitting than occurs with oxygen corrosion and there will be limited corrosion products as much of the ferrous ion will remain in solution and migrate away from the corroding area. The corrosion products will be ferrous and ferric oxides and will appear brown and orange. Soil resistivities are usually low. Soils with chloride concentration >100 ppm are more corrosive.

5.3.1 Moderating acid corrosion

- A CP potential more negative than -850 mV (CCSE) is required to prevent corrosion.
- CP current requirement will be high.

5.4 Microbiological influenced corrosion (MIC)

5.4.1 Mechanism

If the pH is neutral or alkaline (pH >7) and the pipeline is buried in clay or heavy loam soils and there is sulphate present there is the possibility of corrosion by the sulphate-reducing bacteria (SRB)[3]. These soils tend to have low to medium resistivity. The SRB are anaerobic and growth is triggered when the oxygen concentration in the soil falls to a low value and sulphate is present. The SRB obtain energy for survival and growth by oxidizing short chain carboxylic acids to carbon dioxide and water using the oxygen in the sulphate radical. To do this they use unique enzymes. The sulphate is released as hydrosulphide (HS-) ion. Some strains of SRB can also utilize hydrogen and it is these strains that are implicated in corrosion.

The corrosion product formed is a mixture of iron hydroxides, iron sulphides and some iron oxides and is black (mainly iron sulphides) and, if freshly exposed slimy. The morphology of corrosion by SRB depends on the pH. As clays and heavy loams are usually neutral to alkaline the corrosion appears as multiple overlapping hemispherical pits filled with black slime (see Figure 5-6). It is important to check the corrosion product as soon as possible and to determine the presence of sulphide by addition of acid to the corrosion product and noting the evolution of hydrogen sulphide (bad egg odour). Where the corrosion product has been exposed, the iron sulphide will have oxidized and the corrosion product will be orange and brown.

The bacterial corrosion rates increase over time and become roughly linear. The exponent in the penetration equation is in the range 0.9 - 0.95.

5.4.2 Moderating microbiological corrosion

- Check corrosion product for presence of iron sulphide
- Check soil for presence of sulphate-reducing bacteria
- The CP potential must be more negative than -950 mV (CCSE).
- A higher CP current will be required
- It may be necessary to modify the CP system to ensure the potentials at the ground beds are not excessive.

If corrosion is identified by ILI survey and the CP OFF potentials are correct and the coating surveys do not indicate coating damage, then it is likely that the corrosion is occurring under a shielding coating (see Figures 5-7, 5-8, 5-9). Most coatings can shield except fusion bonded epoxy coatings. When corrosion occurs under an FBE coating the coating tends to crack, sooner or later, allowing the CP current to reach the surface and protect the steel. There is some uncertainty about the behaviour of dual layer FBE coatings. CP cannot protect a pipeline under a coating that shields. Figure 5-8 illustrates corrosion that occurred under a heat shrink sleeve.

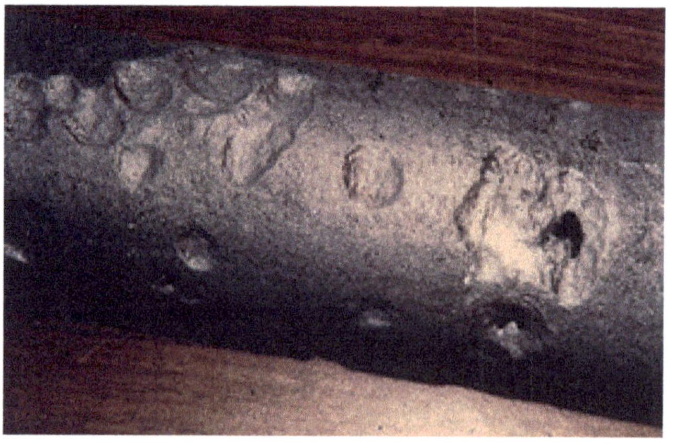

Figure 5-6. External corrosion of water main in clay soil as a result of activity of sulphate-reducing bacteria.

5.5 Shielding

If corrosion is identified by ILI survey and the CP OFF potentials are correct and the coating surveys do not indicate coating damage, then it is likely that the corrosion is occurring under a shielding coating (see Figures 5-7, 5-8, 5-9). Most coatings can shield except fusion bonded epoxy coatings. When corrosion occurs under an FBE coating the coating tends to crack, sooner or later, allowing the CP current to reach the surface and protect the steel. There is some uncertainty about the behaviour of dual layer FBE coatings. CP cannot protect a pipeline under a coating that shields. Figure 5-8 illustrates corrosion that occurred under a heat shrink sleeve.

5.5.1 Moderating shielding corrosion

- Excavate the pipeline
- Validate the depth of corrosion determined by the ILI survey
- Check the pipe surface for cracking.
- If the pipeline is cracked, then estimate the depth of the cracks
- If cracks are shallow, they may be ground out
- If cracks are deep, then either a type B sleeve should be installed, or the section of pipeline cut out and replaced.
- The remainder of pipeline should be checked using EMAT or an ultrasonic crack detection tool.

QR Code 5-3. Scan here for additional information on corrosion caused by SRB.

QR Code 5-4. Scan here for information on microbiological corrosion.

Figure 5-7. Detachment of an integral coating creating a void between coating and pipeline.

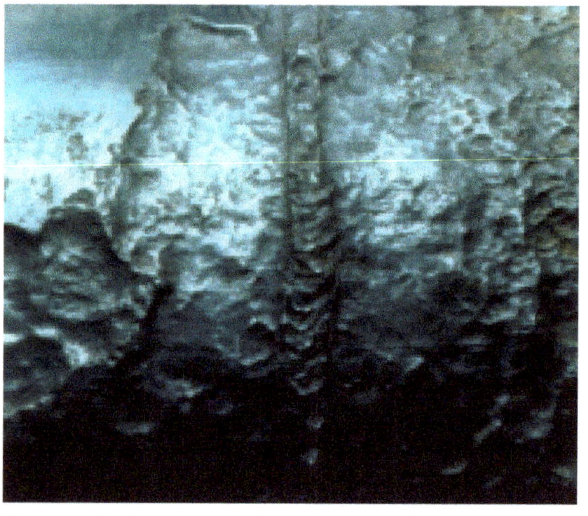

Figure 5-8. Corrosion under a detached field joint coating showing enhanced attach on both sides of the girth weld.

Figure 5-9. Stray current corrosion of a pipeline showing severe localized attack.

5.6 Preferential weld corrosion

5.6.1 Mechanism

Longitudinal, spiral and girth welds are present in pipelines and the metallurgy of the welds and heat affected zones adjacent to the welds is different to the parent material. The difference in potential between weld area and parent pipe creates a hazard of galvanic corrosion. Corrosion at a longitudinal weld presents a high hazard because, in most cases, the highest stress in a pipeline is the hoop stress which is attempting to increase the diameter of the pipeline. Corrosion at a longitudinal weld occurs because coatings can detach from the weld creating a void between coating and pipe adjacent to the weld.

Corrosion can occur within the void forming axial grooves along the longitudinal weld leading to a risk of bursting of the pipeline. Refer to the section on Shielding. Helical (spiral) pipe presents similar problems but the stresses at and adjacent to the spiral welds will be lower than the hoop stress.

Girth welds are protected by field joint coatings. Despite best efforts the quality of these field applied coatings is lower than the factory applied coatings. Water may manage to permeate under the field joint coating and initiate corrosion.

5.6.2 Moderating preferential weld corrosion

- Consider Pearson or DCVG coating survey to identify corrosion at field joints
- Check the inline inspection survey data to see if there is evidence of corrosion at the welds or at the field joints.
- It may be necessary to use a special pig because conventional pigs are less efficient at measuring axial metal loss.
- Excavate the pipeline at representative areas and inspect the weld and field joint area
- Validate the finding of the inline inspection
- Repair, sleeve or replace the affected section of the pipe

5.7 Stray current corrosion

5.7.1 Mechanism

Stray current corrosion occurs when an electrical current leaves the pipeline to return to its source. The stray current may come from a third-party CP systems, AC and DC powered electric railways and trams or high voltage transmission systems. The rate of corrosion may be very high. Pipelines with a good coating will suffer deeper metal loss because the current will flow through fewer defects in the coating (see Figure 5-9).

Relevant documents are NACE RP-0169[4] and ISO 21857[5]. The installation of a new pipeline requires permission to cross from the operators of the other pipelines. Test points should be provided to allow easy checking of interference. There may be limitations placed on the maximum current that can be supplied from the local CP station.

Stray currents from AC and DC system may be infrequent and short term and consequently difficult to detect. It may be necessary to use recording voltmeters to detect the amplitude and duration and time when stray currents occur.

5.7.2 Moderating stray current corrosion

- For adjacent pipelines apply standard tests at test stations to determine if there is interference
- Adjust CP system to minimize interference or, if agreed with other operators, install appropriate resistive bonds or adjust the existing resistive bond.
- Where stray current interference is suspect install voltmeter data logger to identify the nature of the interference and when it occurs. This will help to track down the source of the interference.
- For AC interference from other sources it may be necessary to install an earthing facility or a sacrificial anode to act as an earth. The use of long strip anodes has advantage as the long anode presents a very low resistance to earth.

5.8 High pH carbonate cracking

5.8.1 Mechanism

High pH stress corrosion cracking[6] mainly occurs on gas pipelines downstream of compressor stations when cooling of the gas is not provided and the temperature is in the range 50 - 80°C. The hazard of cracking increases if there is cyclic loading superimposed on the operating pressure. Soil water evaporates and the concentration of carbonates increases resulting in an increase in local alkalinity. Failure of the coating is likely to occur by cathodic disbondment when the pH exceeds approx. 10, allowing the soil water to contact the pipeline surface. The fatigue loading initiates cracks on the steel surface, and these will propagate because of the combination of high static stress with superimposed fatigue loading (see Figure 5-10).

The factors leading to crack initiation are:

- High operating pressure, above 60% SMYS
- Presence of cyclic loading >5% mean stress
- Carbonate and bicarbonate present in the soils
- Temperature above 50°C

The majority of failures have occurred at the lower sections of pipelines, 5 - 7 o'clock position. The CP potentials are usually in the correct protection range though in some cases there may be slight overprotection.

Multiple cracks are formed and are usually axial, forming at 90° to the hoop stress. The cracks are intergranular and narrow. There will be little or no evidence of corrosion and the cracks are not associated with metallurgical features. To date most failures have occurred on pipelines with asphalt and coal tar pipelines but this may only be because these coatings were used on older pipelines. Earliest known time to failure 5 years with average time to failure of after 20 - 30 years in service.

High pH SCC cannot be detected by conventional MFL and UT pigs. Specialized crack detection pigs must be used. EMAT pigs are also able to detect cracks provided the wall thickness is modest.

Pipelines that cannot be inspected using an intelligent tool can be assessed using the guidelines provided in NACE SP-0204.

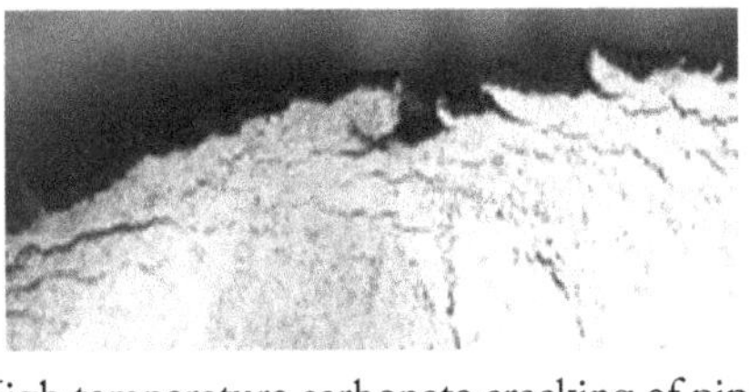

Figure 5-10. High temperature carbonate cracking of pipeline surface showing crack colonies initiating at pitted areas.

5.8.2 Moderating high pH carbonate cracking

- Excavate the pipeline
- Check the pipe surface for cracking.
- If the pipeline is cracked, then measure the depth of the cracks
- If cracks are shallow, they may be ground out and the corrosion coating repaired
- If cracks are deep, then either a type B sleeve should be installed, or the section of pipeline cut out and replaced.

5.9 Near neutral stress corrosion cracking

5.9.1 Mechanism

Near neutral stress corrosion cracking occurs on both gas and oil pipelines[7]. The main factor leading to NNSCC is shielding. Shielding may be a result of the detachment of the coating or the formation of voids under the coating during the application process. Shielding may also occur where creep of coating has occurred creating voids between coating and pipeline surface. Examples of such voids are illustrated in Figure 5-11 and 5-12 for tape or polyethylene over a girth or longitudinal weld or where tape is over-lapped onto tape. Field joints under high insulation field joint coatings present a higher likelihood of shielding.

Water vapour and carbon dioxide can permeate through the coating into the voids to produce a slightly acidic water on the pipeline surface. The carbon dioxide in soils is many times higher than the concentration above ground. CP is not effective because the coating is integral. Corrosion results in deep pitting. Once the corrosion is sufficiently deep the corrosion metal loss will result in higher stress at the bottom of the corroded area and, if the material is locally embrittled, the corrosion may convert to cracking.

Multiple cracks are usually formed in the axial direction at 90° to the hoop stress (see Figure 5-13). Circumferential NNSCC has been observed on pipes that are subject to high stresses as a result of soil movement. Most cases of NNSCC have been found on pipelines protected with tapes, polyethylene and asphalt coatings.

Cracks are wide and transgranular and associated with surface corrosion, usually carbonates but sulphides may be present if the soil contains sulphate-reducing bacteria. Crack growth rate is approx. 0.6 mm/year and some shallow cracks (<1 mm deep) may become dormant. The cracks are often associated with metallurgical features. The pH of the trapped water is pH ~8.

NNSCC cannot be detected by conventional MFL and UT pigs. Specialized crack detection pigs must be used. EMAT pigs are also able to detect cracks provided the wall thickness is modest.

Pipelines that cannot be inspected using an intelligent tool can be assessed using the guidelines provided in NACE SP-0204.

5.9.2 Moderating near neutral pH cracking

- Excavate the pipeline
- Check the pipe surface for cracking.
- Measure the depth of the cracks
- If cracks are shallow, they may be ground out and the corrosion coating repaired
- If cracks are deep, then either a type B sleeve should be installed, or the section of pipeline cut out and replaced.

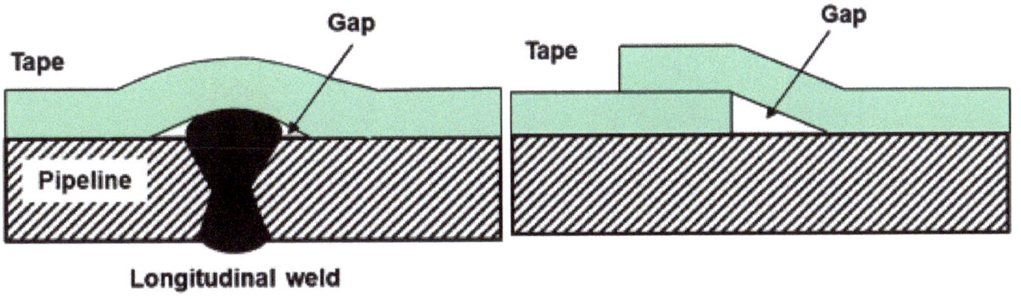

Figure 5-11. Schematics showing voids under a tape wrap coating created by the tape bridging the longitudinal weld or at tape overlaps.

Figure 5-12. Corrosion adjacent to the longitudinal weld as a result of void under a tape coating.

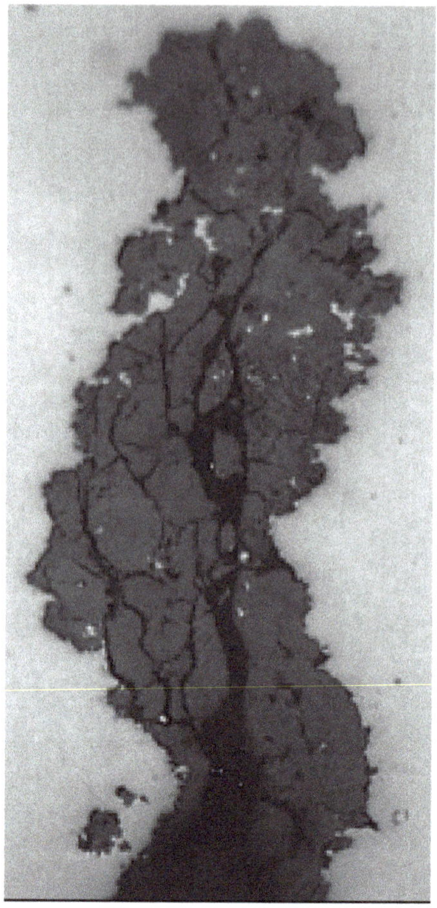

Figure 5-13. Example of near neutral pH carbonate cracking showing wide cracks with corrosion product present.

QR Code 5-5. Scan here for additional information on carbonate cracking.

5.10 Corrosion fatigue

5.10.1 Mechanism

Corrosion fatigue is a mode of cracking in materials under the combined actions of cyclic loading and a corrosive environment. Corrosion fatigue crack growth rates can be substantially higher in the corrosive environment than fatigue crack growth under cyclic loading in a benign environment.

Cracks, which influence the structural integrity of the pipeline, are mainly longitudinally orientated, caused by the predominant stress distribution in the steel. Fatigue cracks can grow both from the internal or the external surface of the wall. Often external fatigue cracks initiate from an external damage to the surface of the pipeline.

Submarine pipelines may suffer fatigue loading at spans if the pipeline span is above a critical length. If corrosion occurs in a sour service pipeline the fatigue life of the pipeline will be affected. There is no agreed value on the reduction in fatigue life, but estimates vary from 30% to 50%.

If a pipeline is overprotected by cathodic protection, then there may be sufficient hydrogen generation on the pipe surface to embrittle the pipe material. Both embrittlement by sour corrosion and excessive CP potentials result in atomic hydrogen migration into the steel. Higher strength steels are more susceptible than lower strength steels. There is limited information of the reduction in fatigue life. API 5L X65 pipeline steels may lose 10 - 30% fatigue life.

Fatigue cracks are becoming increasingly common, mainly due to the increasing accumulated number of pressure cycles in the aging pipeline population.

5.10.2 Moderating corrosion fatigue

- Estimate the loading on the pipeline and calculate the number of cycles to failure
- Allow for reduction in fatigue life if the pipeline is in sour service and corrosion monitoring indicates corrosion is occurring in the pipeline
- Check CP potentials. Sections of the pipeline where the potential is more negative than -1200 mV (CCSE) may be embrittled and fatigue life will be reduced.
- Sour service pipeline may also be embrittled if internal corrosion occurs.

5.11 References

[1] NACE SP-0502 Pipeline External Corrosion Direct Assessment Methodology, NACE International, Houston.

[2] NACE SP-0204 Stress Corrosion Cracking (SSC) Direct Assessment Methodology, NACE International, Houston.

[3] A Practical Manual on Microbiologically Influenced Corrosion, Ed. J.G. Stoeker, NACE International, Houston, 2001

[4] NACE RP-0169 Control of External Corrosion on Underground or Submerged Metallic Piping Systems, NACE International, Houston.

[5] ISO 21857 Petroleum, petrochemical and natural gas industries. Stray current of pipeline system.

[6] Stress Corrosion Cracking of Pipelines, Y. F. Cheng, Wiley, 2012

[7] Stress Corrosion Cracking, Recommended Practice, CEPA, December 2007.

CHAPTER 6

Internal corrosion

Dr Roger King

6.1 Introduction

20 - 40% of pipeline failures are the result of internal corrosion and results from the presence of water in the fluids being transported. The water may be produced with oil or condense from gas from the reservoir as the gas cools. Carbon dioxide (CO_2) and hydrogen sulphide (H_2S) dissolve into the water and reduce the pH. Corrosion occurs as a result of the reaction of the acidic water with the pipeline steel. Oxygen can also enter a pipeline and cause localized corrosion usually in the form of deep pits though this is not common in high pressure pipelines.

The iron goes into solution and hydrogen is evolved. The iron in solution will react with anions in the water to form a range of corrosion products from hydroxides, sulphates, chlorides to oxides. Most rust is brown or black but may be white or red depending on the composition.

Internal corrosion occurs by the following mechanisms:

- Sweet corrosion if CO_2 is present, perhaps with trace H_2S.
- Sour corrosion in the presence of sufficient H_2S
- Pitting corrosion by oxygen entering the pipeline through seals, pump shafts, aerated chemical addition, etc.
- Presence of short chain carboxylic acids, e.g. formic, acetic, propionic acids
- Microbiological influenced corrosion (MIC) when sulphate is present in the water

Corrosion by these mechanisms may be restricted to zones in the pipeline:

- preferential weld corrosion of the longitudinal and girth welds and the heat affected zones;

- top of the line corrosion (TOLC) in hot gas pipelines when there is a high condensation rate of water, this form of corrosion is enhanced by the presence of volatile carboxylic acids;
- corrosion under deposits, wax and scale in the pipeline;
- at bends by flow-enhanced corrosion where the flow velocities of the fluids are high;
- at bends by erosion-corrosion when there are particulates in the fluids.

Pipelines may fail suddenly by cracking:

- by sulphide stress cracking (SSC) or stress orientated hydrogen induced cracking (SOHIC) when stress and H_2S are present;
- by hydrogen induced cracking (HIC) when H_2S is present and pipe metallurgy is inadequate; this form of cracking initially appears as blistering and progresses by cracks linking the blisters resulting in stepwise cracking;
- corrosion fatigue when the pipeline is subject to cyclic stress resulting from changes in pressure and/or temperature or an external force.

Water is necessary for corrosion to occur and internal corrosion may be prevented by removal of water in the fluids. Dried gas transmission pipelines should not suffer internal corrosion unless there is a major upset in the gas processing, but a relative humidity above 60% may result in superficial surface corrosion in a gas pipeline leading to the formation of 'black dust'. Low water cuts in oil pipelines and product pipelines usually prevents internal corrosion provided the flow rate of the hydrocarbon is high enough to prevent water settling out and wetting the pipeline surface.

Where the presence of water cannot be avoided, the corrosion rate may be reduced by the injection of corrosion inhibitor into the fluids. The corrosion rate may be reduced to 5 or 10% of the uninhibited corrosion rate. Microbiological corrosion may be controlled by the injection of biocides.

The effectiveness of chemical treatment is higher if the pipeline is regularly cleaned by housekeeping pigs. The risk of microbiological corrosion will be markedly reduced by frequent pigging.

6.2 Sweet corrosion

6.2.1 Mechanism

Carbon dioxide dissolves into water to form a weak acid and dissociation of this acid results in reduction of the pH of the water. The corrosion process is a sequence of reactions and the overall reaction is iron and carbonic acid converting to iron carbonate and hydrogen[1,2].

The corrosion is determined by the acidity of the water and increases with carbon dioxide concentration, the operating pressure and temperature. Where corrosion occurs, there is the formation of an iron carbonate layer on the pipeline surface. The yellow-brown iron carbonate film (siderite) is unreactive and acts as a diffusion barrier and reduces the corrosion rate. Damaging this carbonate film will result in an increase in corrosion rate. Refer to sections on flow enhanced corrosion and erosion corrosion.

Prediction models are available to estimate the potential corrosion rate which depends on the operating pressure and temperature and the concentration of carbon dioxide. Prediction models are considered reasonable if the CO_2/H_2S ratio is above 500. Flow also has an effect and research suggests that the corrosion rate is related to the Reynold's number to exponent 0.2.

Carbon dioxide corrosion may appear as pitting or axial grooving depending on the flowrate of the fluids. Where a water phase persists, there may be general corrosion over the wetted area. Pitting often initiates at flaws or surface emergent inclusions at the pipeline surface, e.g. sulphide materials which are dissolved in the acidic water to create incipient pits. Seamless pipe is more prone to pitting corrosion than the other forms of pipe fabricated from strip or plate because seamless pipe has a higher concentration of sulphide inclusions on the inner surface; this results from the segregation of the sulphides in the centre of the steel billets that will be formed into pipe.

Corrosion can be identified by inline inspection with magnetic flux leakage (MFL) or ultrasonic testing (UT) tools or by external ultrasonic inspection with UT or radiography. Figure 6-1 gives an example of carbonic acid corrosion that resulted in the failure of a North Sea pipeline. Note the colour of the corrosion product, an isolated pit and the beginning of axial (mesa) corrosion at bottom of the pipe. This pipeline failed by bursting because of the grooving corrosion at the bottom of pipeline resulting from stratification of water and low flow rates.

Figure 6-1. Corrosion by carbonic acid showing isolated pit and the start of mesa (channelling) corrosion along the bottom of the pipeline.

6.2.2 Moderating sweet corrosion

- Use a corrosion prediction model to estimate what the corrosion rate could be based on the operating history of the pipeline.
- Compare the predicted corrosion rate with the actual measured corrosion.
- Compare the predicted corrosion rate with corrosion monitoring data, if available. Ensure that the corrosion rate from weight loss coupons is calculated based on the area where corrosion has occurred and not total coupon area.
- If actual corrosion rate is high and corrosion monitoring data does not reflect this consider an alternative location for the corrosion monitoring access fittings.
- Check the location of the corrosion with the location of longitudinal and girth welds to determine hazard of preferential weld attack.
- Review operating history of the pipeline to identify if there has been a change in operating procedure or fluid composition that may have triggered the corrosion. Often there will have been a change in water cut or a change in morphology of the fluid flow.
- Review length of corrosion against width of corrosion to identify if flow has had an influence on the corrosion process. Metal loss that is longer than it is wide indicates flow effects whilst metal loss that is wider than it is long indicates pooling of water at that location.
- Compare the corrosion location with profile of the pipeline. In most cases corrosion is most severe at low points where water accumulates and long upward slopes where the water runs back down the pipeline.
- Check the history of corrosion inhibition. Has the inhibitor been injected at the correct rate? This may be determined from plant records and cross-checked by inventory of chemical.
- Check the inhibitor injection skid to determine that it is operating correctly and that there are no blockages or bottlenecks preventing correct dosage.
- Review the programme for pigging the pipeline. If pigging has not been done for a long period, then consider cleaning the pipeline.
- If the pipeline is internally clean and the inhibitor has been correctly injected and at the correct dosage rate, then re-evaluate the inhibitor.
- Check for other corrosion mechanisms: microbiological corrosion, top of the line corrosion, preferential weld corrosion.

QR Code 6-1. Scan here for additional information on sweet corrosion.

6.3 Sour corrosion

6.3.1 Mechanism

Sour corrosion occurs when significant hydrogen sulphide is present in the produced fluids. As the concentration of hydrogen sulphide increases iron sulphide will progressively replace iron carbonate as the major corrosion product[3]. Sour corrosion is considered to dominate when the CO_2/H_2S ratio is below 20[4]. For CO_2/H_2S ratios 20 - 500 the corrosion rate prediction models are adequate but should be treated with caution.

The iron sulphide films are protective whilst they are complete and general corrosion rates are low but there is a higher likelihood of pitting corrosion if a new iron sulphide film cannot form at an area where the sulphide film is damaged; this occurs because there is a potential difference between iron sulphide and steel that creates a galvanic cell.

Corrosion product films are black and evolve hydrogen sulphide when treated with acid. The black debris will oxidize to form free sulphur and iron oxides. Deposits from pig receivers thought to contain iron sulphides must be kept wet as they will spontaneously ignite if allowed to dry out.

The protectiveness of the iron sulphide films is related to temperature. Top of the line corrosion may also occur several kilometres downstream of the inlet, see section on Top of Line Corrosion.

Corrosion can be identified by inline inspection using MFL or UT tools or by external ultrasonic inspection. Weight loss coupons will show black iron sulphide films and may show pitting.

6.3.2 Moderating sour corrosion

- Determine the CO_2/H_2S ratio.
- If the ratio is acceptable for CO_2 corrosion, then use a corrosion prediction model to estimate what the corrosion rate could be for the operating history of the pipeline.
- Compare the predicted corrosion rate with the actual corrosion measured.
- Compare the predicted corrosion rate with corrosion monitoring data, if available. Ensure that the evaluation of the weight loss coupons includes measurement of pit depths.
- If actual corrosion rate is high and corrosion monitoring data does not reflect this consider an alternative location for the corrosion monitoring facility.
- Follow the steps given in the sweet corrosion section.

- Check for other corrosion mechanisms: top of the line corrosion, cracking mechanisms.

6.4 Oxygen corrosion

6.4.1 Mechanism

Oxygen is a powerful corrosive agent. Entry of oxygen into a hydrocarbon pipeline is not usual because pipelines operate at high pressure. Oxygen may be introduced inadvertently in the chemicals added to the pipeline, e.g. methanol for hydrate suppression contains high concentrations of dissolved oxygen. Often methanol tanks are blanketed with an inert gas or process gas to exclude oxygen. Dissolved oxygen reacts with the steel to produce ferrous hydroxide. Oxygen may also react with hydrogen sulphide to form free sulphur which is very corrosive.

Water injection pipelines may suffer oxygen corrosion if the water treatment does not remove all oxygen. Corrosion rate is related to oxygen concentration, water flow rate and operating temperature.

Entry of oxygen into a pipeline will cause local pitting close to the point of entry. Pits have relatively smooth sides and can be up to 1-inch (25 mm) diameter. Corrosion products are voluminous and yellowish brown. Oil field corrosion inhibitors may not be effective.

Metal loss can be identified by ILI using MFL or UT tools or by external UT. Weight loss coupons close to the point of entry will show pitting corrosion.

6.4.2 Moderating oxygen corrosion

- If oxygen rrosion is suspect review points of entry of oxygen into the pipeline, e.g. suction inlet of pumps, treatment chemicals.
- Ensure pit depths are measured on weight loss coupons.
- Modify equipment to prevent oxygen ingress.
- Consider blanketing of bulk chemicals with low oxygen nitrogen or inert gas.

QR Code 6-2. Scan here for additional information on sour service.

QR Code 6-3. Scan here for additional information on bacterial corrosion.

QR Code 6-4. Scan here for additional information on top of the line corrosion.

6.5 Organic acid corrosion

6.5.1 Mechanism

Short chain carboxylic acids (sometimes called fatty acids) occur in formation water and concentrations range from 10 - 100,000 ppm. The presence of fatty acids can cause corrosion in pipelines and may interfere with the performance of the corrosion inhibitor. These acids also cause corrosion of the bottoms of crude oil storage tanks.

The dissolved acids dissociate and reduce the pH of the water. The acid water reacts with the pipeline steel to generate hydrogen and cause corrosion. Corrosion appears as generalized corrosion where water collects and persists. Rivulet corrosion may occur if the water phase runs along the bottom of the pipeline similar to sweet corrosion. Corrosion may be more severe at the inlet of the pipeline depending on the concentration of the acid because the acid is consumed in the corrosion process.

The presence of volatile short chain organic acids is implicated in top of the line corrosion. Many bacteria including the sulphate-reducing bacteria utilize fatty acids as a carbon source. Systems containing high concentrations of fatty acids will be at higher risk of microbiological growth and corrosion, if sulphate is present.

Metal loss can be identified by ILI using MFL or UT tools or by external UT.

6.5.2 Moderating organic acid corrosion

Corrosion inhibitors design for reduction of corrosion by CO_2 and H_2S will also reduce corrosion by fatty acids but this may need to be confirmed. The concentration of the inhibitor may need to be increased.

6.6 Microbiological influenced corrosion

6.6.1 Mechanism

A wide range of bacteria can be active and grow in sweet oil pipelines, water injection pipelines and produced water pipelines. The most deleterious bacteria are the sulphate-reducing bacteria (SRB). Seawater injection pipelines are at risk because of the high sulphate and nutrient content of seawater. A typical SRB is shown in Figure 6-2. These bacteria require:

- the presence of water as a persistent layer;
- a food source - usually short chain organic acids;

- sulphate - a minimum of 5 ppm for survival and 50+ ppm for active corrosion;
- presence of a nitrogen source (ammonium ion from scavenger treatment will augment the nitrogen concentration);
- total dissolved solids below approx. 150,000 ppm (the acceptable TDS concentration is related to the nature of the dissolved solids);
- temperature in the range 4 - 80°C, the bacteria may tolerate temperatures outside this range, but activity will be reduced;
- pH in the range 5 to 9, the bacteria may tolerate pH outside this range, but activity will be reduced.
- MIC almost always occurs at the bottom of the pipeline and will be more severe and difficult to correct if there is debris in the pipeline. Regular pigging is necessary to prevent and control MIC.

The bacteria oxidize the organic acids to carbon dioxide and water using the oxygen in the sulphate radical, this results in the production of sulphide. The sulphide reacts with the pipeline to form iron sulphide. A galvanic cell is established between the iron sulphide and adjacent steel. An example of pitting corrosion by SRB is given in Figure 6-3. The bacteria are sustained by their ability to use the hydrogen evolved on the iron sulphide to reduce sulphate to water producing additional sulphide for reaction with the pipeline steel.

Work by Pots[5] proposed that the threat of MIC may be evaluated using appropriate parameters that included concentrations of nutrients, nitrogen, sulphate, flow velocity and frequency of pigging.

The number of bacteria is measured as the number of bacteria per unit volume of water. These bacterial populations need to be adjusted for water flowrate, pipeline diameter and length to determine the hazard that the bacteria present to the pipeline.

6.6.2 Moderating bacterial corrosion

- Consider cleaning the pipeline using progressively aggressive pigging.
- In water injection pipelines drive the final pig with water treated with a high concentration of biocide.
- Check water from pipeline to evaluate success of the treatment.
- For oil pipelines combine final pig run with high dosage biocide slug using a trail pig to keep the biocide slug intact.

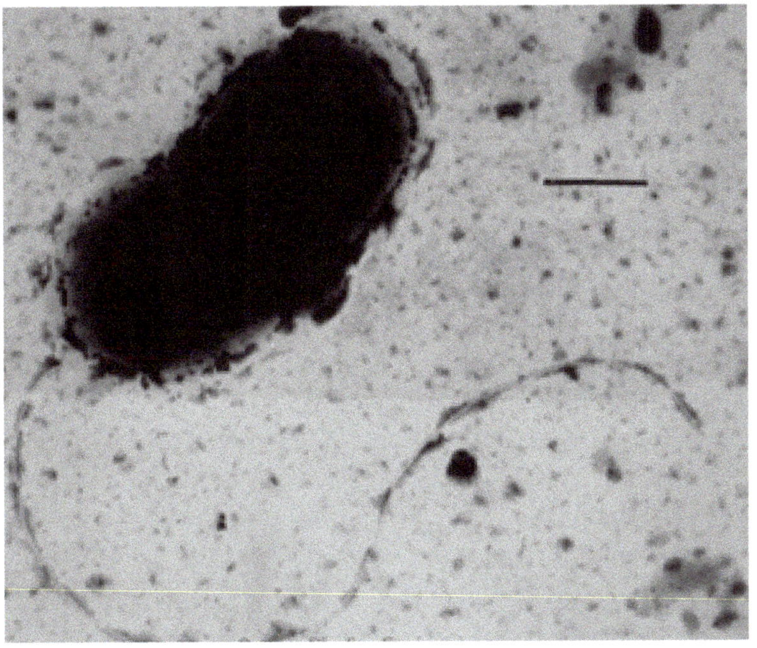

Figure 6-2. SRB isolated from Norwegian waters showing adherent sheath of slime (polysaccharides) and iron sulphide.

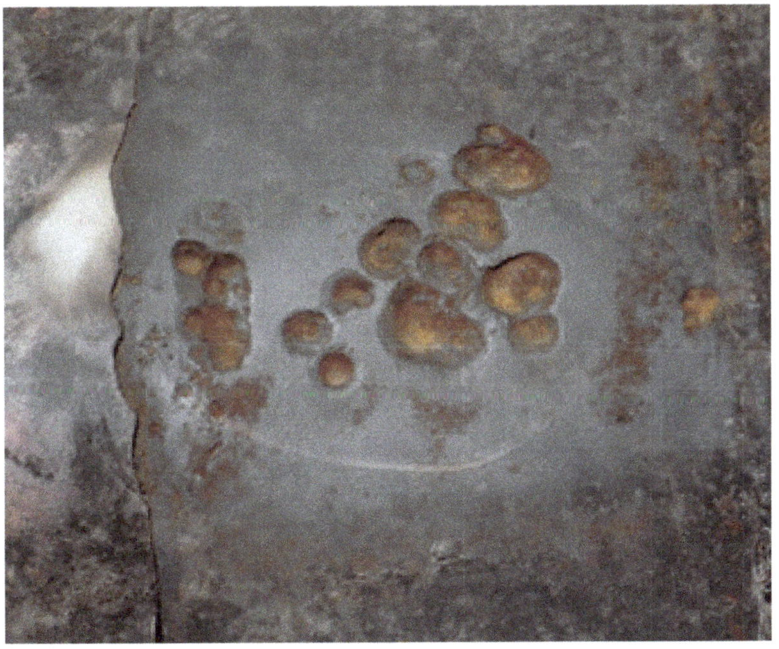

Figure 6-3. Example of localized pitting corrosion by SRB. Depending on flow conditions the localized pitting can link together to form long axial corrosion metal loss resulting in bursting of high pressure pipelines, roughly 50% of water injection pipeline fail by bursting.

6.7 Preferential weld corrosion

6.7.1 Mechanism

Longitudinal, spiral and girth welds are present in pipelines and the metallurgy of the welds and heat affected zones adjacent to the welds is different to the parent material. Corrosion will occur adjacent to these welds because the grain structure of the weld and heat affected zone is coarser; corrosion rate increases with increasing grain size. The change in grain size may also alter the effectiveness of the inhibitor. Examples of PWC are given in Figures 6-4 and 6-5.

Corrosion at a longitudinal weld presents a high hazard because, in almost all cases, the highest stress in a pipeline is the hoop stress which is attempting to increase the diameter of the pipeline. Corrosion at a longitudinal weld may occur because water condensing from the crown of the pipeline is caught on the longitudinal weld.

Helical (spiral) pipe presents similar problems with condensing water following the weld down to the bottom of the pipeline. The nominal calculated stress at and adjacent to the spiral welds will be lower than the hoop stress but actual stress will include residual stresses and may be higher overall.

Girth welds of submarine pipelines are protected by field joint coatings, but concrete will not usually be present. The field joints may be slightly cooler than the main pipeline leading to a higher internal condensation rate. Sections of pipeline underneath sacrificial anodes may also be cooler.

Preferential weld corrosion can be detected by inline inspection surveys using MFL and UT tools or external ultrasonic inspection.

Preferential weld corrosion can be identified by ILI with MFL or UT tools. May also be identified by external UT; angle probes may be necessary.

6.7.2 Moderating preferential weld corrosion

- It is impractical to repair every weld in a pipeline but may be possible to reduce progression of the corrosion. Review the efficiency of the inhibitor and the dosage rate used.
- Increase the frequency of inspection

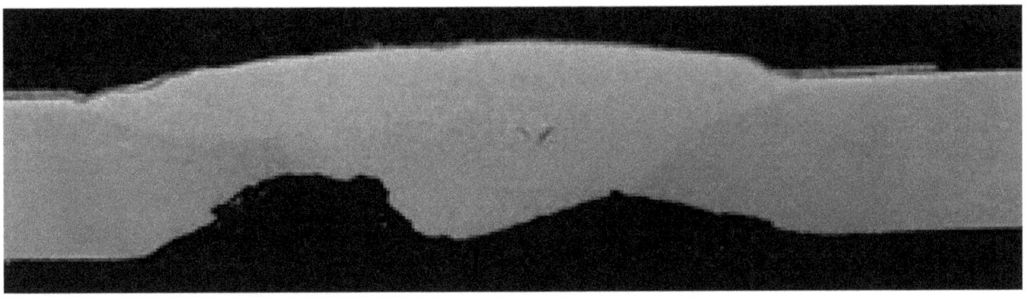

Figure 6-4. Preferential weld corrosion showing corrosion of the heat affected zones adjacent to the weld.

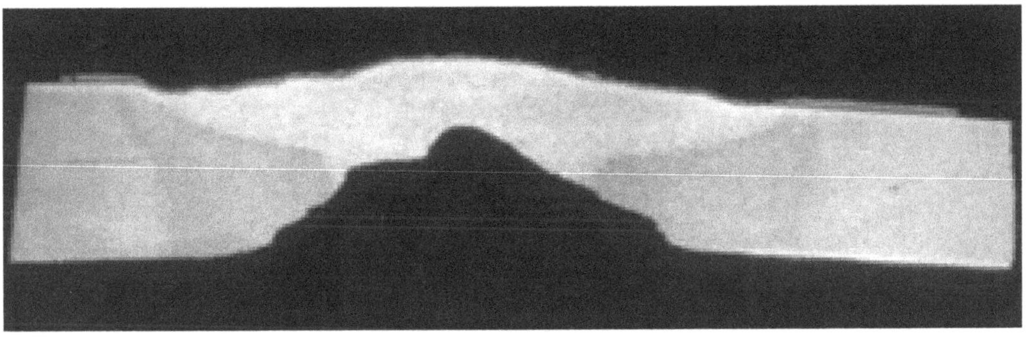

Figure 6-5. Preferential weld corrosion showing corrosion of the weld.

6.8 Top of the line corrosion

6.8.1 Mechanism

All wet gas pipelines suffer some corrosion of the upper surface of the pipeline but excessive corrosion at the top of the pipeline is described as top of the line corrosion (TOLC)[6]. TOLC can occur in gas and gas-condensate pipelines that operate at elevated temperature and where flow is stratified. Hot gas entering the pipeline will cool and the condensed water will form droplets on the upper surface of the pipeline and water will run down the sides of the pipeline to the bottom of the pipeline and add to the stratified water layer. An example of TOLC is shown in Figure 6-6.

In most pipelines the condensation rate of the water is low to modest. Though corrosion occurs iron carbonate and/or iron sulphide films form and reduce the rate of corrosion to approximately 10% of the predicted corrosion rate in free condensed water. The condensing water will be low pH because the water is free of salts that would moderate the pH.

If the external ambient temperature is low, for example pipelines in cold climates and submarine pipelines, the condensation rate will be high. If the condensation rate exceeds 0.25 g/ m^2/s then the water cannot build up sufficient dissolved ferrous ion for iron carbonate films to form. As a consequence, corrosion rates are high. TOLC can occur in a pipeline with a gas inlet temperature of approx. 50°C or above, depending on the external ambient temperature.

For submarine pipelines the gas ambient temperature will usually fall to seabed temperature within ~1 km from the platform though this depends on the volume of condensate present in the transported fluids, the external coatings on the pipeline and the depth of burial of the pipeline in the seabed. There are no values for the length of pipeline at risk of TOLC for onshore pipelines.

When estimating the length of pipeline that is at risk of TOLC ensure to account for the presence of valves and other pipeline furniture which may provide a heat sink or heat conservation item that will distort the temperature profile.

Conventional corrosion inhibition will not protect pipelines with high water condensation rates. The inhibitor is absorbed into the water and rapidly removed from the gas stream. Shortly downstream of the inlet to the pipeline there will be inadequate inhibitor to protect the pipeline.

The presence of volatile short chain carboxylic acids will increase the hazard of TOLC. Formic and acetic acids are absorbed into the condensing water and react with ferrous ions to form soluble corrosion products. By reduction of available ferrous ion, the formation of iron carbonate is prevented.

TOLC also occurs in wet sour gas pipelines at distance downstream from the inlet. This form of corrosion is not related to temperature but the lack of protectiveness of the iron sulphide films[7,8].

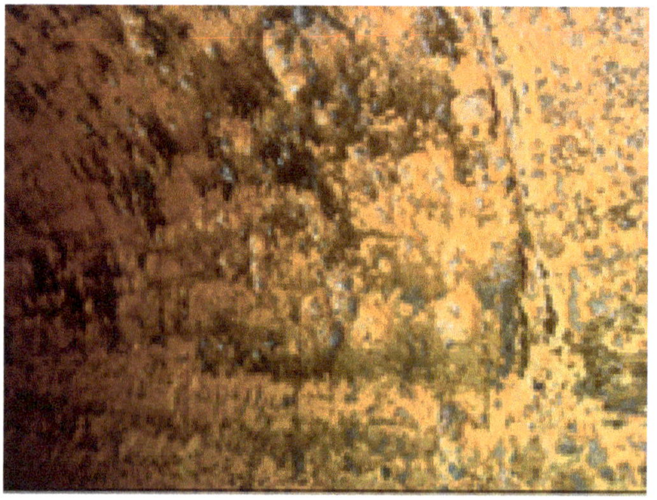

Figure 6-6. Localized corrosion attack as a result of top of the line corrosion in a gas pipeline.

6.8.2 Moderating TOLC

- TOLC appears in ILI surveys as extensive metal loss on the upper surface of the pipeline. There is less corrosion in the bottom of the pipeline because of the presence of corrosion inhibitor in the stratified water phase.
- Calculate the temperature profile along the pipeline from the inlet. From the temperature profile calculate the condensation rate. If the condensation rate exceeds the critical value, then TOLC may be the corrosion mechanism.
- If the condensation rate is below or close to the critical condensation rate, check formation water for presence and concentration of short chain carboxylic acids, in particular, formic and acetic acids.
- The impact of TOLC can be prevented by reducing the temperature drop along the pipeline. This may be achieved by:

  a) increasing gas flow rate such that flow is annular mist flow which will ensure that inhibited water is around the full circumference of the pipeline;
  b) install cooling system to reduce the gas temperature at the inlet of the pipeline;
  c) insulating the pipeline to lessen the steepness of the temperature profile;
  d) for submarine pipelines by burying the pipeline under seabed sediment. Mattresses and rock dump are not considered adequate but use of geotextiles combined with mattress and rock dump may be.

- The impact of TOLC may be reduced by:

  a) batch dosing the pipeline with corrosion inhibitor
  b) using an inhibitor dispersal pig to collect water at the bottom of the pipeline and spray the inhibited water onto the upper surface of the pipeline.

6.9 Under deposit corrosion

6.9.1 Mechanism

Deposits in pipelines may include millscale exfoliated from the surface of the pipeline, waxes, scales, solids produced from the reservoir and corrosion products. Continuous wax and uncracked scale will prevent or markedly reduce corrosion. Water will penetrate through porous deposits to wet the pipeline surface beneath the deposits. Carbon dioxide and hydrogen sulphide migrate through deposits whilst corrosion inhibitors do not as they are absorbed onto the deposits. Though the corrosion rate will be low beneath the deposits the rates will be high in comparison to the clean areas of the pipeline because of the limited corrosion inhibitor present at the pipeline surface beneath the deposits.

Wax deposits may not be homogeneous. For flow assurance, regular pigging is necessary to remove excess wax. A water layer may form on the exposed pipeline surface and then be covered with a wax layer. Carbon dioxide and hydrogen sulphide will permeate through the wax, but corrosion inhibitor permeation will be limited.

Pipelines that are regularly pigged for flow assurance may need to have corrosion inhibitors that quickly reform the protective inhibitor layer on the pipeline surface, or the dosage rate of inhibitor should be increased for a period to aid renewal of the inhibitor film.

6.9.2 Moderating under deposit corrosion

- Evaluate volume and location of debris by use of a calliper or instrumented pig. Instrumented pigs are standard cup or bi-di pigs fitted with accelerometers and data loggers that identify the location of debris.
- Clean the pipeline by pigging being careful not to be too aggressive with initial pigs.
- If the pipeline cannot be pigged, then corrosion will need to be controlled by corrosion inhibition. Evaluate the option of a penetrative corrosion inhibitor.
- Gas flushing has been trialled for water pipelines, but data are limited. Inert gas is introduced into the pipeline as a gas slug and the outlet pressure is reduced. The gas slug moves along the pipeline expanding and accelerating. A sufficiently high velocity may be achieved to detach and fluidize the deposits.
- Collect debris from pigging for estimate of volume and nature.
- After pigging ensure pipeline surface is inhibited by increasing corrosion inhibitor dosage rate threefold for 24 hours or as necessary to ensure there is the re-formation of an effective inhibitor film on the surface of the pipeline.

6.10 Flow enhanced corrosion

6.10.1 Mechanism

Corrosion product films act as diffusion layers and reduce corrosion. The films have limited adhesion to the pipeline surface and may be removed by the shear force on the pipeline surface imposed by high fluid velocity flow. Corrosion inhibitor films may also be removed by high shear forces and there are limits to the shear force that a specific inhibitor can tolerate. Flow effects are amplified at bends and tees in pipelines and the amplification depends on the acuity of the bend. Flow amplification varies from 1.25 - 1.8. Flow enhanced corrosion may also occur downstream of probes inserted into the pipeline and at tees and wyes [9] (see Figures 6-7, 6-8).

Flow velocities in pipelines may be designed to conform with API-14E (this practice is considered conservative) but allowance should be made for bends.

Flow enhanced corrosion may be identified by ILI surveys using MFL and UT tools and external UT at bends and other areas of higher flow velocity.

Figure 6-7. Flow enhanced corrosion induced by presence of a probe inserted into the pipeline. Flow is from left to right.

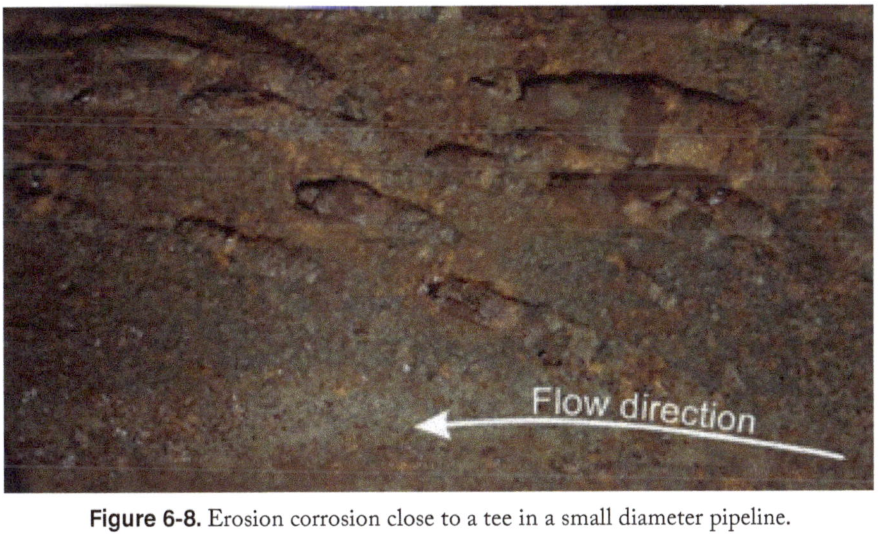

Figure 6-8. Erosion corrosion close to a tee in a small diameter pipeline.

6.10.2 Moderating flow enhanced corrosion

- Calculate the flow velocities along the pipelines. Gas pipelines may have marked differences in flow rate at outlet compared to inlet.
- Use API RP-14E as first pass to identify if enhanced flow rate may contribute to corrosion. Make allowance for bends and tees in the pipeline.
- Check ability of the corrosion inhibitor to tolerate the calculated shear forces. Review selection of the inhibitor if the inhibitor is found sensitive to flow.

6.11 Erosion and erosion corrosion

6.11.1 Mechanism

Erosion results from the impingement of solid particles on the pipeline surface [10]. The extent of damage is related to the density and velocity of the particles and the angle at which they strike the surface. In most cases erosion damage occurs because repeated impacts cause work hardening and embrittlement of the surface with subsequent brittle failure. Erosion damage is more extensive at bend and tees.

Erosion corrosion is the result of the combination of removal of corrosion product film by impingement by particles with consequent reformation of corrosion product film followed by removal of this film by repeated impact. Erosion corrosion can cause rapid failure.

6.11.2 Moderating erosion corrosion

- Investigate reduction in volume of particles in the flow by modification to the production wells, e.g. by provision of downhole sand screens or gravel packs.
- Consider reduction in flow rate
- Consider an inhibitor programme that can provide a rapid re-formation of inhibitor film on the pipeline surface. This approach will reduce erosion corrosion damage but not prevent it.

6.12 Sulphide stress cracking (SSC)

6.12.1 Mechanism

SSC occurs when there is a combination of hydrogen sulphide, stress and a susceptible material. In the NACE document (NACE MR-0175)[11] issued before 1997 the critical partial pressure of hydrogen sulphide was 0.05 psia (0.345 kPa) and a total pressure above a pressure dependent on the service of the pipeline, either gas or multiphase. Produced water pipelines are treated as multiphase pipelines. After 1997 the document relates hazard of SSC to the combination of partial pressure of hydrogen sulphide and system pH[12].

When corrosion occurs, atomic hydrogen is generated. Most combines to form molecular hydrogen but some will enter into the steel. Atomic hydrogen bonds to iron sulphide films delaying the combination of atomic hydrogen to hydrogen gas and a higher flux of atomic hydrogen enters the pipeline steel. Hard material is highly susceptible embrittlement; and when the embrittled steel is subject to a high stress the material may fail by brittle fracture. The crack will progress into normal material and be arrested but the process will repeat. Cracks progress discontinuously with the rate of cracking accelerating as the crack lengthens. Cracks progress through wall at 90 degrees to the hoop stress. An example of SSC is given in Figure 6-9.

To ensure a carbon steel pipeline is not susceptible there is a limit on the steel hardness. Parent pipe, welds and heat affected zones must have hardness below Rockwell C 22 or Vickers hardness (10) 248 (often rounded to 250). The majority of pipelines constructed after 1975 will conform with the requirements of the NACE document.

Concern will only arise when service has changed, i.e. reservoir souring[13] has occurred or an alternative hydrocarbon is being considered or where the pipeline has been damaged and cold worked, e.g. dents.

Figure 6-9. Sulphide stress crack initiated in the heat affected zone.

6.12.2 Avoiding or moderating SSC

- SSC can be avoided by ensuring the pipe material and the welding procedures comply with NACE MR-0175/ISO 15156 and NACE TM-0177.
- Where spare, similar pipe is available it is possible to do routine tests on the material to determine the window within which the material can be used. This will be required if the nature of the transported fluids has altered.
- Where there is concern, test affected areas with portable hardness tester.
- On submarine pipelines that were originally intended for sweet service the welds of sacrificial anode pigtails to the pipeline may be harder than allowed.
- There is no accepted method to predict the rate of propagation of SSC.
- Periodic inspection using angle probe ultrasonic inspection or time of flight diffraction ultrasonic inspection will give an indication of the rate of propagation of SSC.
- An inhibitor programme will reduce the rate of corrosion and the rate of propagation of SSC but cannot halt propagation.

6.13 Hydrogen induced cracking (HIC)

6.13.1 Mechanism

The hazard of hydrogen induced cracking (HIC) is related to the composition and metallurgical structure of the steel. The presence of platelets of manganese sulphide determines the resistance to, or susceptibility of the steel to cracking. When corrosion occurs, atomic hydrogen is generated when the hydrogen ion (H+) is discharged on the surface by combination with an electron. Atomic hydrogen bonds to iron sulphide films delaying the combination of atomic hydrogen to hydrogen gas. A higher flux of atomic hydrogen enters the pipeline steel. If atomic hydrogen enters a manganese sulphide inclusion it will combine with another atomic hydrogen to form molecular hydrogen.

The hydrogen will be trapped in the steel matrix. Pressure build-up over time results in high-pressure hydrogen within the lamination, resulting in blistering of the steel (see Figures 6-10, 6-11). Tearing or cracking of the steel may occur resulting in connections between blisters. This advanced morphology is termed stepwise cracking (see Figure 6-12).

For pipelines fabricated from plate or strip the laminations are centre wall. For seamless pipe the blistering and cracks will be close to the inner surface of the pipe, typically 10 - 20%. Should a crack propagate to the inner surface of the pipeline the hydrogen will be voided and the blisters and cracks will fill with the transported fluids.

Blistering and cracking may then propagate through the outer ligature of the pipe wall. Failure of the pipeline is usually by bursting. The inner ligature of blisters close to the inner surface of the pipeline may be work-hardened during blister formation and this ligature may be susceptible to SSC. Pipe constructed before 1997 may not be resistant to HIC because the NACE document issued before 1997 did not consider HIC though there were documents related to HIC testing available. HIC is avoided by ensuring the material is adequately resistant to this form of cracking and testing to NACE TM-0284. If the hydrogen sulphide concentration in the fluids in a pipeline are increasing as a result of reservoir souring or other reason, then the resistance of the material to HIC will need to be established. This can be done by routine tests on spare, similar pipe if available.

Indication of resistance may also be obtained by chemical analysis and metallurgical examination of the steel, but this requires destructive testing. Pipe that is resistant to HIC has a sulphur content below 0.03% (usually 0.003%) and will have been calcium treated for shape modification. Metallurgical examination will indicate the presence of spherical manganese inclusions rather than platelets and the presence of calcium can be checked by XRD and a scanning electron microscope.

6.13.2 Avoiding and moderating HIC

- HIC may be avoided by testing pipeline material to ensure the material is resistant to HIC. Tests are described in NACE MR-0175/ISO 15156 and NACE TM-0284.
- Validate inline inspection indications of HIC using time of flight diffraction ultrasonic inspection.
- The time to failure by HIC is long compared to SSC. Inhibition may be effective at reducing the corrosion rate and consequently the volume of atomic hydrogen migrating through the pipe wall, but inhibition cannot halt all propagation.
- There is no known method to predict the propagation rate of HIC.
- Periodic inspection will be required to identify the rate of propagation of the HIC and whether it has converted to step-wise cracking.
- Measurement of hydrogen permeation through the pipe wall may be useful to evaluate the effectiveness of the corrosion inhibitor but this can only be done on a section of pipe that is not blistered.

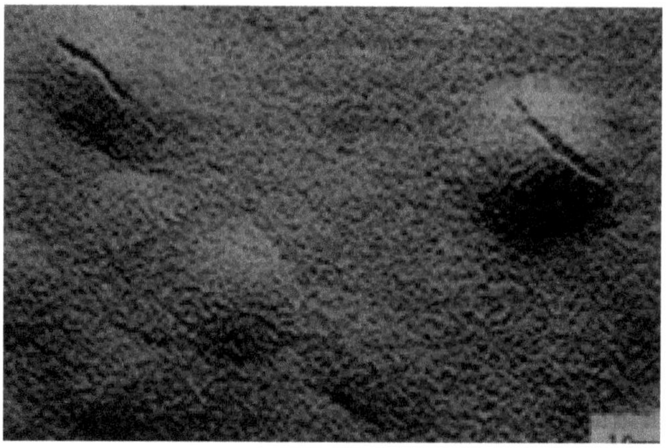

Figure 6-10. HIC surface blisters in a pressure vessel.

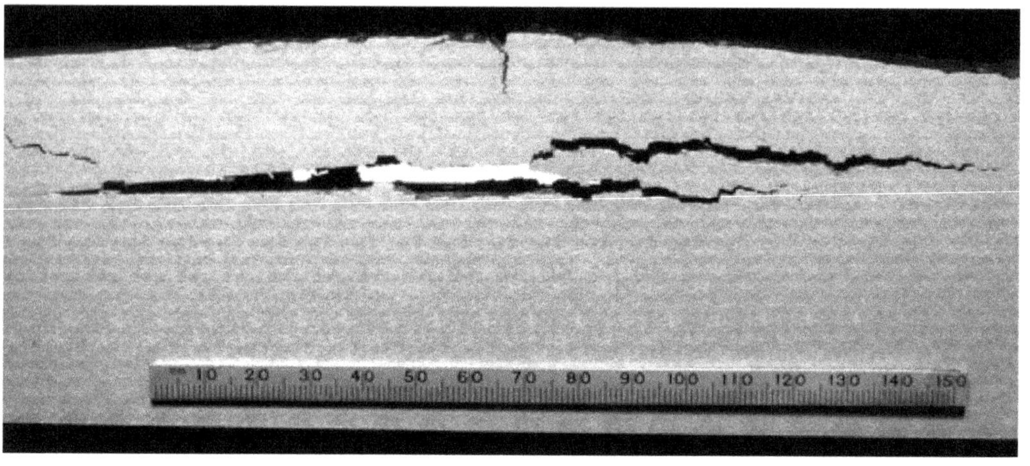

Figure 6-11. Hydrogen blister in pipeline showing sulphide stress crack initiated in the external ligament of the pipeline as a result of work hardening of the ligament as the blister formed.

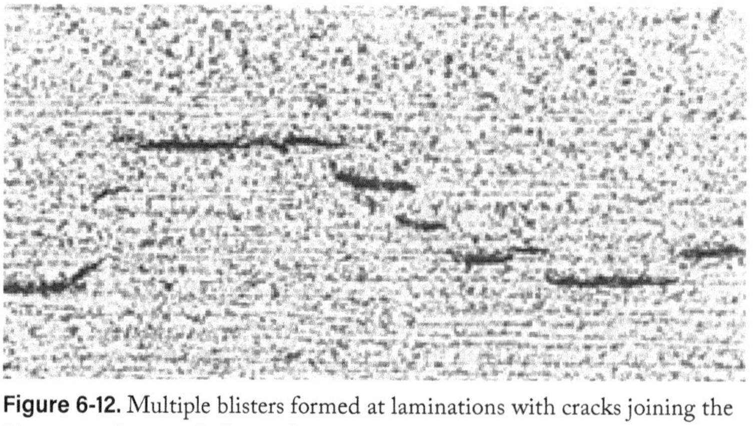

Figure 6-12. Multiple blisters formed at laminations with cracks joining the blisters results in early form of stepwise cracking.

6.14 Stress orientated hydrogen induced cracking (SOHIC)

6.14.1 Mechanism

Stress orientated hydrogen induced cracking (SOHIC) is the combination of hydrogen induced cracking (HIC) and sulphide stress cracking (SSC). The horizontal blisters are linked by the vertical sulphide stress cracks (see Figure 6-13).

The mechanism of SOHIC is not fully understood. One explanation is that SOHIC occurs in steels that are strengthened by use of micro-alloying elements (titanium, niobium/columbium, vanadium). During welding the distribution of the micro-alloying elements is altered with a resultant formation of strong and weak bands through the wall thickness of the pipeline adjacent to the weld. Under stress the softer layers may tear in a ductile manner resulting in voids that can fill with hydrogen resulting in blister formation. The additional stress causes work hardening of the stronger band and these are cracked by SSC linking the blisters in a vertical array.

SOHIC is not detected by conventional MFL and UT tools. SOHIC can be identified by crack detection tools.

6.14.2 Avoiding SOHIC

- SOHIC may be avoided by testing pipeline material and the welds.
- If SOHIC is suspected then the susceptibility to SOHIC may be evaluated using spare, similar pipe if available. The girth welds will need to be completed following the weld procedure used for installation of the pipeline.
- Until 2016 there was no standard test to determine the resistance of a pipeline to SOHIC. A new test, similar to the test for SCC and incorporating a torsional stress, has been developed[14, 15].
- Full ring tests are also suitable for evaluation of susceptibility to SOHIC.
- Areas identified as possible SOHIC should be validated using time of flight diffraction ultrasonic testing. Phased array ultrasonic inspection may also be effective.
- Reduction of corrosion rate reduces the evolution of atomic hydrogen and reduces the rate of permeation of atomic hydrogen. A quality corrosion inhibition programme will delay propagation of SOHIC but is unlikely to halt propagation.
- There are no known methods to predict the rate of propagation of SOHIC.

6.15 Corrosion fatigue

6.15.1 Mechanism

Corrosion fatigue occurs if a pipeline is subject to cyclic stress. The stress may arise from variations in operating pressure and/or temperature, an external force, for example loading by road traffic or oscillations of a submarine pipeline by vortex shedding at a span.

Pumps and compressors impose cyclic loading. The pressure variations will be dampened by friction and turbulence and will generally only be significant for a short distance downstream of the pump or compressor station. An example is given in Figure 6-14.

The fatigue loading repeatedly fractures the corrosion products on the pipeline surface exposing bare metal that corrodes to form new corrosion product. Corrosion reactions may also result in hydrogen embrittlement of the fatigue crack tip, reducing ductility and increasing the mechanical damage caused by the stress cycles.

The fatigue life will be reduced by up to 30% if the pipeline steel is embrittled by hydrogen charging. Sour corrosion and over protection by the cathodic protection system can cause embrittlement (see section on external corrosion).

6.15.2 Moderating corrosion fatigue

- Check operation of the pipeline to determine if there is pressure or thermal cycling.
- Calculate loads and determine fatigue life. Adjust expected fatigue life to account for sour service.
- Run a geopig to determine if there is a deviation of the pipeline from the planned route.
- Excavate pipeline to determine if cracks are present. Shallow external cracks may be ground out.

QR Code 6-5. Scan here for additional information on corrosion inhibitors.

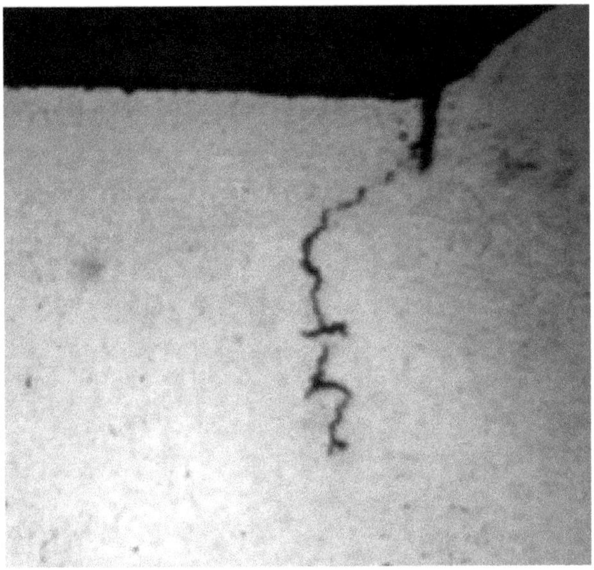

Figure 6-13. Near vertical array of hydrogen blisters linked by cracks in a heat affected zone (weld is to right of crack).

Figure 6-14. Longitudinal fatigue crack possibly initiated at the weld of the small diameter pipe on the surface of the spool piece.

6.16 References

[1] Influence of Liquid Flow Velocity on CO_2 Corrosion: A Semi-Empirical Model, C. de Waard & U. Lotz, Paper 128, Corrosion95, 1995, NACE International, Houston.

[2] CO_2 Corrosion Control in Oil and Gas Production: Design Considerations, Working Party Report No 23, European Federation of Corrosion, Eds MM.B. Kermani & L.M. Smith, Inst. Materials, 1997

[3] Predicting Corrosion in Slightly Sour Environments, S.N. Smith, Materials Performance, August 2002

[4] Discussion of the History and Relevance of the CO_2/H_2S Ratio, S.N. Smith, Paper 11065, Corrosion 2011, NACE International, Houston.

[5] Improvements on de Waard-Milliams Corrosion Prediction and Applications to Corrosion Management, B.F.M. Pots, R.C. John, I.J. Rippon, M.J.J. Simon Thomas, S.D. Kapusta, M.M. Girgis & T. Whitham, Paper 02235, Corrosion 2002, NACE International 2002.

[6] Top of the line corrosion, M. Singer, Chapter 16, 'Trends in Oil and Gas Corrosion Research and Technologies, http://dx.doi.org/10.1016/B978-0-08-101105-8.00016-4, Elsevier, 2017

[7] Sour Top of the Line Corrosion in the Presence of Acetic Acid, K. singer, A. Camacho, B. Brown, S. Nesic, Paper 10100, Corrosion 2010, NACE International, Houston.

[8] Top of Line Corrosion Mechanism for Sour Wet Gas Pipeline, D.V. Pugh, S.L. Asher, J.Cai, W.J. Sisak, J.L. Pacheco, F.C. Ibrahim, E.J. Wright, A. Dhokte, S. Venaid and D. Robson, Paper 09285, Corrosion 2009, NACE International, Houston.

[9] DNV RP 0501: Erosive Wear in Piping Systems, 2011.

[10] Study of Factors Affecting CO_2 Corrosion and Inhibitor Effectiveness Using a Multi-Phase Flowloop, D.D. Salama & B.N. Brown, Paper 09476, Corrosion 2009, NACE International, Houston.

[11] NACE MR-0175: Materials for Sufide Stress Cracking and Stress Corrosion Cracking Resistance in Sour Oilfield Environments, NACE International, Houston.

[12] ISO 16165/NACE MR-0175, Petroleum and Natural Gas Industries - Materials for use in H_2S-containing environments in oil and gas production, Parts 1 - 3.

[13] Petroleum Microbiology, Ed. B. Ollivier & M. Magot, ASM Press, Washington, 2005

[14] Corrosion Testing of Pipelines, C.M. Fowler, Exova/Element Report.

[15] ISO 17945/NACE MR-0103: Petroleum and Natural Gas Industries - Metallic materials resistant to sulfide stress cracking in corrosive petroleum refining environments.

CHAPTER 7

External cracking

Dr Phil Hopkins and Dr Roger King

7.1 Introduction

Pipelines are the main mode of oil and gas transportation due to economics and safety. Their safety record is primarily due to pipelines being regulated by governments, and designed, constructed, and operated using standards. The most popular standards are from the American Society of Mechanical Engineers[1, 2], but other national standards are also popular (for example, [3 to 5]).

Our very old pipelines (say, before the 1940s) were built to any design, using any method: this was not good for safety, the environment, or security of supply. This can result in older pipelines containing defects, and additionally they may not have been inspected/tested to modern standards which would detect defects before a pipeline is put into service.

All large diameter, long distance oil and gas pipelines operate in hostile environments (underground, or under the sea) and these environments can cause the pipeline to become defective; for example, pipelines can corrode and contain large corrosion defects (see Figure 7-1).

Pipeline operators are aware of the possibility of defects being present in their pipelines. These operators regularly inspect and survey their pipelines to both prevent and detect defects, before they threaten pipeline safety. These defects can include:

- corrosion, on the inside of the pipeline (caused by the product in the pipeline), or the outside (caused by moisture in the surrounding soil (see Figure 7-1), or surrounding seawater);
- dents and gouges, caused by impacts from earth moving equipment (onshore pipelines), or anchors (subsea pipelines).

A particularly severe defect is a crack (see Figure 7-2). There are several reasons a pipeline can contain cracks including:

- cracking due to the environment (corrosion can cause cracking in the pipeline);
- cracking created during the manufacture of the line pipe, or during pipeline construction;
- cracking caused by the operation of the pipeline (repeated stressing on a pipeline can lead to cracking caused by 'fatigue').

This Chapter explains what cracking is, why pipelines can contain cracks, and how we can manage cracks in pipelines. The focus of this Chapter is cracking occurring on the external surface of the pipeline. Cracking may also initiate and grow on the inside surface of a pipeline when it transports 'wet' fluids (fluids with a water content) containing hydrogen sulphide: this form of internal cracking is discussed in the Chapter on internal corrosion.

7.1.1 What is a crack?

A crack is a very sharp defect: it is a 'planar' (two-dimensional) defect. ASME B31.8S[6] states that a crack is a... 'very narrow, elongated defect caused by mechanical splitting into two parts'.

We now often hear the term 'crack-like'. The USA's National Transportation Safety Board (NTSB) has defined a 'crack-like' defect as[7]: '... "crack-like" is used to describe a feature that has the shape of a crack but where the material on opposite faces of the feature was never fused or bonded' (see Figure 7-3). API 579 states[8]: 'Crack-like flaws are planar flaws that are predominantly characterized by a length and depth, with a sharp root radius'.

Crack or crack-like? We inspect our pipelines to detect defects such as cracks, but inspection equipment only measures reflections, changes, etc. (see Figure 7-4). The inspection cannot tell us if the reflector is definitely a crack. We rely on interpretation, judgements based on defect location, reflected/changed signal, etc. This introduces human judgement and hence human error.

It is not easy to say with total confidence that the defect you have detected is a crack, unless you can actually see part of the crack. Clearly, the crack-like defects in the pipeline girth weld in Figure 7-3 are on the inside of the pipe, and not visible from the outside. An inspector would not be able to see these crack-like defects, and hence would need to use some type of inspection methods (e.g., ultrasonics (see Figure 7-4)) to locate and size these two defects. He/she would not be able to say, without some doubt, that these two defects were cracks, or crack-like defects.

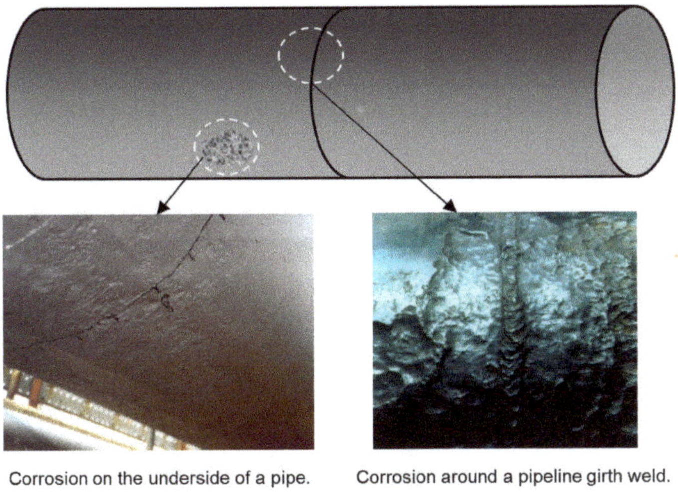

Figure 7-1. Corrosion in pipelines (photographs copyright and courtesy of Rosen Ltd,UK).

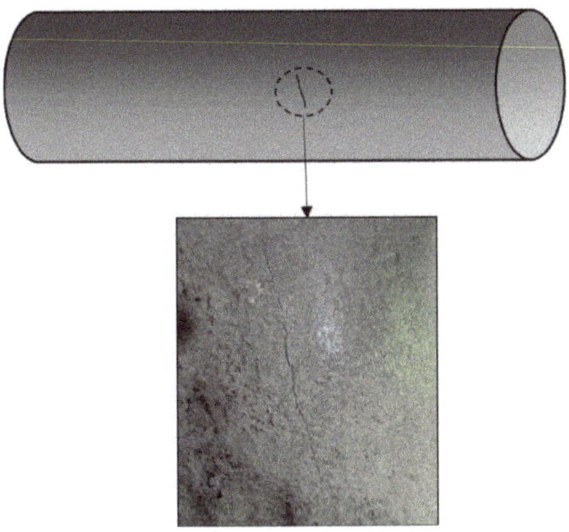

Figure 7-2. Crack in pipeline (photograph copyright and courtesy of Penspen Ltd, UK).

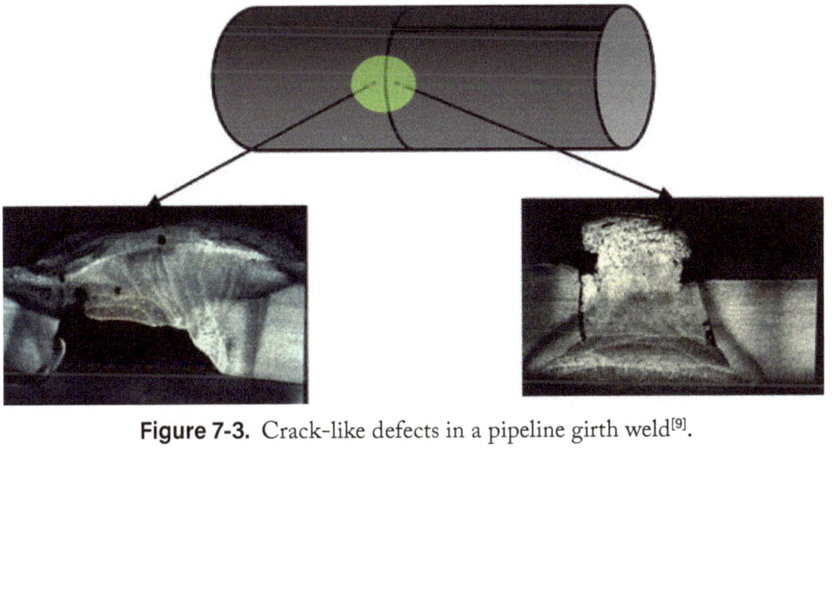

Figure 7-3. Crack-like defects in a pipeline girth weld[9].

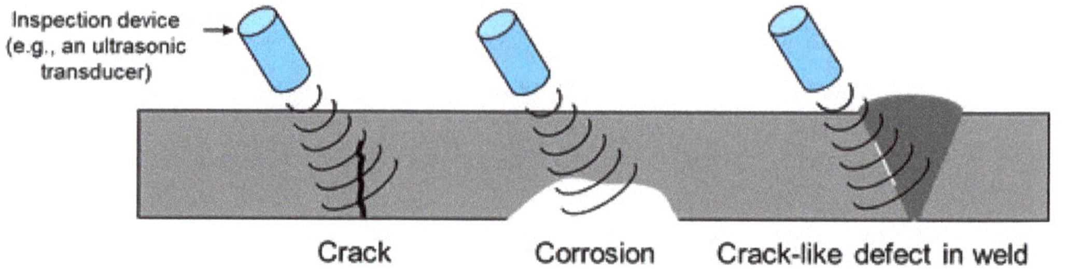

Figure 7-4. It can be difficult to discriminate between cracks, crack-like defects, and other defects.

Doubts about discriminating between cracks and crack-like defects have led to standards assigning likelihoods to their discrimination; for example[10]:

- 'likely crack' - high certainty that the indication called by inspection vendors correlates to a crack-like defect (e.g., stress corrosion cracking, hook cracks, etc.).
- 'possible crack' - reduced certainty of the indication being an actual crack, or, the operator cannot determine with certainty that the indication is not a crack-like defect.
- 'unlikely crack' - high certainty to correlate to a non-crack-like feature or imperfection (e.g., corrosion, misalignment, trim features, etc.).

7.1.2 Why are cracks in pipelines?

Line pipe manufacture and pipeline construction. Cracks can be created during the manufacture of our line pipe, or its construction. These cracks are inadvertently 'built into' a pipe, but are usually too small to cause pipe failure[10], or are usually detected:

- during inspections at the mill making the line pipe;
- during the hydrostatic test performed at the mill before the line pipe is released for delivery; or,
- during the pipeline's pre-service pressure test.

If they are not detected (for example, the pre-service pressure test is not conducted, or not conducted to a high level) they can cause failure of the pipeline during service.

Operation. Cracks can also be created during the pipeline operation; for example:

> They can be caused by corrosion of the line pipe (for example, 'stress corrosion cracks' (SCC – see later); or, cyclic loadings on the pipeline (fatigue cracks – see later).

7.1.3 Always understand the cause of cracking

BSI 7910[11] states: 'Cracks can be indicative of poor design features, workmanship or inspection, or stress corrosion, fatigue or corrosion fatigue, and therefore the cause should be ascertained before proceeding with an assessment.'

A crack is an indication that something has gone wrong, or something is going wrong... therefore, standards do not allow them. ASME B31.4 states[1]: 'Verified cracks except shallow crater cracks or star cracks in girth welds shall be considered defects and removed or repaired unless an engineering evaluation shows that they pose no risk to pipeline integrity...'.

The 'engineering evaluation' can be conducted using standards such as References 8 and 11, but caution is needed with these evaluations, as cracks are:

- difficult to detect and size;
- grow quickly if the environmental conditions are favourable (e.g., if corrosion is present, or the structure is fatigued); and,
- this growth can be difficult to predict and calculate.

All these considerations lead to uncertainties in their assessment. Accordingly, crack assessments should be carried out by experts[12], within the control of a pipeline integrity management system (for example, [6, 13 to 16]).

QR Code 7-1. Scan here for more on images of cracks in pipelines.

7.2 Why are cracks a problem in structures?

Why are cracks such a big problem? The presence of a defect, such as a crack (see Figure 7-5):

- reduces the structure's load bearing area in the remaining ligament below the crack; and,
- its presence also leads to a concentration and elevation of stresses at its tip.

This reduction in load bearing area, and stress elevation, can lead to failure at stresses below the material's yield strength.

The material's toughness (defined as a material's resistance to a crack or crack propagation) resists the presence of a crack and its growth through a structure, so we need high toughness materials to resist the presence of cracks. A crack has very high stresses at its tip. This is because the stresses that would have been carried by the structure in the absence of a crack concentrate around the crack (see Figure 7-6). The concentration raises the local stress to above the yield strength, and starts extensive straining (deformation) at the tip of the crack:

- Brittle (low toughness) materials cannot deform and have low resistance to cracks. They can fail at high stresses but cannot deform. This means that high local stresses at the crack tip cannot be relaxed by straining, and hence the brittle material will fail quickly at a crack.
- Ductile (high toughness) materials deform and relax the stresses at a crack tip and will resist a crack. They will also allow the crack to blunt (reducing the stress concentration).

A crack is sharp with high stresses at its tip (see Figure 7-7) leading to fracture, or tearing, at its tip. This tearing can continue through the ligament, preventing the stress in the ligament reaching the yield strength, and this can lead to a failure stress lower than the material's yield strength.

The high stresses at the (sharp) crack tip go above the material's yield strength, and can cause failure, but many defects in our structures are not sharp (they are 'blunt'; for example, corrosion). Blunt defects do not create such high stresses at their tip and fail at much higher loads. The blunt defect is effectively increasing the material's toughness.

We now know that toughness is the ability of a material to withstand the presence of a defect, such as a crack. The higher the toughness, the more resistance to crack growth, and hence the higher the failure stress of the crack. Toughness will be an essential part of our crack assessment (see later).

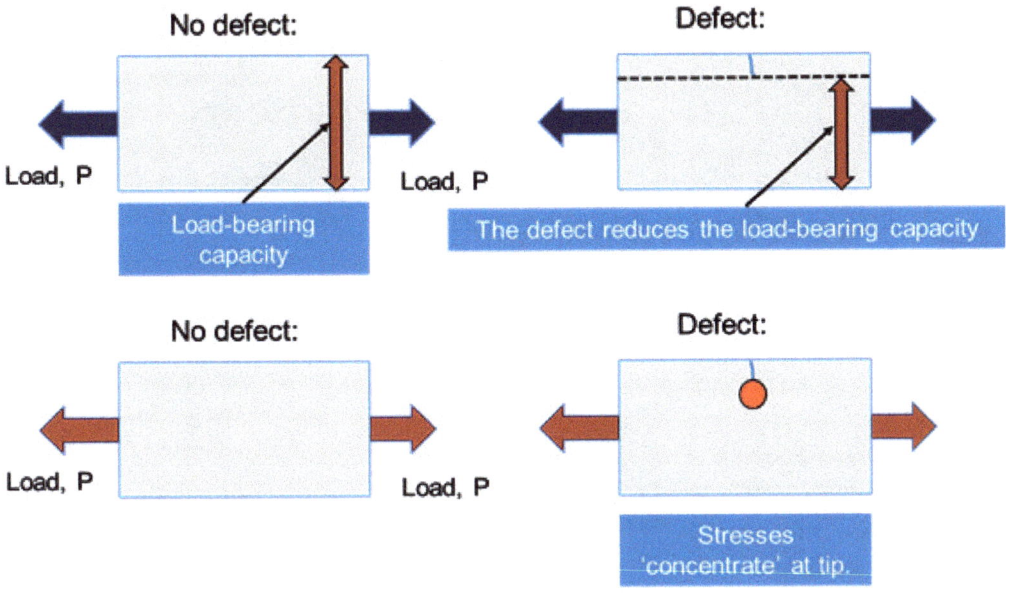

Figure 7-5. Reduction in load-bearing capacity (top) and concentration of stresses (bottom) caused by a crack.

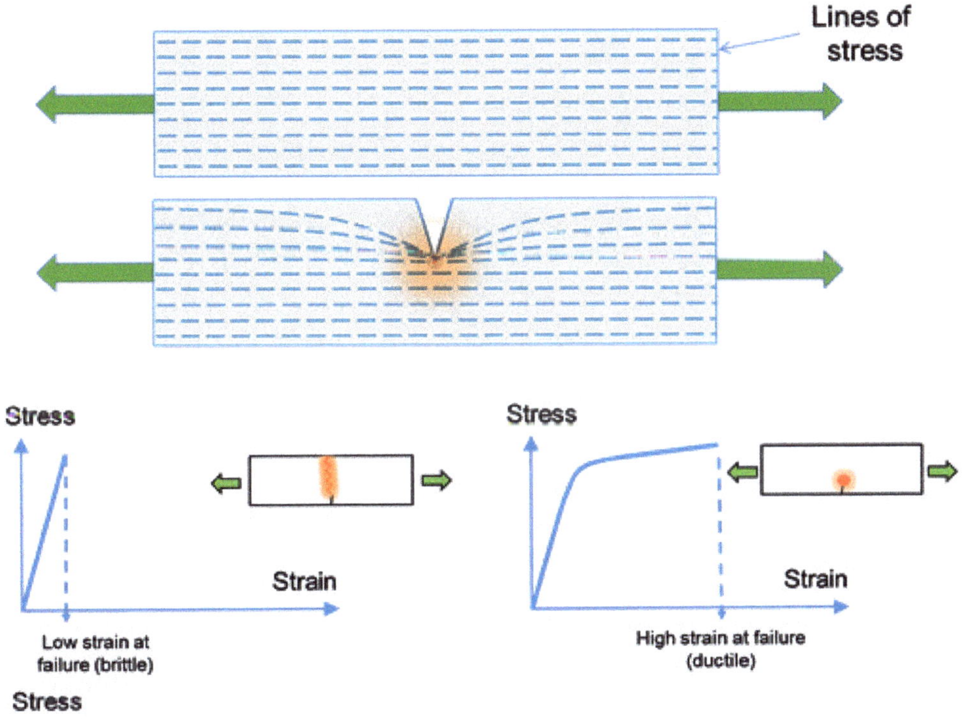

Figure 7-6. Cracks locally raise stress levels.

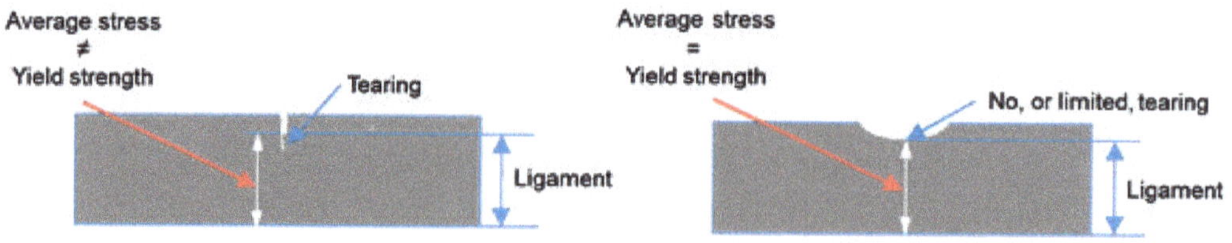

Figure 7-7. Cracks can grow when subjected to stresses.

7.2.1 What cracks are in pipelines?

Pipeline can crack due to a variety of reasons[10]; for example:

- 'environmentally assisted cracking' such as: stress corrosion cracking, sulphide stress cracking (SSC), hydrogen-induced cracking (HIC), or stress-oriented hydrogen-induced cracking (SOHIC);
- manufacturing defects associated with the pipe body or seam weld, such as: hook cracks, lack-of-fusion, mismatched plate edges, in electric resistance welded (ERW) line pipe;
- 'mechanical damage' caused by impact of the pipeline by foreign objects, such as earth moving equipment, and resulting in gouges that can contain cracking;
- construction and fabrication defects.

All these cracks can extend further during operation as a result of pressure-cycle-induced fatigue, with any eventual failure being caused by the fatigue extending the original crack.

This section will focus on two common types of cracking found in pipelines:

- fatigue cracking; and,
- stress corrosion cracking.

QR Code 7-2. Scan here for more information on cracks in pipelines.

Fatigue cracks. Very few structures are subjected to a single ('static') stress during service. Most structures are subjected to varying, or 'cyclic', stresses during operation. When we subject a structure to a cyclic (repeated) stress we say it is being 'fatigued'. Cracks can be both created and extended by fatigue, if the cyclic stresses and number of stress cycles are sufficiently high.

Fatigue failures take time, and this time to failure is primarily a function of the magnitude of the stress range, and the number of cycles at this stress range. A corrosive environment will also accelerate this fatigue crack growth (this is 'corrosion fatigue' – see later).

Fatigue has been known and understood for many years: many structures are fatigued, and many structures fail; for example, the first commercial jet airplane (the Comet) failed in the early 1950s due to fatigue around its square-shaped windows. A combination of poor design and bad manufacturing led to many deaths.

Pipelines are subjected to cyclic stresses due to[17]:

- internal pressure variations;
- changes in temperatures;
- external loads (e.g., traffic loading, or movement on a seabed from currents);
- etc.

The internal pressure in a pipeline can vary, due to changes in inputs, or changes in customer demands, although the internal pressure does not vary in gas pipelines as much as in liquid lines[10]. This means that fatigue management is particularly important in liquid lines. Incidents of failures in gas pipelines caused by fatigue (cyclic pressure) are very few in numbers. This is because[18]:

- the operation of a gas pipeline is usually continuous product shipping, without the use of batching operations, resulting in few large amplitude internal pressure cycles;
- natural gas is compressible, meaning changes in the pressure level that may occur at the discharge of a compressor station are damped, and do not propagate a significant distance along the pipeline.

It should be emphasized that fatigue is generally not considered in most pipeline designs, principally because the number of stress cycles that occur in the life of a pipeline are typically fewer than required to initiate a fatigue-related failure in the pipe wall[3]. When fatigue failures do occur, they occur at some pre-existing flaw (such as a dent) or design fault (such as a poorly designed weld): 'There are no known cases of fatigue failures due to pressure cycle effects in gas or liquid pipelines in the absence of some sort of significant initial flaw, damage, or ill-considered design feature that concentrates stresses locally' [19].

This quote is supported by data from CONCAWE[20], an organisation that reviews failures in Western European onshore liquid pipelines. They note very few pipeline failures from deterioration due to fatigue. Fatigue failures do sometimes occur when pipelines have suffered some construction fault, or subsequent damage such as dents. These are relatively infrequent causes of spillage.

This means that when we detect defects such as cracks and dents, we may need to conduct a fatigue analysis on these defects. Standards such as References 8 and 11 can be used for these analyses.

Corrosion fatigue. Fatigue becomes a greater hazard when a pipeline is subjected to cyclic stress (from changes in pressure and/or temperature, or an external force) is in corrosive soil and the cathodic protection system is not providing adequate protection. Both onshore and offshore pipelines can suffer from this 'corrosion fatigue', and a good example of corrosion fatigue is given in Reference 21.

Corrosion pitting may often be the initiator of the fatigue cracks and the rate of growth of the cracks is higher when the environment is corrosive. The cracking can be identified using specialized in-line inspection tools (see later).

Figure 7-8 provides a flowchart for identifying pipelines threatened by corrosion fatigue. Near neutral stress corrosion cracking (see later) is often regarded as corrosion fatigue because constant loading rarely leads to crack growth.

QR Code 7-3. Scan here for more on fatigue in pipelines.

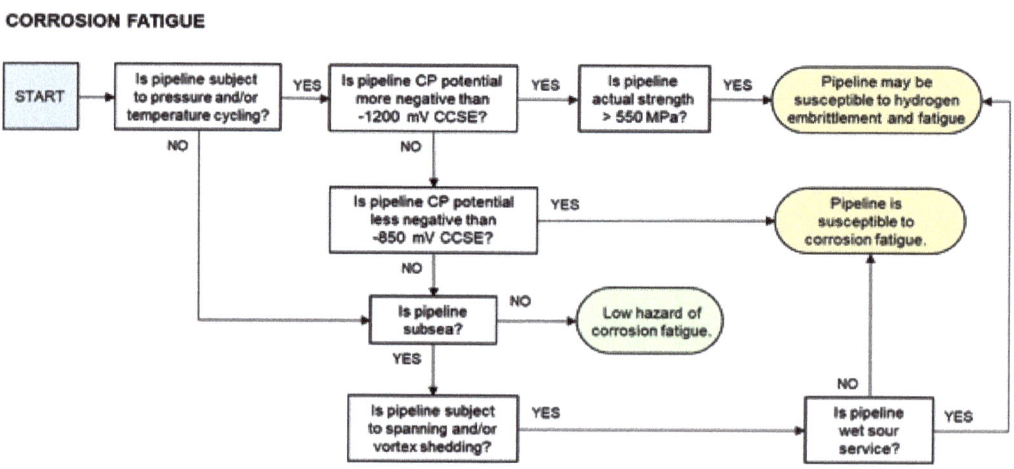

Figure 7-8. Screening criteria for identifying corrosion fatigue hazard.

Controlling fatigue. Fatigue is controlled/reduced by:

- removing any 'starter' defect/stress concentration by inspection and repair, or pressure testing;
- reducing stress ranges;
- reducing number of cycles of fatigue.

In a corrosive/erosive environment, reducing the corrosion/erosion will increase fatigue life.

7.2.2 Stress corrosion cracking

'Stress corrosion cracking', SCC[22, 23], is 'environmental cracking' (cracking caused by the environment surrounding the structure) and occurs in many industries. External SCC in pipelines has been well known for many years. The first documented case of SCC causing a pipeline failure was in 1965 in Natchitoches, Louisiana. This rupture caused a gas release and fire resulting in 17 fatalities.

ASME B31.8S[6] defines SCC as a: '… form of environmental attack of the metal involving an interaction of a local corrosive environment and tensile stresses in the metal, resulting in formation and growth of cracks.'

The steel cracks due to the presence of hydrogen (from a corrosion process created by a breakdown or disbonding of the external coating on the pipeline) and can cause pipeline failures. SCC accounts for only about 1% of all USA pipeline failures. 8% of product loss incidents in the last 40 years, on European pipelines operating above 15 bar (13 incidents), have been as a result of external SCC[23]. These are relatively low incident rates, but SCC in a pipeline is difficult to detect, and can cause ruptures. It can affect both gas and liquid lines, but it has never been reported in subsea pipelines[24]. SCC can crack a material without any obvious deformation or material deterioration (although pitting corrosion is associated with the cracking). SCC always propagates in a direction perpendicular to the direction of the principal stress, typically the hoop stress[22]. This means that SCC is usually in the axial direction, and in highly stressed pipelines. Cyclic stresses assist formation/growth.

Circumferential SCC is sometimes present in a pipeline, and is associated with:

- areas of local high stress concentration (e.g., a dent); or,
- high external stresses (e.g., mountainous regions).

Regardless of orientation, the local stresses are also high; e.g., the cracks are associated with dents, at/around welds, or, in corrosion pits.

The SCC occurs at any location in the pipeline, usually in clusters ('colonies'). Fracture faces are covered in 'magnetite' (Fe_3O_4) and 'carbonate' (a salt of carbonic acid) films.

7.2.3 Types of SCC

We have two types of external SCC (Table 7-1):

- 'high pH SCC' (occurs in a high pH environment of ≥9);
- 'near-neutral SCC' (occurs in a pH environment of between 5 and 7.5).

It is the pH value at the tip of the SCC that is important, not the pH of the soil around the pipe.

The electrolyte (wet soil) around an onshore pipeline contains CO_2 from the decay of organic matter. The CO_2 can dissolve in water and form carbonic acid (H_2CO_3), which then can dissociate to form hydrogen (H^+), bicarbonate (HCO_3^-), and carbonate (CO_3^{2-}) ions.

The pipeline's cathodic protection (CP) plays a key role in creating the pH value at the pipe's surface. API 1176 states[10]: 'The balance between the level of cathodic protection reaching the surface and the CO_2 partial pressure in the ground water is critical in determining the pH at the steel surface, and hence, whether high-pH stress corrosion cracking, near-neutral-pH stress corrosion cracking, or neither of these, occurs.'

Table 7-1. Ranking system for stress corrosion cracking[10].

Factor	High pH carbonate cracking	Near neutral stress corrosion cracking
Pipe temperature	A	C
Pipeline age	A	A
Soil properties	B	A
Maximum stress	A	A
Stress fluctuations	B	A
Stress concentrators	B	A
Surface condition	A	A
Pipe coating	A	A
Girth weld coating	A	A
Pipe manufacturer	C	B
Cathodic protection related parameters	B	B
A - factor with major influence. B - indicates a factor with some effect. C - factor with minor effect.		

API 1176 states: 'SCC can only occur where coating is disbonded or damaged and where inadequate CP is available at the pipe surface'.

Accordingly, the principal factor contributing to high pH SCC is coating condition, whereas the principal factor contributing to near-neutral SCC is a breakdown in the pipe coating and the CP.

pH

'pH' is a measure of the acidity of a liquid. Water has a pH of 7 and is 'neutral'. A lower (<7) pH level indicates increasing 'acidity', while high pH levels (>7) indicate increasingly 'basic' or 'alkaline' solutions. pH is defined as the negative log (base 10) of the hydrogen ion concentration.

High pH SCC. High pH SCC (sometimes called 'high pH carbonate stress corrosion cracking') can occur in an environment involving an interaction between hydroxyl ions (produced by the cathode reaction in the CP system) interacting with CO_2 in the soil (from the decay of organic matter, the activity of plant roots and bacterial activity). The CP current causes the pH of the electrolyte beneath disbonded coatings to increase (between pH 8 and 11). CO_2 dissolves in the elevated pH electrolyte, resulting in the generation of the concentrated carbonate- bicarbonate electrolyte.

Failures typically occur within 15 – 20 years of commissioning a pipeline (but as short as 6 years, and as long as 40 years)[23]. The cracking mechanism typically occurs at stress levels >60% of SMYS, and requires a cyclic component to the stress, and is accelerated at elevated temperature.

There is no documented experience of high pH SCC occurring under FBE, two- and three-layer extruded PE, and liquid coatings, including epoxy, epoxy urethane and polyurethane (PU)[23].

Near-neutral SCC. Near-neutral SCC occurs in an environment of[23]:

- a groundwater containing dissolved CO_2 (from the decay of organic matter and geochemical reactions in the soil); and,
- little, if any, CP current reaching the pipe surface over a prolonged period, either because of the presence of a shielding coating, or rock or other material close to the pipeline, a highly resistive soil, or ineffective CP.

It initiates at locations on the outside surface, where there is already pitting or general corrosion, in an environment having a pH between 5.5 and 7.5.

It occurs under disbonded coal tar enamel, asphalt and tape coatings where the surface under the disbonded coating:

- is being shielded from CP (near neutral pH SCC can only occur when there is an inadequate level of CP on the pipe surface[23]); and,
- shows active corrosion.

It typically occurs at stress levels around 70% of SMYS and is accelerated by cyclic loading. Temperature is not implicated in the crack initiation or propagation of near neutral pH SCC.

SCC threat assessment. Figure 7-9 provides a flowchart for identifying pipelines threatened by high pH SCC, and Figure 7-10 provides a flowchart for identifying pipelines threatened by near neutral SCC. See Table 7-2.

7.2.4 Controlling cracks

Cracks are normally controlled by:

- manufacture: quality control of both the line pipe and welding, including inspection and testing;
- operation: control of operating conditions (e.g., minimise cyclic stresses);
- inspection: during fabrication of the pipeline and during operation;
- pressure testing: testing the pipeline before (and sometimes during) service to remove large cracks.

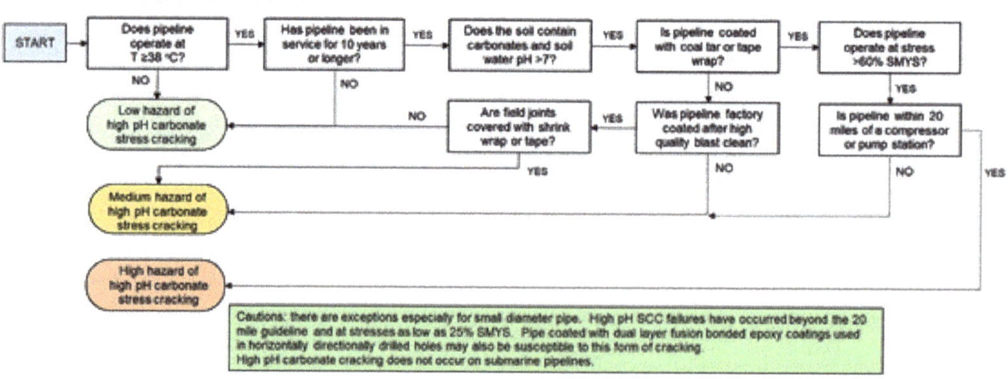

Figure 7-9. Screening criteria for high pH stress corrosion cracking.

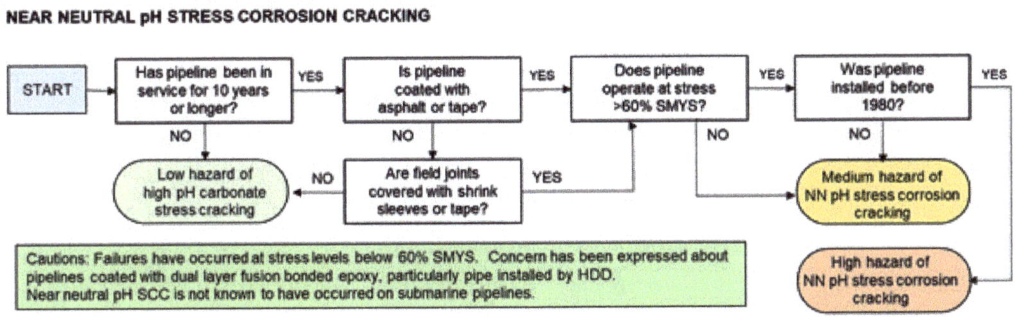

Figure 7-10. Screening criteria for near neutral stress corrosion cracking.

7.3 Crack management

How do we manage a cracked pipeline? There are many standards to help with the management of cracks (for example, References 6, 10, 13). These standards help us manage cracks in pipelines by:

- presenting suitable methods for assessing the condition of the cracked pipeline;
- establishing criteria for in-line inspection (ILI) results;
- helping calculate the remaining lives of cracks remaining in the pipeline; and,
- giving appropriate hydrostatic test levels and duration.

API 1176[10] recommends:

- selecting methods for assessing the condition of the cracking in liquid and gas pipelines;
- establishing response criteria for ILI results and determining pressure reductions where action is delayed beyond calculated timelines;
- determining appropriate hydrostatic test levels and duration;
- calculating remaining lives of anomalies that may remain in the system, and re-inspect/test or repair; and,
- developing preventive and mitigative measures for the cracking.

Serious cracks require 'immediate' response; i.e., usually excavation and repair and/or restriction in operating pressures to maintain safety margins. Serious cracks are very deep (e.g. >70% wall thickness). Lesser cracks (e.g. >50% wall thickness) have to be addressed within 365 days.

Other cracks may require re-inspection, or retest, or might remain in the pipeline, if shown to be insignificant by an engineering critical assessment.

7.4 Pipeline condition assessment for cracks

We must regularly assess the condition of our pipeline. Pressure testing, and internal inspection using 'smart pigs' are methods of assessing the condition of an in-service pipeline (see Figure 7-11).

Other assessment methods include 'direct assessment' (e.g., [25, 26]). 'Direct assessment' is an integrity assessment used to determine the pipeline's integrity; for example, external corrosion direct assessment can be indirect assessment using a coating survey, followed by excavation of the pipeline and direct measurement.

Cracked pipelines are usually assessed using[10]:

- in-line inspection (ILI) tools; or,
- pressure testing.

A pressure test can be used instead of ILI where:

- ILI is not an option; or,
- where inspection tools might be incapable of detecting the cracks.

Pressure testing can also be used to validate the results of inspection tools.

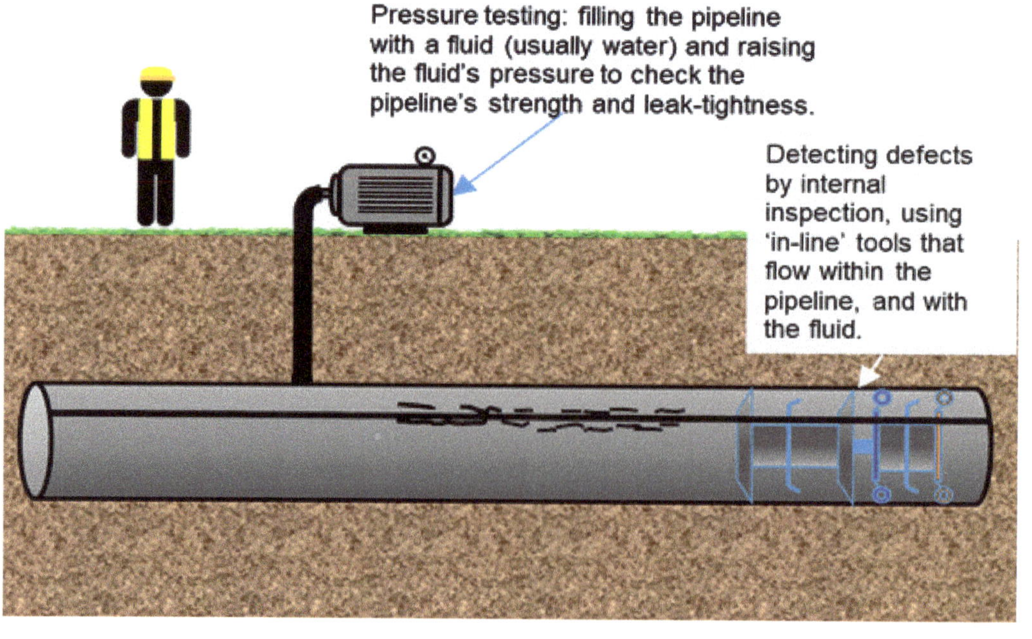

Figure 7-11. The condition of a pipeline can be determined by pressure testing, or in-line inspection tools.

7.4.1 Pressure testing

Generally, pipe that has been inspected in the mill, and has successfully undergone a post-construction pressure test to a minimum level of 1.25 times the maximum operating pressure (MOP) does not contain flaws that result in failures later in service, in the absence of a growth mechanism[10].

We can conduct an in-service pressure test on a pipeline containing cracking: many standards/regulations require this test on cracked pipelines. The test will fail large cracks and allow a safe period of operation following the test.

ASME B31.8 [2] gives the following guidance (see Figure 7-12):

- hydrotest (i.e., pressure test using water as the pressurizing medium) to a pressure to give a hoop stress of ≥100% of the specified minimum yield strength (SMYS) of the line pipe;
- hold the pressure long enough for the test pressure to stabilize (in most cases 30 minutes or less); then, follow this with…
- a leak test at a test pressure of at least 1.10 times the maximum allowable operating pressure (MAOP) of the pipeline.

QR Code 7-4. Scan here for more on stresses in pipelines.

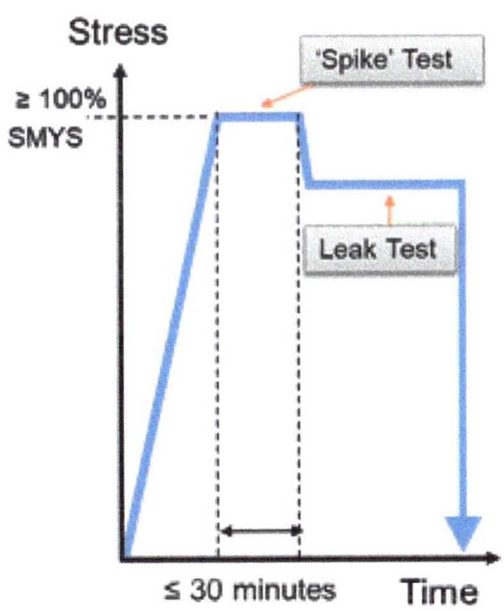

Figure 7-12. Example of pressure testing parameters for crack management.

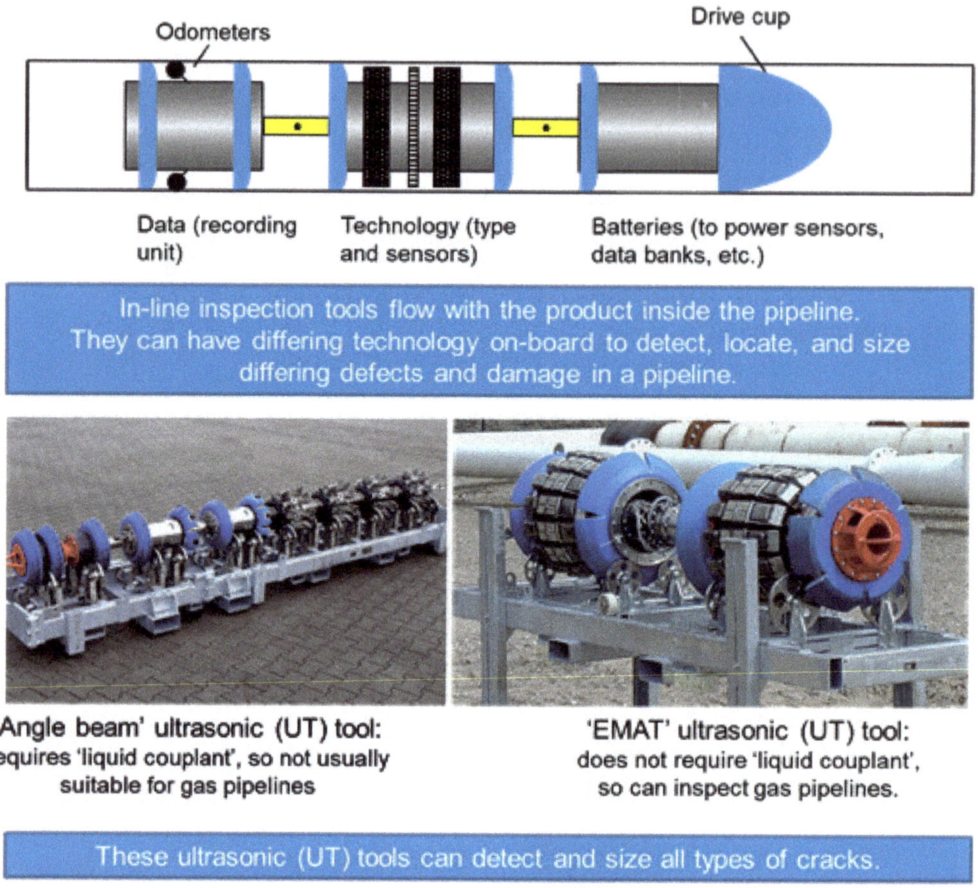

Figure 7-13. Examples of in-line inspection (ILI) tools that can detect and size cracks in pipelines (photographs copyright and courtesy of Rosen Group).

7.4.2 In-line inspection

We have several in-line inspection tools that can detect cracks, or 'crack-like' defects; for example, Figure 7-13, and the capabilities of a variety of ILI tools for crack detection are presented in Table 7-2[10].

Table 7-2. Detection and sizing capabilities of in-line inspection tools[10].

Type of axial crack	Location	Magnetic Flux Leakage (MFL) Tools		Ultrasonics (UT) Tools	
		Transverse field inspection MFL	Spiral MFL	Angle Beam UT	Electro-magnetic acoustic transducers (EMAT)
Stress corrosion cracking	Pipe	D*	D*	S	S
Fatigue cracks	Pipe and Seam Weld	D*	D*	S	S
Fatigue cracks in dents	Pipe	D*	D*	D*, S*	D*, S*
Hook cracks	Seam Weld	D*, S*	D*, S*	S*	D, S*

Notes:
'D' is detection capability as specified by vendor, with limited or no sizing capability.
'S' is detection capability with sizing capability specified by vendor.
D or S* indicate that a qualification on ability to detect or detect and size is needed.*

7.4.3 Crack assessment

You do not have to repair all cracks in pipelines. Cracks detected in pipelines can be assessed. API 579[8] and BSI 7910[11] present methodologies for the assessment of many types of defects, including cracks in pipelines.

It should be emphasised that crack assessment using these standards is not simply a calculation. It requires and relies on (see Figure 7-14):

- a pipeline/product/configuration suitable for inspection/testing;
- pipeline records (e.g. material properties);
- an accurate and reliable condition assessment method (e.g., in-line inspection (ILI)), followed by robust data analysis;
- engineering assessment of the crack;
- field verification and repair, mitigation or re-assessment.

This is a lot of required information that is often not available. Indeed, API 1176 states[10]: 'When managing cracks, a lack of adequate data for making critical integrity decisions can make the use of assumptions necessary.'

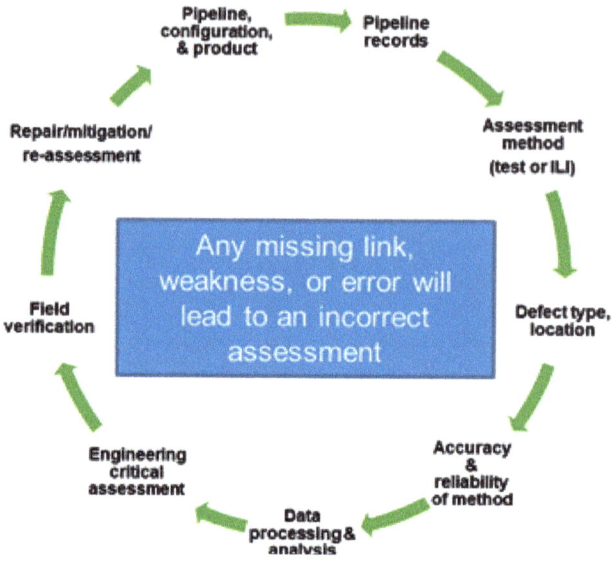

Figure 7-14. Crack assessment is not simply an inspection and a calculation.

7.5 Summary

Cracks are sharp defects which indicate something is going/gone wrong in our pipeline. We treat cracks with great caution as cracks are:

- difficult to detect and size;
- grow quickly if the environmental conditions are favourable (e.g., if corrosion is present, or the structure is fatigued); and,
- this growth can be difficult to predict and calculate.

These cracks can be created during service (e.g., by corrosion or fatigue), or may have been in the pipeline since it was constructed (e.g., line pipe cracks, or cracks in welds).

We can detect cracks by pressure testing the pipeline, or we can use in-line inspection tools. These tools use various technologies to detect and size cracks but the preferred technology is ultrasonics. It should be noted though, that cracks are the most difficult defect to detect and size, because they are two-dimensional, have irregular shape, and can be located in welds.

If we detect cracks in our pipeline, we can assess the significance using standards such as API 579 and BS 7910. These assessments should be conducted or supervised by subject matter experts.

References

[1] Anon., 'Pipeline Transportation Systems for Liquids and Slurries', American Society of Mechanical Engineers. New York, USA. ASME B31.4- 2016. 2016.

[2] Anon., 'Gas Transmission and Distribution Piping Systems', American Society of Mechanical Engineers. New York, USA. ASME B31.8-2018. 2018.

[3] Anon., 'Pipelines - gas and liquid petroleum - design and construction', AS 2885.1-2012. Australian Standards. Australia, 2012.

[4] Anon., 'Oil and gas pipeline systems', CSA Z662-15, Canadian Standards Association. Canada. 2015.

[5] Anon., 'Pipeline systems – Part 1: Steel pipelines on land – Code of practice', British Standards Institution. UK. Published Document. PD 8010- 1:2015. 2015.

[6] Anon., 'Managing System Integrity of Gas Pipelines', ASME B31.8S-2018. American Society of Mechanical Engineers. New York, USA. 2018.

[7] Anon., 'Pacific Gas and Electric Company Natural Gas Transmission Pipeline Rupture and Fire San Bruno, California September 9, 2010', National Transportation Safety Board. USA. Accident Report NTSB/PAR-11/01 PB2011-916501. August 30th, 2011.

[8] Anon., 'Fitness-For-Service', Third Edition. American Petroleum Institute. API 579-1 / ASME FFS-1. 2016.

[9] P Hopkins, 'Some Experiences of Applying the Pressure Systems Regulations to Older Systems', IMechE Paper C454/006/93, Institute of Mechanical Engineers, London, 1993.

[10] Anon., 'Recommended Practice for Assessment and Management of Cracking in Pipelines', American Petroleum Institute. API 1176. July, 2016.

[11] Anon., 'Guide to methods for assessing the acceptability of flaws in metallic structures', BS 7910:2013+A1:2015. British Standards Institution. UK. 2013.

[12] M Unger, P Hopkins, 'What is a 'Subject Matter Expert'?', The Journal of Pipeline Engineering, Volume 16, No. 4, Fourth Quarter. 2017.

[13] Anon., 'Managing System Integrity for Hazardous Liquid Pipelines', American Petroleum Institute. API Recommended Practice 1160. Third Edition, February, 2019.

[14] Anon., 'Integrity Management of Submarine Pipeline Systems'. Recommended Practice. DNV-RP-F116. Det Norske Veritas. Norway. October 2017.

[15] Anon., 'Pipeline systems Part 4: Steel pipelines on land and subsea pipelines – Code of practice for integrity management', BSI PD 8010-4:2012. British Standards Institution. UK. 2012.

[16] Anon., 'Pipeline Safety Management Systems', American Petroleum Institute. USA. ANSI/API Recommended Practice 1173. July 2015.

[17] A Cosham, P Hopkins, 'Time for Fatigue'. Pipeline Technology Conference. Ostend, Belgium. Tiratsoo Technical. 2015.

[18] V Semiga et al., 'Fatigue Considerations for Natural Gas Transmission Pipelines', Report for INGAA, USA. BMT Fleet Technology Limited. Report 30348.FR (Rev. 02). 30 June 2016.

[19] M J Rosenfeld, J F Kiefner, 'Basics of Metal Fatigue in Natural Gas Pipeline Systems – A Primer for Gas Pipeline Operators', Pipeline Research Council International, Report No. L52270. 2006.

[20] M Cech et al., 'Performance of European cross-country oil pipelines. Statistical summary of reported spillages in 2016 and since 1971', Report 6/18. CONCAWE, Brussels. 2018.

[21] Anon., 'Enbridge Incorporated. Hazardous Liquid Pipeline Rupture and Release. Marshall, Michigan. July 25, 2010', National Transportation Safety Board. NTSB/PAR-12/01. PB2012-916501. July, 2012.

[22] Anon., 'CEPA Recommended Practices for Managing Near-neutral pH Stress Corrosion Cracking', 3rd Edition. Canadian Energy Pipeline Association. Canada. May 2015.

[23] Anon., 'Industry Good Practice Guide: Near Neutral pH and High pH Stress Corrosion Cracking', UK Onshore Pipeline Operators' Association (UKOPA). UKOPA/GP/009. Edition 1. 2016.

[24] R Fessler, 'Pipeline Corrosion: Final Report', Michael Baker Jr., Inc. Report for U.S. Department of Transportation Pipeline and Hazardous Materials Safety Administration. November 2008.

[25] Anon., 'Pipeline External Corrosion Direct Assessment Methodology', ANSI/NACE SP0502-2010, NACE International. USA. 2010.

[26] Anon., 'Stress Corrosion Cracking (SCC) Direct Assessment Methodology', SP0204-2015-SG, NACE International. USA. 2015.

CHAPTER 8

Mechanical damage

Dr Phil Hopkins

8.1 Introduction

There are many threats to pipeline safety. Threats such as corrosion, ground movement, and material defects can cause failures, but the major cause of pipeline failures in most developed countries is 'external interference'.

Pipelines can be impacted by excavating machines (onshore), anchors and trawl boards (subsea), etc.: this impact is called 'external interference'. This external interference causes 'mechanical damage' in the pipe wall such as a gouge (or scratch, groove, or notch), or a dent, and is sometimes called 'third-party damage' as most (but not all) of the damage is caused by third-parties.

This Chapter explains the causes and effects of mechanical damage in pipelines, and how to protect a pipeline from this type of damage.

8.2 Mechanical damage

High pressure, transmission pipelines are usually buried underground, or on a seabed. This means they are invisible from aboveground or sea level. This burial and sea cover gives considerable protection to a pipeline, but also puts them at risk from damage from excavating machines, ships, etc., whose owners may not be aware of their presence.

There are millions of excavations near pipelines every year, and thousands of ships passing near subsea pipelines (see Figure 8-1).

Some plant working near a pipeline can present a threat; for example, the use of back actors ('back hoes') and diggers ('excavators' or 'front hoes') account for approximately 54% of all damage incidents, and 28% of pipeline failures[1]. This is a relatively low threat (data show a 3% chance of failure if the pipeline is damaged) from these mainly small machines, but the use of power drills accounted for <4% of all recorded damage incidents, but approximately 28% of pipeline failures[1]. This is a high threat

(43% chance of failure if the pipeline is damaged). The size and type of machinery working near a pipeline significantly affects its chances of causing a failure.

Impacts on a pipeline are usually called 'external interference', and the resulting damage (puncturing, gouging, and denting (see Figure 8-2)) is called 'mechanical damage'. A puncture is when the pipe wall is completely penetrated. ASME B31.8S[2] defines a gouge as a 'mechanically induced metal-loss that causes localized elongated grooves or cavities in a metal pipeline', and a dent as 'permanent deformation of the circular cross-section of the pipe that produces a decrease in the diameter and is concave inward'.

Most (>80%) onshore pipelines contain many dents, but they do not contain any other defect in the dent (e.g., a gouge)[3]. Pipeline failures due to these 'plain' dents (dents that vary smoothly, do not contain creases, and have no surface damage to the pipe wall) are not a significant portion of reported failures[4]. This indicates that 'plain' dents are not a major threat to pipelines, but other types of mechanical damage (e.g., dents with gouges) are a major threat.

Plain dents may not be a major threat to a pipeline, but other types of damage (e.g., a dent containing a gouge) can cause pipeline failures[5] (see Figure 8-3) and casualties[6] (see Table 8-1).

Mechanical damage usually fails a pipeline immediately, typically when construction equipment punctures the pipe and produces a leak at the time of damage[8], but the damage can also fail sometime after it has been introduced ('delayed failure')[8, 9, 10]. The damage can gradually grow by, for example, fatigue. The majority (~80%) of mechanical damage incidents fail immediately and cause the majority of casualties[8].

Table 8-1. Causes of casualties and fatalities in USA transmission pipelines (1994-2013)[7].

Cause of Pipeline Failure	Natural Gas Pipelines		Liquid Pipelines	
	Casualty	Fatality	Casualty	Fatality
Corrosion	13	6	1	18
Excavation damage	15	49	12	37
Incorrect operation (e.g., human error during pipeline operation)	0	9	9	16
Material/weld/equipment failure	8	70	4	12
Natural force (e.g., flooding or lightning)	0	2	0	1
Other outside force (e.g., vandalism or sabotage)	0	13	4	5
All other causes (e.g., unknown, or not fitting into the above causes)	5	46	10	43

Figure 8-1. Work on and around a pipeline can lead to damage on the pipeline (photographs courtesy and copyright of Penspen Ltd).

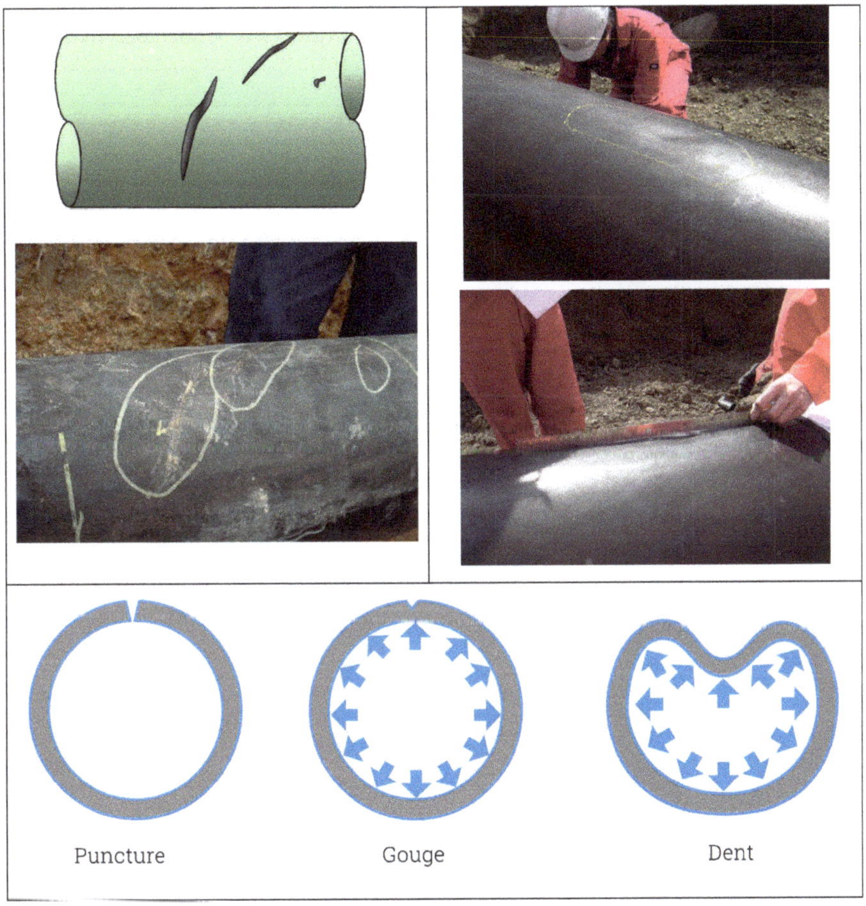

Figure 8-2. Mechanical damage in pipelines (top left gouge, top right dent) (photographs courtesy and copyright of Penspen Ltd).

8.2.1 Gouges

A gouge is surface damage to a pipeline. It is usually caused by foreign objects removing part of the pipe wall. Gouges in pipelines are usually short and shallow, but are treated with caution because:

- the gouge could be associated with a dent (see Figure 8-1);
- the gouge can be associated with a 'hard layer' and cracking;
- the gouge can subsequently crack due to hydrogen cracking.

Pipeline standards (e.g., [11, 12]) advise that gouges, grooves, and notches should be prevented, eliminated, or repaired, as they '… have been found to be an important cause of pipeline failures…'[11]. This cautious approach is in recognition of [13]:

- the known severity of mechanical damage; and,
- the fact that a residual dent may have been overlooked around a gouge.

Denting associated with a gouge. An impact on a pipeline can be very high energy: excavating equipment or ships' trawl gear/boards can impose many tonnes of force on the pipeline. Consequently, gouges may be associated with denting.

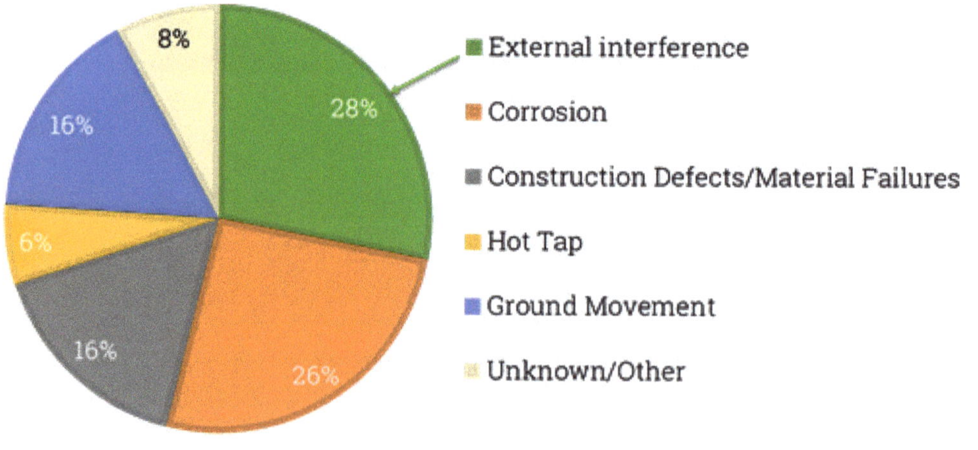

Figure 8-3. Onshore Western European gas pipelines failure causes (2007- 2016).

The impact can dent and gouge the pipeline, and after impact and removal of the impactor, the dent will attempt to (see Figure 8-4):

- 'springback', due to the elasticity in the pipe wall; and then,
- 're-round', due to the internal pressure in the pipeline.

Ductility and Hardness

'Ductility' is[2] '...[a] measure of the capability of a material to be deformed plastically before fracturing.' 'Hardness' means the resistance of a metal to localised plastic deformation.

Cracks can form at the base of a gouge in the dent, during this springback and re-rounding of the dent. This cracking can be difficult to detect. Similarly, the denting of the pipe may or may not be apparent, due to the springback and re-rounding[11].

A combined dent and gouge is the most severe form of mechanical damage. This combined defect fails at stresses well below that of a gouge, or a dent, alone. Standards (e.g., [12]) require this type of defect to be removed, repaired or evaluated.

Gouges and the 'hard layer'. An impact on the line pipe will crush the steel, leading to 'work hardening' or 'cold work'. This reduces the line pipe's ductility. Additionally, an impact on a pipeline will also create high energy; some of this energy is heat, and this heat can be very intense. The heat can change the microstructure of the metal, and as the metal subsequently cools, it becomes brittle (low toughness).

Figure 8-4. A combined gouge and dent (top image is gouging in dents) (photographs courtesy and copyright of Penspen Ltd).

This combination of crushing and heating causes[9]:

- plastic flow;
- metal transfer; and,
- re-melting due to the heat of friction.

The result is a 'hard layer' (low ductility and low toughness) around the gouge (see Figure 8-5).

The gouge is surrounded by a hard, deformed layer, and this layer cracks easily; additionally, any 'springback' or 're-rounding' of the pipe will crack the deformed layer (see Figure 8-6) [14, 15].

The cracking can extend due to fatigue, usually caused by the fluctuating pressure in the pipeline.

Gouges and hydrogen cracking. An impact on a pipeline can easily remove coating on the pipeline, allowing the environment to contact the line pipe. This will lead to corrosion, but the cathodic protection on the pipeline should prevent this.

We know that corrosion takes place when:

- electrons are removed from the metal at the anode area on the pipe surface; and,
- consumed by the reaction at the cathode area on the pipe surface with oxygen or hydrogen.

When electrical current leaves the anode to enter the electrolyte, small particles of iron are dissolved into solution, causing pitting at the anode (see Figure 8-7). Wherever the current enters the cathode, hydrogen gas is formed on the surface, and the cathode is preserved and protected from corrosion (see Figure 8-7).

Cathodic protection (CP) should prevent corrosion, but it does produce hydrogen at the cathode, particularly with onshore pipeline CP systems and environment. The hard layer around the gouge is susceptible to 'hydrogen cracking'.

Hydrogen ('stress') cracking needs the following three conditions for it to occur:

- a steel of high hardness and high strength (i.e., the gouge's hard layer);
- sufficient sustained tensile stresses (a pipeline is always under stress from the internal pressure); and,
- a source of atomic hydrogen (the atomic hydrogen can come from the CP reaction).

This hydrogen cracking can continue to grow and extend as more and more hydrogen is generated. Its growth can also be assisted by fatigue.

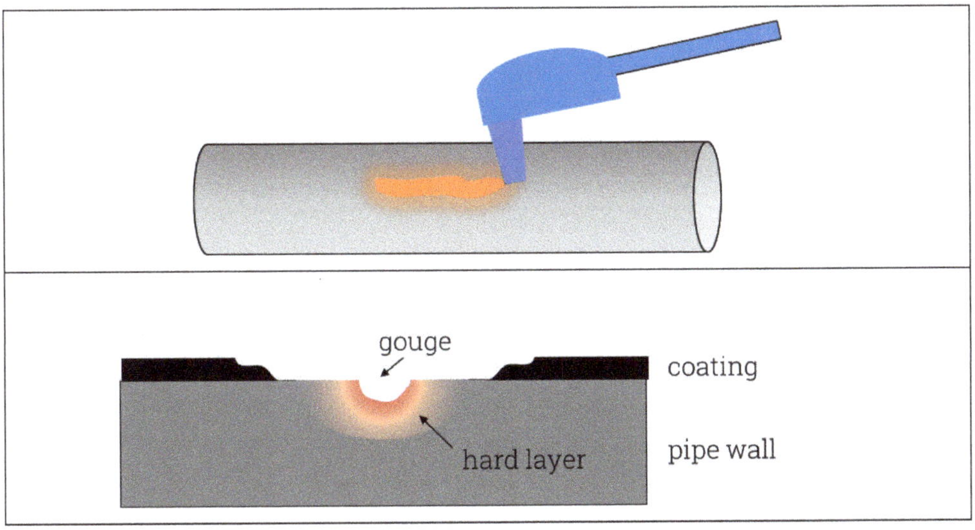

Figure 8-5. Impacts can create high temperatures, and plastic deformation around a gouge.

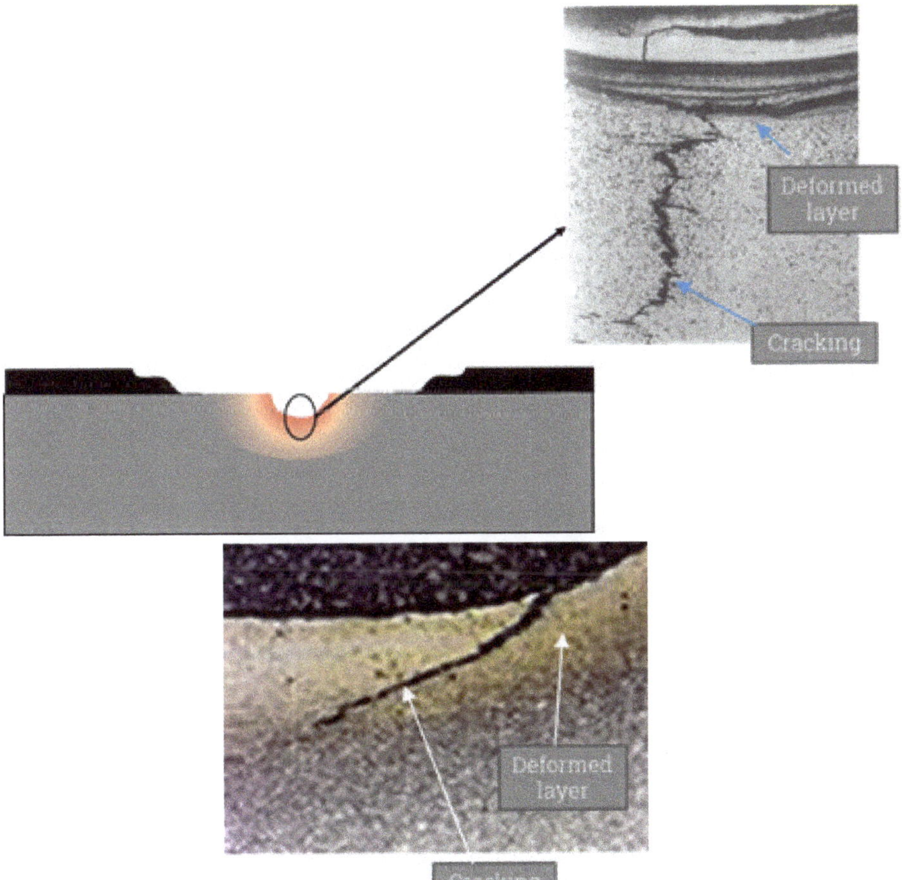

Figure 8-6. Cracking in hard deformed layer around a gouge[14].

Inspection of gouges. Gouges should be inspected for the presence of denting around the gouge, and surface cracking in the gouge. The cracking can be caused by the energy and heat of impact, but it can also by caused by an associated dent springing back and re-rounding, or by fatigue or hydrogen.

This associated cracking is severe and requires repair.

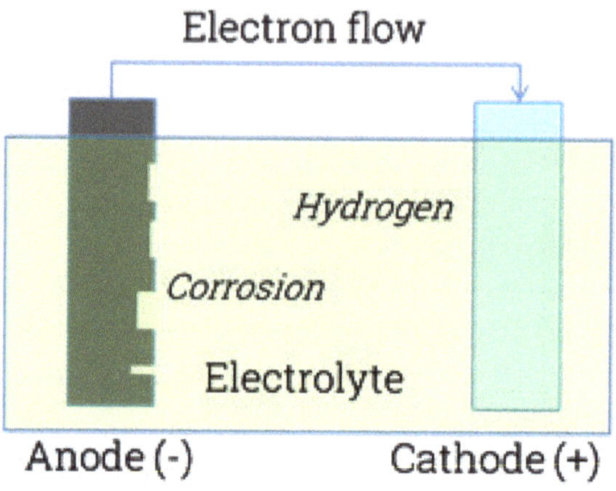

Figure 8-7. Cathodic protection principle.

QR Code 8-1. Scan here for more photos of damage to pipelines.

QR Code 8-2. Scan here for more information on the hard layer around a gouge.

QR Code 8-3. Scan here for more information on how to assess the significance of gouges in pipelines.

8.2.2 Denting

Line pipe is not a perfect circular tube; the pipe will have 'out-of-roundness' or 'ovality'; i.e., deformation that is elliptical in shape. A 'dent' is a local change in the pipe surface contour[16]. Dents are caused by:

- transportation/storage/installation;
- pipe settlement during or after construction;
- 'external interference';
- etc.

Most (>80%) onshore pipelines contain many dents, and most of these dents are at the bottom of the pipe (having being caused by the pipeline originally being laid on rocky ground, and rocks denting the pipe) and do not contain any other defect (e.g., a gouge)[3].

In-line inspection vehicles find many of these plain dents in pipelines which have not failed for many years. This indicates that plain dents are not a major threat to pipelines.

Dents are usually characterised by their depth, H (see Figure 8-8). This depth is usually expressed as a percentage of the pipeline's outside diameter, D (e.g., 6% D). The length and width of the dent affects the stress and strain distribution in the dent, and, its geometric stiffness, but these effects are secondary compared to the dent depth when considering the dent's 'static strength' and 'fatigue life'[15].

Static strength and fatigue strength of dents. Pipelines need to withstand all the loads imposed on them. Usually, the main load comes from the internal pressure, but other loads including thermal loads (when carrying hot products), bending loads (for example, spanning on subsea pipelines), and external pressure, can add to this main load.

Damage in a pipeline must be able to withstand these loads, and this is called the 'static strength' of the damage – its ability to withstand the maximum load imposed on the pipeline.

The loads on a pipeline are never constant; for example, pipelines can be heavily pressure cycled, particularly pipelines carrying liquids[17]. Under cyclic pressure loading there will be large cyclic stress and strains localised in the dent. The pressure cycles will make the dent move in and out, bending the line pipe, and this can lead to fatigue cracking (see Figure 8-9).

Cyclic pressure fatigue tests on plain dents in pipes indicate that plain dents reduce the fatigue life compared to plain circular pipe. Therefore, a dent in a pipeline has to be able to withstand all the cyclic loadings imposed on the pipeline. This is termed the 'fatigue strength'. There are models available to calculate the fatigue strength of a dent in a pipeline[20 to 24].

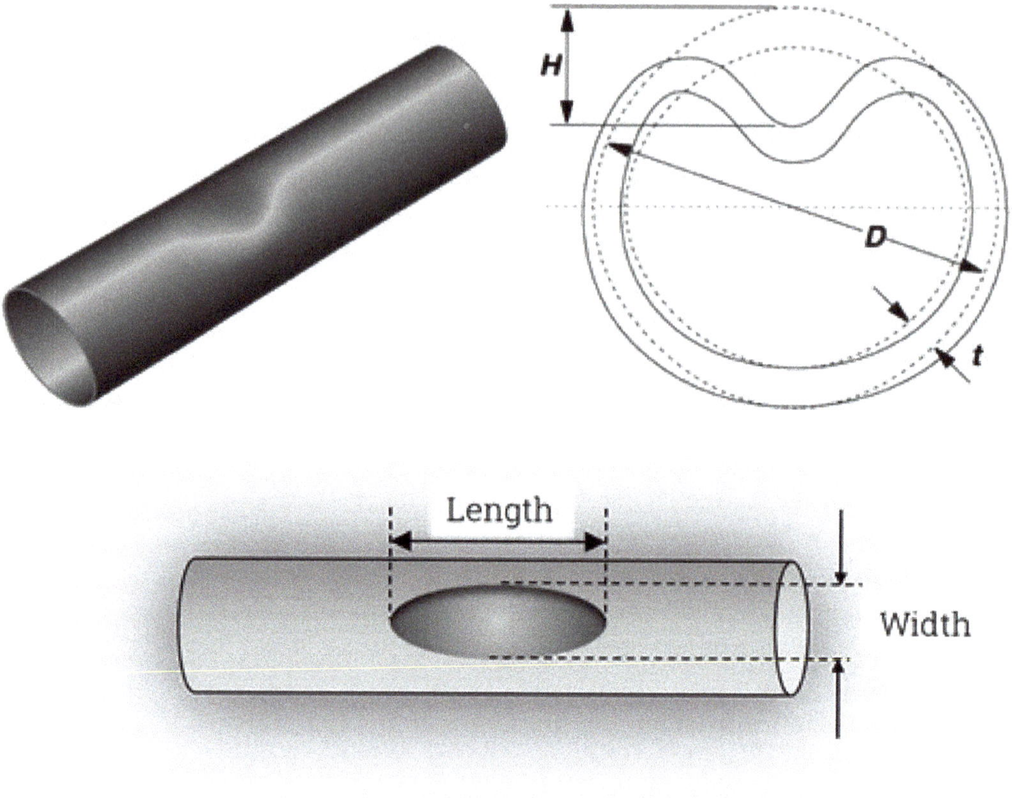

Figure 8-8. Dents are usually characterised by their depth, H.

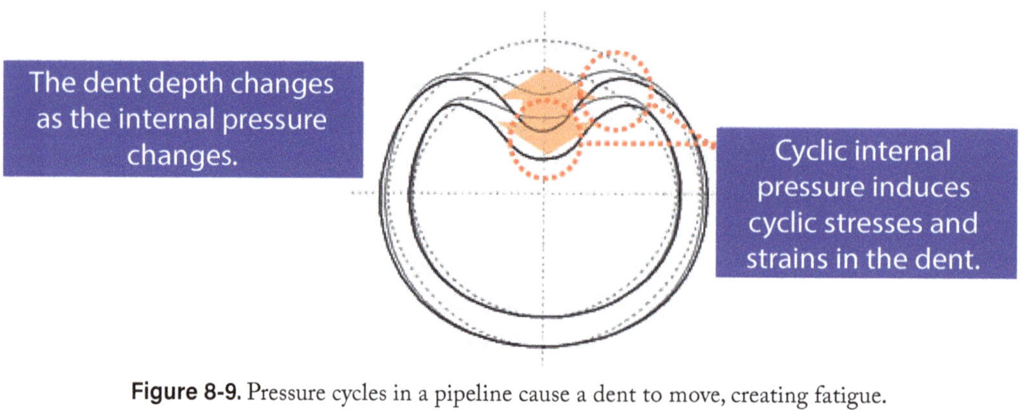

Figure 8-9. Pressure cycles in a pipeline cause a dent to move, creating fatigue.

Dent acceptance levels in standards. Dents are common in pipelines; therefore, standards give acceptance levels; for example, Table 8-2[12].

Most damage in pipelines is detected by in-line inspection vehicles; therefore, standards give guidance on responses; for example[18], Table 8-3.

Table 8-2. Dent acceptance levels during operation[12].

Dents requiring removal or repair or an engineering evaluation
Dents with a depth greater than 6% of the nominal pipe diameter [0.250 in. (6.4 mm) in depth for a pipe diameter NPS 4 and smaller].
Dents containing gouges, grooving, scratches, cracking or other stress raisers.
Dents containing metal loss resulting from corrosion or grinding where less than the minimum wall thickness permitted by the component specification remains.
Dents that affect pipe curvature at a girth weld or a longitudinal seam weld.

Table 8-3. Dents requiring immediate responses when detected in pipelines[18].

'Anomalies' requiring immediate response in all pipeline segments
Any dent containing cracking.
Any dent above the 4 and 8 o'clock position that contains stress raisers (gouges, notches, scratches), or corrosion (unless an assessment shows it to be of no risk).
Any dent above the 4 and 8 o'clock position with a depth greater than 6% of the nominal pipe diameter (unless an assessment shows it to be of no risk).

Plain dents. Many experiments have shown that plain dents do not affect the strength of pipelines when the pipeline is subjected to static pressure. The plain dent introduces high localised stresses (causing yielding and wall thinning) in the dented area, but these high stresses and strains are accommodated by the strength and ductility of the line pipe.

On pressurisation, the dent attempts to move outward, allowing the pipe to regain its original circular shape, through:

- 'spring back' (the elasticity of the steel); and,
- 're-rounding' (due to internal pressure).

The dent will not reduce the static strength of the pipe provided nothing restricts the movement or acts as a stress concentration (e.g., a gouge or a kink), but note that large dents, with the associated ovality, will reduce the pipeline's buckling/collapse capacity.

Experiments have shown dents of depth ≤ 7% D are acceptable under 'static' (no cycling) pressure[15]. This ≤ 7% D is a technical limit[19], but these deep dents:

- may restrict product flow;
- can prevent the passage of pigs along the pipeline; and,
- create high strains that may disbond coatings.

These considerations meant dents acceptance limits have historically been set at 6% D, to ensure the dents do not affect operation, in-line inspection, or coatings[9, 13].

ASME B31.8 also 'allows' plain dents to remain in a pipeline:

- if the dent is ≤ 6% pipe diameter; or,
- providing strains in the dent are ≤ 6% strain.

The standard gives equations to calculate these strains, but the length, width, and depth of the dent must be known (see Figure 8-8). The 6% strain limit only applies to the static strength of dents – cyclic pressure loading which can lead to fatigue cracking in a dent is not considered.

The 6% strain was selected as a reasonable limit because[9]:

- strains up to 3% occur during cold field bending of line pipe during pipeline construction;
- coatings, such as coal tar, may disbond at strains of 4% to 5%;
- cracks have been observed in pipeline buckles at strains that are a large fraction of the uniaxial elongation; and,
- 12% is a strain when line pipe may crack.

Dents containing welds. A dent may be on, or coincide, with a weld (see Figure 8-10). Pipeline welds are usually stronger ('over- matched') than their 'parent' material (line pipe), and the fatigue life of a weld is always lower than the parent material; therefore, a dent with a weld will have a lower fatigue life than a dent in 'parent' material.

Experiments have shown that dented seam welds can exhibit both very low static failure stresses under internal pressure loading (as low as 7% SMYS) and very short fatigue lives. Dents on 'good quality'[23], ductile, welds, free from major defects, have high failure stresses and fatigue lives; however, in some impacts, the large stress and strains associated with the denting process cause cracking in the weld, Figure 8-10. This cracking can lead to very low failure stresses and short fatigue lives.

API 1160[18], ASME B31.8S[2], and the USA's Pipeline Safety Regulations for gas and liquid lines (CFR 192 and 195) have differing guidance on dents affecting longitudinal and girth welds. ASME B31.8S and the CFRs consider dents of depth ≤2% pipe diameter containing a 'good quality' weld, with its coating intact, are insignificant, but the strain level in the deformation must not exceed 4%[13, 23]. When a pipeline is pressure cycled, its fatigue life must be assessed using a suitable S-N curve for the weld, and the increased stresses in the dent.

UKOPA[23, 24] considers dents containing a 'good quality' weld, with its coating intact, are insignificant if the depth is ≤ 4% pipe diameter, and the strain level in the deformation does not exceed 4%.

Dent failure

In full scale tests ('full scale test' means a laboratory or field test on a section of line pipe) of plain dents, failure of the pipe has not generally occurred in the area originally containing the dent, but has occurred at a weld or an area of local weakness elsewhere in the pipe body.

QR Code 8-4. Scan here to learn more about calculating the fatigue life of a dent.

QR Code 8-5. Scan here to learn about assessing dents based on strain.

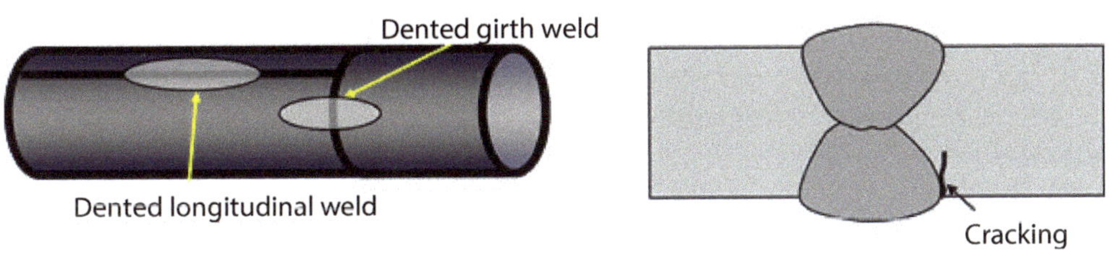

Figure 8-10. Dents can coincide with welds.

Dents caused by rocks. Pipelines can be dented by rocks that rest under them. These dents are 'constrained' (they cannot move outwards due to the presence of the rock). Dents created by impact can move outwards; therefore, these are 'unconstrained' dents.

The overburden on a pipeline can cause the rock to puncture the pipeline, but it has been concluded[25]: "Constrained dents less than 6 percent in depth are not a problem, unless they involve a sharp object which might cause a puncture or they interfere with operations such as pigging."

> **Failures from Dents caused by Rocks**
>
> Rock dent failures are rare, and all recorded failures have been leaks (not ruptures)[26, 27].
>
> The failure is usually a circumferentially-orientated crack created by the rock puncturing the pipe[26, 27].

Dents created by rocks may have long term issues[9]:

- coating damage;
- shielding from the cathodic protection;
- corrosion;
- stress corrosion cracking (SCC) or hydrogen cracking; and,
- punctures due to continued settlement (if rock is sharp and overburden load is high).

Another publication[28] warns that plain dents caused by rocks that have been excavated and that have had the indenting object removed (allowing the pipe circular cross section to partially re-round under internal pressure) are susceptible to fatigue from pressure cycles acting on the unrestrained residual dent.

Dents containing corrosion. Tests on corrosion in rock dents have shown the dent and corrosion to have high static failure strength, due to the dent preventing the corrosion 'bulging'[9, 13]. ASME B31.8[11] states that corrosion in dents can be assessed using standard methods[29], assuming the dent is acceptable. Corrosion protection will still be needed around the dent to prevent further corrosion and possible stress corrosion cracking.

Dents containing gouges. Most onshore pipelines contain many dents, but pipeline failures due to 'plain' dents (dents with no surface damage) are not a significant portion of reported failures[3, 4]; however, dents containing gouges do fail pipelines.

The large strains created as a dent moves outwards (under the action of the internal pressure) can cause tearing of any defect within the dent. The tearing goes through the remaining ligament of the pipe wall and is resisted by the line pipe's toughness. Hence, we need good line pipe ductility and toughness to resist a gouge in a dent[17].

Dents with gouges record low failure stresses, as the line pipe cannot withstand the high strains around the gouge[17], and they record very poor fatigue lives[15] (see Table

8-4). This had led to pipeline standards requiring immediate action if they are detected in a pipeline; for example[12]: '… dents containing gouges, grooving, scratches, cracking, or other stress risers … shall be removed or repaired unless an engineering evaluation can demonstrate that other mitigative action… will reduce the risk to an acceptable level.'

Table 8-4. Example of fatigue lives of gouges and dents[17].

Dent depth (% pipe diameter)	Weld	'Gouge' depth (% wall thickness)	Fatigue life (cycles)
no dent	*no weld*	20	>145,500
4	*no weld*	*no gouge*	<6,930
4	weld in dent	*no gouge*	<789
4	*no weld*	20	<199

Dents in subsea pipelines. Pipelines exposed at river crossings, lake crossings, subsea, etc., can be damaged by:

- fishing trawl doors;
- fishing trawl beams;
- ships' anchors;
- dropped objects;
- etc.

These impacts can:

- damage any concrete coating, and the corrosion coating on the pipeline;
- cause the pipeline to be dented, and bent and deformed; and,
- create high 'locked-in' stresses, both compressive and tensile.

Consequently, the assessment of dents in subsea pipelines can be very complex, and all loads (including external pressure) need to be considered[30].

QR Code 8-6. Scan here to learn about corrosion-assessment.

8.2.3 Theft, vandalism, sabotage

There are many different types of 'third-parties', but the most dangerous are those third-parties who deliberately damage or steal from a pipeline. These criminals incur huge repair costs, environmental clean-up costs, loss of product and production, but most tragically, cause many fatalities and casualties. This section briefly reviews these third-party problems.

Vandalism. Pipelines are sometimes vandalised; for example, an above ground pipeline was vandalised (bullets penetrated the pipeline and caused a major leak) in Alaska, USA, in 2001[31], resulting in a $17 million clean-up bill, and $8 million lost royalties and taxes.

Vandalism is not a major problem in pipelines around the world, mainly because pipelines are not easily accessed, and most pipelines are large structures that are difficult to damage.

Sabotage. Historically, there has been sabotage on a number of pipelines around the world, usually in regions with ongoing unrest or conflict; for example, Colombia's Caño Limón oil pipeline was attacked hundreds of times by the National Liberation Army, causing the loss of millions of barrels of crude oil[32]. In 2001 alone, sections of the pipeline were blown up 170 times, putting it out of action for 266 days[32].

Protecting pipelines from sabotage, particularly when the saboteurs are armed, is difficult. Burying pipelines is an obvious good method for protection, but many of the pipelines in hostile environments are above ground.

CCTV, isolation valves, pipeline-wide computerised sensor systems, (to detect pressure changes), etc., have been used to improve pipeline security, but with limited success[33].

A better method, but with significant cost and dangers, is 24-hour manned patrols[33]. Employing local communities, or even local militia, to patrol the pipeline can be beneficial, as it can[33]: reduce local violence; increase local employment; and, 'redistribute' the wealth from the oil industry.

Theft. Pipeline vandalism and sabotage are problems, but by far the greatest problem today is theft; for example, Nigeria is losing almost 15% of its oil production through theft on pipelines[34]. At current prices this would mean a financial loss of almost $3 billion/year.

It is a big business, with quick rewards: stealing 10,000 litres of crude from one pipe gives a profit of about US $4,600, with the 'tapping' (drilling into the pipe and making a connection) taking about 30 minutes[34, 35]. It is a growing business: fuel theft from pipelines in Mexico has evolved from small time criminal activity to a sophisticated operation linked to Mexico's main criminal groups (the drug cartels) who see it as a 'side business'[34, 35].

These thefts are now becoming common throughout Western Europe, and are increasing every year[36, 37]; although Africa, South America, and Mexico remain the areas most affected. There were 2,500 'illegal taps' in Mexico in the first nine months of 2014, stealing $1 billion in fuel[33]. In 2017 there were 10,363 illegal taps detected in state-owned pipelines (average of approximately 28 illegal taps every day)[38]. These thefts cost the pipelines' operator (Pemex) > $1 billion in losses in 2016. The estimated cost for 2017 is even higher[38]. The cause of theft is complex, and linked to social (e.g., poverty, and lack of basic services) and political (e.g., corruption) factors[33]; therefore, reducing the threat of theft is very complex. Any approach will require police or para-military assistance. One approach to reduce theft is:

- Patrol pipelines at areas of high risk (e.g., near villages or local criminals).
- Work with police to destroy the 'organised' crime (many of the thefts are for a criminal with an organisation selling to customers). This will require special detective work.
- Review internal staffing (often criminals are working with pipeline staff).

8.3 Protecting pipelines

8.3.1 Protection

Pipeline operators should have a damage protection program[10] that includes:

- up-to-date maps of the system;
- procedures and guidelines, and trained staff, to react to a One-Call system, where such a system is in place;
- marker maintenance along the right-of-way;
- establishing and continuing a public awareness program, with land occupants, excavators, and contractors;
- periodic ground/air/sea surveillance;
- erection of physical barriers (or deeper depth of cover) on and around a pipeline in a high-risk area;
- recordings of all impacts and near misses (this helps determine which damage prevention techniques are the most cost-effective).

8.3.2 Protection methods

In the USA, a pipeline owner will normally mark where the pipeline is (by markers, paint, etc.) prior to any excavation work in its vicinity, but most damage is still caused when excavators are working near a pipeline[39]:

- without either knowing there is a pipeline there; or,
- not notifying the pipeline owner of their operations.

Hence, pipelines need protection against this damage. The protection will need to be significant, as machines of up to 32 tonnes (operating weight) can work around an onshore pipeline[40, 41], and ships' anchors can be very heavy (for example 25,000 kg (25 tonnes, or 27 tons)).

Pipelines can be inspected, and modern in-line inspection vehicles can detect and size mechanical damage. Most mechanical damage incidents occur immediately; therefore, inspection methods ('reactive' methods) will only reduce the failures/consequences by a small percentage[10]. It is better to protect a pipeline against damage – be 'proactive'.

There are various methods that can be used to protect a pipeline (see Table 8-5).

Table 8-5. Example of methods of protecting a pipeline.

Method	Method
'Awareness' methods (internet sites, personal visits to landowner/shipping, radio alerts, leaflets, etc.).	Patrolling and surveillance (walking, aerial, sea).
Monitoring of building developments around pipelines, and liaison with local authority planners to establish early stages of any proposed building work around the pipeline.	Barriers over/around a pipeline (casings, concrete slabs, mattresses (for subsea pipelines), rock dumping (for subsea pipelines), etc.).
'One-call' systems (telephone systems to call before any onshore excavation. A call centre alerts excavators of the presence of an underground pipeline/utility and alerts the pipeline/utility owner).	Depth of cover (the deeper the pipe is buried the more difficult it is to reach to damage).
Special pipeline coatings or special backfills.	Marker posts (indicate where a pipeline is).
Marker tape (tape can be buried and placed over a pipeline and will become visible during an excavation).	Thicker pipe wall (the thicker the pipe wall, the more it resists denting and puncture).
On-line monitoring (equipment can detect machinery approaching or hitting a pipeline).	

8.4 Summary

Mechanical damage (e.g., gouges, or dents) is a major cause of pipeline failures, and casualties. Consequently, pipelines need to be protected against this damage.

Gouges are treated with caution due to the possibility of associated denting and cracking.

Dents containing no surface damage (such as gouging) fail at high stresses but can be a fatigue problem.

Dents containing welds can have low failure stresses and fatigue lives.

Dents containing gouges are a severe form of damage and should be repaired.

Pipeline vandalism and sabotage are problems, but by far the greatest (and increasing) problem today is theft of product from pipelines.

QR Code 8-7. Scan here to learn about pipeline protection.

8.5 References

[1] Anon., 'An Assessment of Measures in Use for Gas Pipelines to Mitigate Against Damage Caused by Third Party Activity'. Contract Research Report 372/2001. HSE Books. Her Majesty's Stationary Office, UK. www.hse.gov.uk. 2001.

[2] Anon., 'Managing System Integrity of Gas Pipelines'. American Society of Mechanical Engineers. ASME B31.8S - 2018. 2018.

[3] S J Dawson et al., 'Emerging Techniques for Enhanced Assessment and Analysis of Dents', International Pipeline Conference, Calgary, Canada. 2006. ASME. Paper IPC 2006-10264. Also, in Global Pipeline Monthly, Vol 4. Issue 5. July 2008.

[4] Anon., 'Integrity Management Program Delivery Order DTRS56-02-D-70036: Dent Study', Final Report. M Baker Jr. Inc. November 2004.

[5] Anon., 'Gas Pipeline Incidents: 10th Report of the European Gas Pipeline Incident Data Group (period 1970 - 2016)', EGIG. Doc. number VA 17.R.0395. March 2018.

[6] 'Underground pipeline corrosion: Detection, analysis and prevention', Editor: M E Orazem. Elsevier, Woodhead Publishing, UK. 2014.

[7] http://primis.phmsa.dot.gov/comm/reports/safety/SigPSIDet_1994_2013_US.html#_ngtrans

[8] Anon., 'Mechanical Damage Final Report', Michael Baker Jnr Inc., DoE PHMSA OPS, April 2009.

[9] M J Rosenfeld, 'Factors to Consider when Evaluating Damage in Pipelines', Oil & Gas Journal Article, September 9, 2002.

[10] J F Kiefner, 'Effectiveness of Various Means of Preventing Pipeline Failures from Mechanical Damage', Gas Research Institute Report GRI-99/0050. August 26, 1999.

[11] Anon., 'Gas Transmission and Distribution Piping Systems', American Society of Mechanical Engineers. ASME B31.8-2018. 2018.

[12] Anon., 'Pipeline Transportation Systems for Liquids and Slurries', American Society of Mechanical Engineers. ASME B31.4-2016. 2016.

[13] M J Rosenfeld et al., 'Basis of New Criteria in ASME B31.8 for Prioritisation and Repair of Mechanical Damage', 4th International Pipeline Conference, American Society of Mechanical Engineers. Calgary, Canada. October 2002.

[14] P Hopkins et al., 'Recent Studies of Significance of Mechanical Damage in Pipelines', AGA and EPRG Research Seminar, San Francisco, USA, September 1983.

[15] A Cosham, P Hopkins, 'The Effect of Dents in Pipelines – Guidance in the Pipeline Defect Assessment Manual (PDAM)', Proceedings ICPVT-10. July 7-10, 2003, Vienna, Austria.

[16] Anon., 'In-line Inspection Systems Qualification Standard', American Petroleum Institute. API Standard 1163. Second Edition. August, 2018.

[17] C Alexander, 'Evaluating damage to on- and offshore pipelines using data acquired using ILI', Journal of Pipeline Engineering, 1st Quarter, 2009. p. 35-46.

[18] Anon., 'Managing System Integrity for Hazardous Liquid Pipelines', American Petroleum Institute. API Recommended Practice 1160. Third Edition. February, 2019.

[19] Anon., 'Fitness-For-Service', API 579-1 / ASME FFS-1, American Petroleum Institute. 2016 Edition, June 2016.

[20] J R Fowler et al., 'Cyclic pressure fatigue life of pipelines with plain dents, dents with gouges, and dents with welds'. AGA Pipeline Research Committee, Report PR-201-927 and PR-201-9324, June, 1994.

[21] I Corder, P Chatain; 'EPRG Recommendations for the Assessment of the Resistance of Pipelines to External Damage', Proceedings of the EPRG/ PRC 10th Biennial Joint Technical Meeting on Line Pipe Research, Cambridge, UK, April 1995

[22] Anon., 'UK Onshore Pipeline Operators' Association – Industry Good Practice Guide, Managing Pipeline Dents'. OPA/GP/004. The United Kingdom Onshore Pipeline Operators' Association. http://www.ukopa.co.uk/documents/UKOPA-GPG004.pdf

[23] A Lockey, R Palmer-Jones, N Jackson, R Ellis, 'Shallow Dents: Updates to the UKOPA Dent Management Strategy', 10th International Pipeline Conference, Calgary; IPC2014-33576; 2014.

[24] Anon., 'Managing Pipeline Dents', UK Onshore Pipeline Operators' Association. Industry. Good Practice Guide. UKOPA/GP/004. Edition 1. 2016.

[25] C R Alexander, J F Kiefner, 'Effects of Smooth and Rock Dents on Liquid Petroleum Pipelines', API Publication 1156, First Edition, November 1997.

[26] J Kiefner, K Leewis, 'Pipeline Defect Assessment – A Review & Comparison of Commonly Used Methods', PRCI Report No. L52314. May 2011.

[27] J F Kiefner, 'Evaluating the Stability of Manufacturing and Construction Defects in Natural Gas Pipelines', US DOT, Final Report No. 05-12R, 2007.

[28] Anon., 'Pipeline Safety Management Systems', American Petroleum Institute. ANSI/API Recommended Practice 1173. First Edition, July, 2015.

[29] Anon., 'Manual for Determining the Remaining Strength of Corroded Pipelines, A Supplement to ASME B31 Code for Pressure Piping, ASME B31G-2012, The American Society of Mechanical Engineers, New York, USA. 2012.

[30] K A Macdonald et al., 'Assessing mechanical damage in offshore pipelines – Two case studies', Engineering Failure Analysis 14 (2007) 1667–1679. Elsevier.

[31] Anon., 'Joint after-action report for the TAPS bullet hole response, October, 2001', Office of Pipeline Safety Report. RSPA. February, 2002.

[32] http://www.offshore-technology.com/features/feature1165/ 23/08/2007.

[33] H-C Imig et al., 'Mitigating Risks', World Pipelines, Vol 10, No. 3. March 2010. p. 109.

[34] R Landstorfer, 'Hot Tapping and Product Theft from Pipelines. A Way to Detect and locate these Spots during Normal operation', Pipeline Technology Journal. Issue 1. P. 26. 2017.

[35] https://www.bloomberg.com/news/articles/2017-01-12/drug-cartels-are-looting-mexico-s-gas-stations.

[36] P Davis, 'Management of Illegal Tappings', United Kingdom Onshore Pipeline operators association. UKOPA Newsletter. June, 2016. http://www. ukopa.co.uk/wp-content/uploads/2016/06/June-2016-Newsletter-Final-Version.pdf

[37] M. Cech et al., 'Performance of European cross-country oil pipelines. Statistical summary of reported spillages in 2015 and since 1971', CONCAWE Report 7/17. Brussels. July, 2017.

[38] A Celovic, 'Pipeline Vandalism/Theft Plaguing Mexico', Pipeline Technology Journal. 22nd February, 2018. https://www.pipeline-journal.net/ news/pipeline-vandalism-theft-plaguing-mexico

[39] A Muradali et al., 'Effectiveness of New Prevention Technologies for Mechanical Damage', Gas Research Institute Report No. 8410. Project L027. June 2004.

[40] P Billet, G Pognonec, 'Two mechanical protections for natural gas transmission pipelines', IGRC 2009. International Gas Union. Paris, France. 2009.

[41] P Vercamer et al., 'Overview and Prospects on Prevention and Protection Measures Related to Pipeline Integrity', International Gas Union, WTG 2009. Paper 00763.

CHAPTER 9

Submarine pipelines

Nigel Curson

9.1 Introduction

This chapter describes the threats to submarine pipelines and how they can be assessed and managed. In this context submarine pipelines include: offshore pipelines, marine outfalls, nearshore pipelines/shore approaches, shore crossings, risers (vertical sections at offshore installations) and river crossings.

9.1.1 Submarine pipelines

Onshore pipelines are pipelines laid on or in land, and can include sections laid under rivers, lakes, and inland watercourses.

An offshore or submarine pipeline is a pipeline laid in maritime waters and estuaries, seaward of (below) the ordinary high-water mark, Figure 9-1.

9.1.2 Constructing submarine pipelines

Submarine pipelines are installed using a lay barge, a reel ship, or constructed and pulled into place either on the surface, on the bottom or mid-depth as a single pipeline or a bundle of pipelines (Figure 9-2). In addition to this, some sections of submarine pipelines are lifted into place; for example, with tie-in spools that connect subsea sections together, risers (Figure 9-3), and short sections such as used in river crossings.

Lay barge construction is by far the most common method of constructing pipelines offshore (and in swamps and marshes). Lay barge methods were developed in the Gulf of Mexico from the 1940s, so they are an established method of construction.

A lay barge contains all the equipment needed to weld together the pipe[1] and form a pipeline. It will have a number of welding 'stations', an inspection station, and coating stations, and can construct several kilometres of pipeline per day.

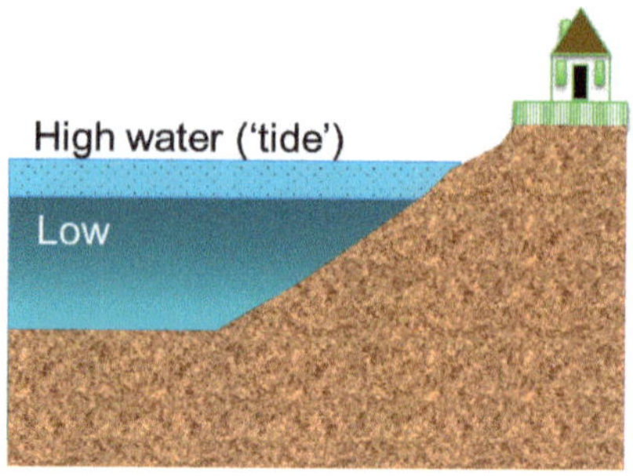

Figure 9-1. Definition of a submarine pipeline.

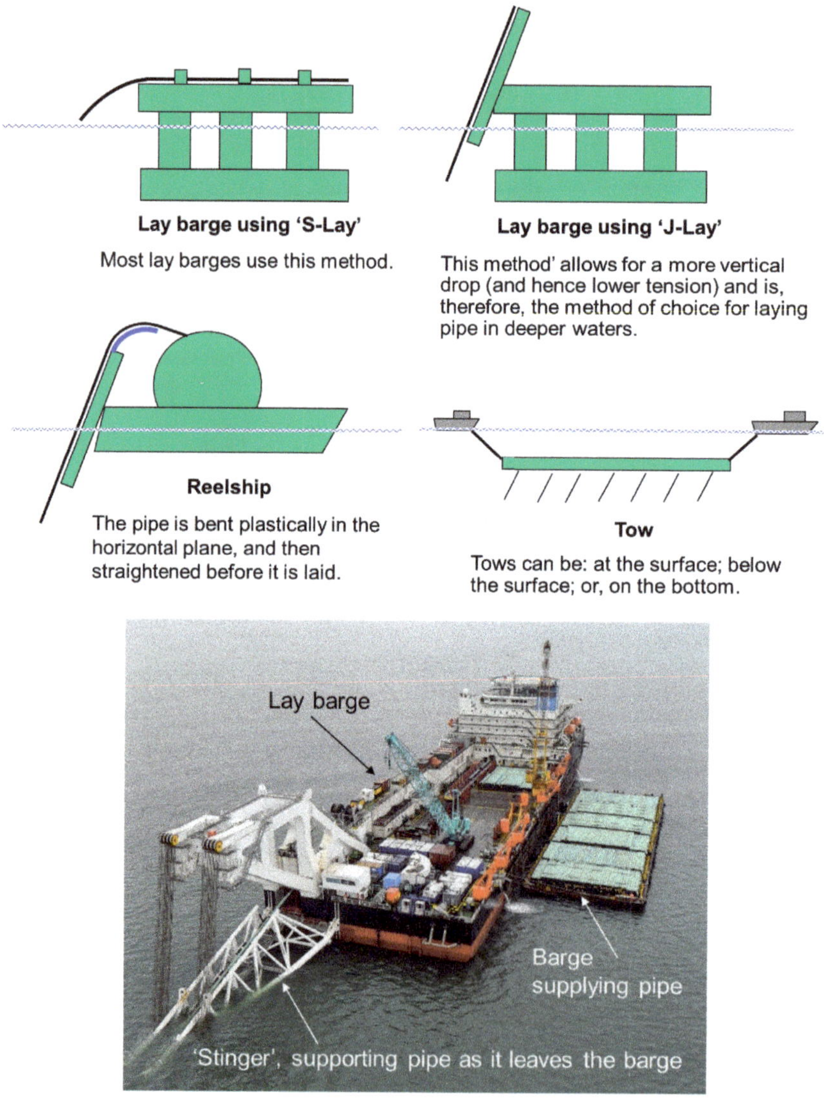

Figure 9-2. Examples of submarine construction methods.

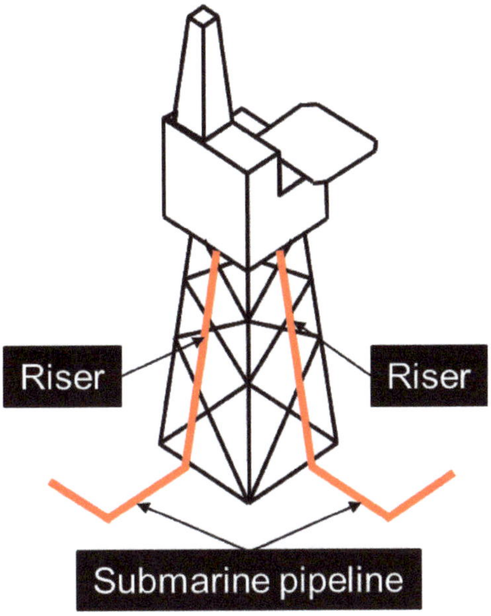

Figure 9-3. Risers on an offshore platform, and submarine pipelines on the seabed.

9.1.3 Threats to onshore pipelines vs submarine pipelines

Threats to submarine pipeline are very different to onshore pipelines due to a very different environment, different construction methods and different possibilities for third-party threats, but the consequences in terms of structural damage and the assessment are identical.

The cost of intervention with marine pipelines is typically high, due to the cost of marine equipment, divers, and safety consequences; therefore, there is an emphasis on designing submarine pipelines to be reliable and to need as little intervention as possible. For the same reasons, the analysis of threats tends to be increasingly more complex, with reducing conservatism.

QR Code 9-1. Scan here for a comparison between threats to onshore and submarine pipelines.

9.2 Threats

Submarine pipelines have specific design and operation issues, particularly due to:

- Construction and installation
- External pressure
- Stability
- Spans
- Buckling

The following sections identify the principal threats to submarine pipelines under the headings of: external pressure, pressure and temperature, waves, tides and current, seabed/coast, third-party, dropped objects, ice, and debris. There is guidance on managing these threats in the literature – for example, see References [2] through [16].

9.2.1 External pressure

Collapse

A large external pressure tends to make a pipeline lose its roundness and collapse. A perfectly round pipeline loaded by a steadily increasing internal pressure would remain circular and collapse suddenly when the stress in the wall reaches the material's yield strength. In reality, pipelines are not perfectly circular, but are oval to some extent. Ovality is controlled by the specification used to procure the pipe ('line pipe') used to construct the pipeline[1].

When an out-of-round pipeline is subject to external pressure, the out of roundness progressively increases until the circumferential bending stress reaches yield strength and the pipe collapses, Figure 9-4. Resistance to collapse is considered in the design of the pipeline and is rarely an issue for an operating pipeline, because it is only a significant factor for deep water, and bending resistance in the pipe wall tends to not be significantly degraded due to corrosion during the life of the pipeline. Collapse can be managed by design, material specification, and if necessary providing operational restraints.

Methods for assessment are included in Reference [9].

Propagating buckle

A compressive load on a structure can cause it to buckle. Pipelines, like other slender structures, can buckle if the axial compression goes beyond a certain level. A pipeline under high axial forces can bend and then buckle.

Both onshore and submarine pipelines can buckle. This can be caused by loading on a pipeline due to high construction loads, operational loads, differential settlement, soil movement (e.g., mining activities), external pressure, and thermal loading. Buckling is usually associated with submarine pipelines, a common cause being high temperatures and pressures. A pipeline containing a pressurized fluid at a temperature greater than the installation temperature will have a compressive load in the cross-section as a whole, even though there still may be a tensile load in the wall of the pipe.

If a pipeline is bent far beyond its yield strength and into the plastic range under external pressure, the mode of buckling can change. The buckle is no longer a local wrinkle on the compression side, but spreads out and moves along the pipe, flattening it into a dumb-bell cross-section.

This process is called 'buckle propagation'. It only starts if the external pressure is greater than an initiation pressure that depends on the size of the bending buckle. Once it has started, however, a buckle can continue to propagate, even into lengths of pipe which are not being bent, provided the external pressure is greater than the propagation pressure.

External pressure

Submarine pipelines do have additional stresses to withstand; for example, these pipelines have high external pressures. We can illustrate this pressure by considering divers: for every 10 metres of depth, a diver experiences an additional 1 atm of pressure due to the weight of the water surrounding them. At 20 metres, the diver experiences a total pressure of 3 atm (1 atm from atmospheric pressure plus an additional 2 atm from the weight of the water). The same applies to a pipeline.

The highest risk of buckle propagation is during construction where the pipeline is subject to potentially high bending loads combined with external pressure. Pipelines are normally designed to resist propagating buckles by ensuring the propagating pressure is above the maximum external pressure. For deep water pipelines this is not always cost-effective because it results in excessive wall thickness. Under these circumstances, thicker sections of pipe, or buckle arresters, can be installed to limit the extent of the propagating buckle if it does occur.

Resistance to a propagating buckle is considered in the pipeline design and is rarely an issue for an operating pipeline, because it is only a significant factor for deep water pipelines, and bending resistance in the wall tends to not be significantly degraded due to corrosion during the life of the pipeline. Propagating buckles can be managed by design, material specification and if necessary providing operational restraints. Methods for assessment are included in Reference [7].

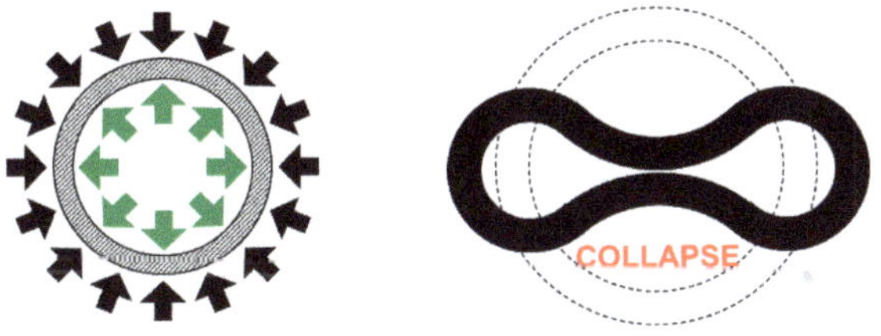

Figure 9-4. Collapse of pipe caused by external pressure.

9.2.2 Pressure and temperature

Pipelines that operate at high temperatures and/or pressures produce high axial loads in restrained sections of the pipeline and potentially large expansion at unrestrained sections, including the ends. Expansion can be accommodated by using suitably designed and installed expansion loops/offsets.

High axial loads will cause a pipeline laid directly on to the seabed to buckle laterally, Figure 9-5. If the pipeline is trenched and/or covered it will be restrained laterally and therefore will tend to buckle vertically. Buckling is a threat to a pipeline, because it can cause local buckling. Local buckling appears as wrinkling on the compressive side of the cross section of the pipeline and can lead to ovalisation, high through thickness strains, and cracking. Buckling can also produce high through-thickness bending strain, fracture, fatigue due to cyclic loads, and be further damaged by fishing equipment.

The effects of high temperature and high pressure can be mitigated by the cooling of wellhead fluids, or by using a High Integrity Pressure Protection Systems (HIPPS). Cooling systems can be provided by subsea spool, tube or coils heat exchangers. Another method of managing the effects of temperature and pressure is to introduce intermediate expansion offsets. This effectively splits the pipeline into a series of shorter pipelines, with a lower axial force in each. If the axial force in the pipeline is sufficiently high to cause buckling this needs to be considered very carefully in the design, construction and continued operation of the pipeline.

QR Code 9-2. Scan here for more information on HIPPS.

Lateral buckling

A pipeline laid onto the seabed may buckle laterally. Lateral buckling under these circumstances is random in nature, with several different buckling mode shapes dependent on local restraint, Figure 9-6.

Each buckle alleviates compressive axial strain at the location of the buckle; however, the number and buckle mode shape are not predictable, as is the resulting strain. This means uncontrolled buckling may or may not result in unacceptable strains.

If the pipeline does not need to be buried for protection, then controlled lateral buckling can be used to avoid unacceptable strains. This is done by building in imperfections that later become lateral buckles. A pipeline can be laid in a zigzag, or S-shape, or on supports or sleepers. The sleepers initially cause vertical buckling, but gravity then transforms the vertical buckling into a horizontal buckle.

Another method of controlling lateral buckling is to use intermittent rock dumping, Figure 9-7. The intermittent rock dumping limits the size of lateral buckles by preventing feed in from one buckle to the other. The choice of method depends on construction equipment available, material available, the soil conditions and water depth.

Methods for designing and assessing lateral buckling can be found in [2].

Local and Global Buckles

'Local buckling' is a local, gross deformation of the pipe cross section. Local buckling is buckling over a short length of pipeline causing gross changes in cross-section (e.g., wrinkling).

A locally buckled pipeline cannot withstand an increased bending moment in the pipeline. This could lead to pipeline collapse and loss of containment.

'Global' buckles in a pipeline represent a large scale displacement of a section of the pipeline.

Global buckling occurs over a long length of pipeline, usually several pipe joints.

Upheaval buckling

A pipeline subjected to high axial strains which cannot move laterally will move vertically, Figures 9-5 and 9-8. The positions where it moves vertically will be determined by the axial force and vertical imperfections in the seabed.

Imperfections can originate from the natural seabed profile, from hard material falling out of the side of the trench during trenching, or from imperfections produced during pipeline welding. Upheaval buckling is controlled by limiting the size of vertical imperfections or by providing cover. Cover provides a downward force which resists vertical force produced by the buckle.

Cover can be natural seabed material that has been swept back into the trench using backfill ploughing, natural backfill, or rock dumping. The amount of cover required can be assessed using finite element analysis, or other suitable analytical methods, combined with a soil resistance modelling. Cover requirements can be based on a typical idealised conservative imperfection shape, or the exact as- laid pipeline profile measured using a remotely operated vehicle (ROV). Typical idealised imperfection shapes are 0.3m high and based on a centrally supported prop. An idealised imperfection is often used for estimating or preliminary design, but this approach is often too conservative for detailed design, or final construction.

An assessment methodology is provided in Reference [3].

Strain localization

High axial loads and bending moments can result in the localising of strain in sections of the pipeline that have reduced cross-sectional area/moment of inertia, and/or a lower yield strength. This can manifest itself as very high axial or bending wrinkles, or local buckles. This can happen at pipeline field joints and welds. This is something that needs consideration during the design process.

Pipeline walking

Pipelines that operate at high temperatures and/or high pressures can exhibit the phenomenon of 'walking' or 'ratcheting'. This is the axial movement of a short pipeline section caused by asymmetric load combinations from start-ups and shutdowns.

A long pipeline that has buckled at several places along its length will behave as several short pipelines, with each segment having the propensity to walk independently. Walking on flat seabed usually proceeds from the hot end of the pipeline to the cold end.

Advice and assessment methodologies are provided in Reference [4].

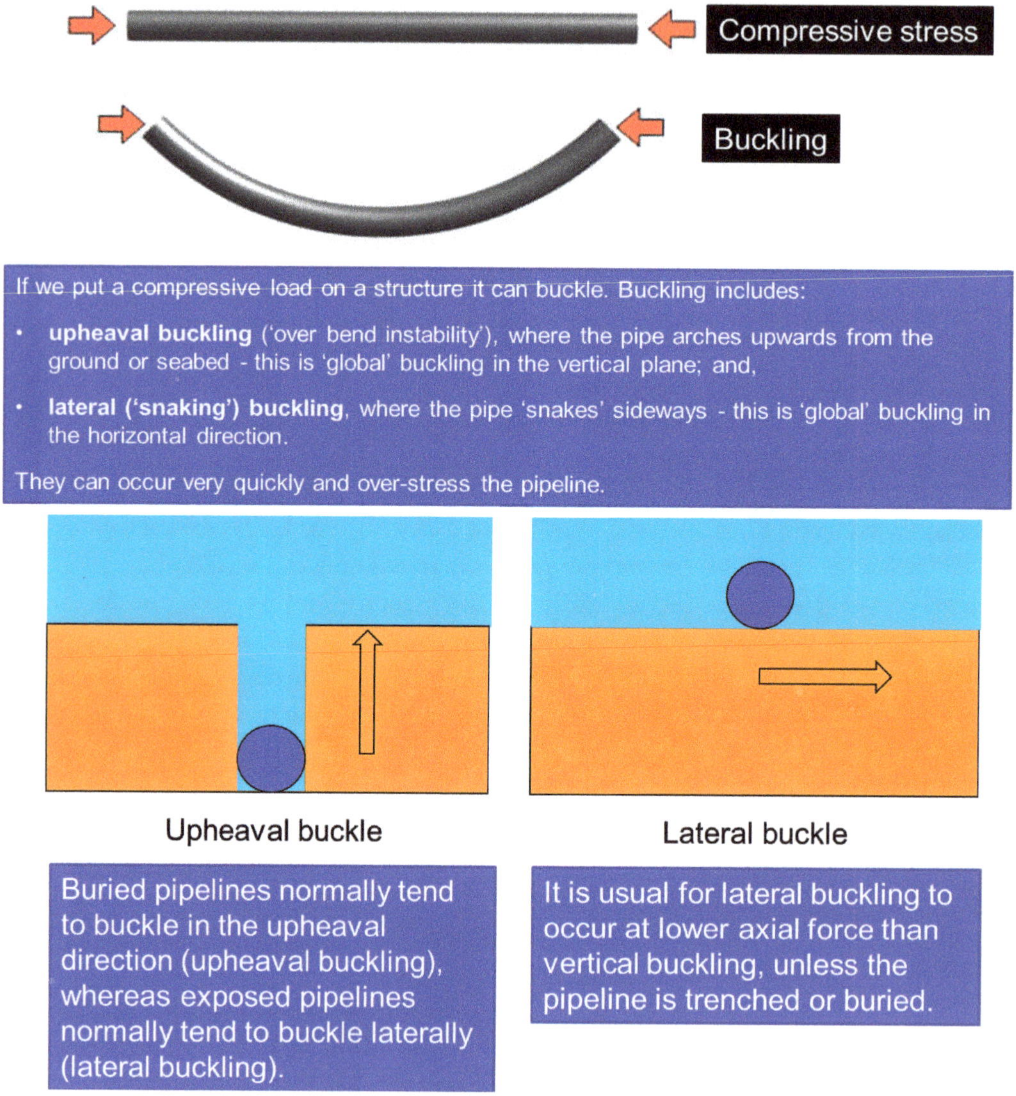

Figure 9-5. Upheaval and lateral buckling.

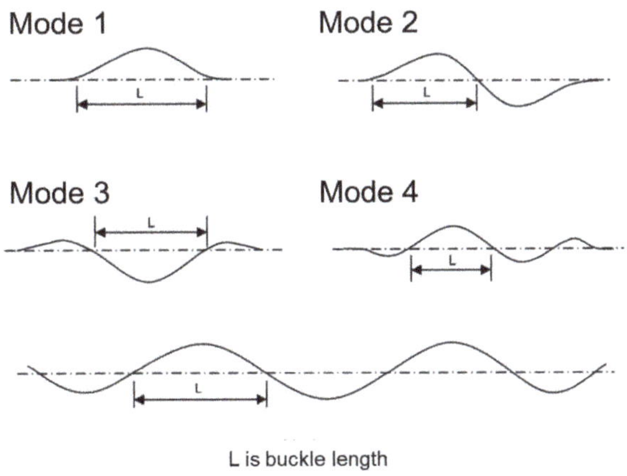

Lateral buckling occurs because of:

- an initial lateral out-of-straightness; and,
- an axial load due to temperature and pressure.

The initial out of straightness is often caused during pipe installation, but can also be caused by movement of the pipe during operation.

Figure 9-6. Buckling mode shapes.

Figure 9-7. Rock dumping over submarine pipeline.

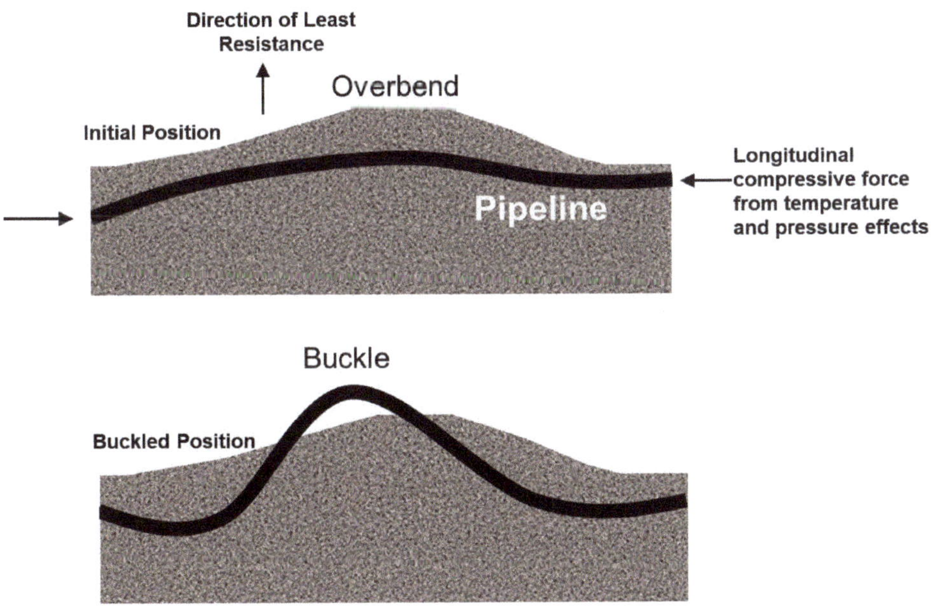

Figure 9-8. Upheaval buckling.

9.2.3 Waves, tides and current

Waves and current exert forces on the pipeline which cause instability and or threats to unsupported sections or spans.

Pipeline stability

Pipeline stability refers to the prevention of unacceptable pipeline movement due to waves and current. The hydrodynamic stability of a pipeline is critical in the design of submarine pipelines and risers, and analyses are required to ensure the stability of both pipelines and risers during construction and operation.

A lack of stability is a threat to a pipeline because it can cause:

- Unacceptable bending stresses
- Cyclic stress leading to fatigue
- Unacceptable bending strain at fixed points such as risers and crossings
- Movement to an extent where the pipeline interacts with other structures or pipelines and it can move the pipeline from its designated route or outside of the protection of a trench

Stability analysis considers the balance of forces in the pipeline, which are (Figure 9-9):

- Weight (steel, concrete and contents)
- Buoyancy (water displaced)
- Hydrodynamic loads
- Seabed friction

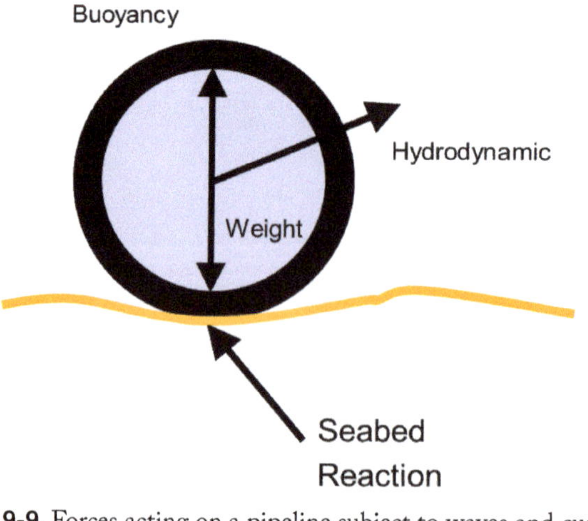

Figure 9-9. Forces acting on a pipeline subject to waves and current.

Force on the pipe is traditionally calculated using Morison's equation[17] in an unsteady flow, but more comprehensive guidelines are now available[5]. Stability can be improved by adding weight in the form of concrete, additional wall thickness, clamped weights, providing a shielding trench, cover, or mechanical anchors.

More advice and an assessment methodology are provided in Reference [5].

Pipeline spanning

A free span is a section of unsupported pipeline, Figure 9-10. This can be on the seabed or riverbed or between clamps in a riser. Pipeline spans on the seabed are often idealised by considering a simple symmetric span with shoulders span height and length (Figure 9-10), but in practice shapes are much more complex, with possibilities of interaction with adjacent spans.

Spans are created by:

- Construction on an uneven seabed
- Pock marks (holes in the seabed)
- Tie-ins to subsea structures or platforms where tie-in points are above the surface
- Crossings of other pipelines or cables or,
- During operation by seabed erosion.

To a lesser extent spans can also be caused by changes in local seabed properties, and pressure and temperature effects (pipeline movement).

Pipeline spans represent threats to a pipeline by three distinct mechanisms:

- Excessive bending due to external applied hydrodynamic and/or self-weight loading (this requires a static analysis)
- Long term fatigue damage by vortex induced vibrations (viv) (this requires a dynamic analysis) and,
- Crossover and hooking from fishing equipment.

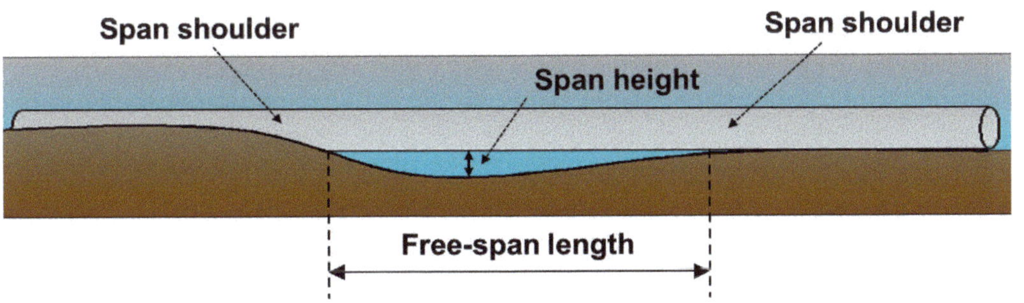

Figure 9-10. Pipeline span.

Static analysis considers deflection under a combination of self-weight and the axial force in the pipeline. The span may also be subject to environmental forces induced by wave and current action (hydrodynamic loads). Maximum bending moments can be calculated using beam-column theory considering pipeline self-weight, bending, effective axial force in the free span, pressure, temperature, and residual lay tension. The effect of tension generated by the sag of the pipeline, external hydrodynamic load, and direct current and wave action (using Morison's equations), need to be considered.

Fishing equipment crossover and hooking can be considered by conducting a static analysis.

Dynamic analysis considers vibration due to the shedding of vortices, or vortex induced vibration (VIV). As water flows over the pipeline a series of vortices are formed. As each vortex is shed, it exerts a small vertical force (cross-flow) and a smaller horizontal force (in-line). In-line at low flow velocities produces small amplitude of oscillation, and cross-flow at higher flow velocities, produces large amplitude of oscillation. If the shedding frequency comes close to the natural frequencies of the span, the span amplitude of oscillation increases significantly. This can cause significant fatigue damage to the pipeline.

An initial assessment might include a static comparison with stress and dynamic analysis based on choosing a span length that avoids the natural frequency. A more sophisticated analysis might consider maximum allowable strain in the static analysis, and fatigue due to vibration for the dynamic analysis.

It is necessary to consider threats from spans during construction and operation, although some vibration can be allowed during the construction phase since the duration is typically quite short.

Spans can be managed by monitoring and rectifying if required. The first preference, of course, is to do nothing, since subsea intervention is expensive. Where the seabed is subject to sediment transportation it is possible for spans and pipeline exposures to correct themselves. When scour creates a span, there is a tendency for the length to increase until the span touches down in the centre and forms two shorter spans. Mobile sediment then fills the gaps and covers the exposed section. The net effect is progressive lowering of the pipeline so that the number of spans decreases with time. Relying on this mechanism requires analysis to demonstrate that the seabed is mobile and that spans will develop and re-fill over the lifetime of the pipeline and monitoring. If this mechanism cannot be relied upon, then there are various other methods of span rectification.

Advice on spans and their assessment is provided in Reference [6].

Risers

Risers are vertical sections of pipe used to connect the pipeline on the seabed to process equipment on a platform (Figure 9-3). They can be dynamic or rigid:

- Rigid risers tend to be fixed at the top with a deadweight support and then guided along the length to resist hydrodynamic loads and allow for movement due to self-weight, temperature, and jacket sway due to environmental loads
- Dynamic risers are flexible or rigid pipe in various configurations including steep s, lazy s, steep wave, lazy wave, still catenary risers, or hybrid.

Risers are subject to environmental loads, vortex shedding, wave slamming, self-weight, buoyancy, internal pressure, external pressure, temperature, and settlement loads. The transportation and installation loads also need to be considered. Risers are typically analysed using static and dynamic stress analysis programs.

QR Code 9-3. Scan here for more information on waves, tides and currents.

QR Code 9-4. Scan here for more information on pipeline stability.

9.2.4 Seabed/coastal

This section considers threats from coastal erosion, scour, settlement of associated structures, sand waves, liquefaction, seabed slope and geological faults. Gathering data is a critical part of the initial phase of any pipeline project. Advice with respect to the gathering of geotechnical data can be obtained from [7].

Coastal erosion

Coastal erosion is particularly important for shore crossings. Pipelines typically have a design life of 25 years but have historically operated for a lot longer. During this period, significant coastal changes can occur with loss of beach material, deposition of beach material, and loss of cliffs. Under these circumstances it is usual to bury the pipeline taking account of beach levels and cliff profiles in the future. Where this is not practical shore crossings must rely on methods such as Horizontal Directional Drilling (HDD), microboring, or tunnels. Advice and assessment methods are provided in References [8] and [9].

Scour

Scour is the removal of granular material from the seabed by hydrodynamic forces. It typically occurs around obstructions on the seabed which cause local increases in velocity. Examples of where this occurs are around seabed structures, piles, and pipelines. It can be beneficial, in that it is part of the process of self-burial of a pipeline, but equally it can expose a pipeline and cause spans, or completely undermine a pipeline if it is inappropriately protected. Scour can be managed by understanding the propensity, considering carefully any aspects of the design that can exacerbate it and, if required, considering mitigation. Scour is a time-dependent process, so design considerations can lead to a requirement for continuous maintenance. Mitigation can be in the form of trenching, artificial seaweed, and/or concrete mattresses, or carefully placed gravel.

Settlement of associated structures

Structures that rely of gravity to fix them to the seabed as opposed to piles, depending on the soil conditions will be subject to settlement. This applies to large gravity-based platforms but can also apply to much smaller structures such as pipeline crossings, manifolds, and riser bases. In some cases this can be quite substantial. Settlement can produce bending and, depending on the extent, can be a threat to the pipeline. This threat can be mitigated by careful design and consideration of tie-in points, particularly where mechanical connectors are used since they can be less tolerant to bending loads.

Sand waves/megaripples

Megaripples are relatively large-scale seabed features caused by sediment transportation. Depending on the diameter of the pipe, it may not be possible to lay directly across megaripples due to the resulting high bending stresses and unacceptable spans produced.

Under these circumstances the seabed must be prepared before the pipeline can be laid. This is referred to as 'pre-sweeping'. From an operational perspective, megaripples translate slowly across the seabed, exposing some pipeline sections and burying others. If possible, pipelines should be laid parallel to megaripples at low points, so that burial is more likely than exposure.

A methodology for assessment is provided in Reference [10].

Soil liquefaction

Liquefaction is where soil under, or around, a pipeline behaves as a liquid. Depending on the pipeline's specific gravity this can cause the pipeline to sink or float. It is specific to granular materials and is caused by:

- An increase in pour pressure due to wave action
- An internal pressure differential across the soil
- Vibration due to construction activities or,
- An earthquake.

It is important to consider it as part of the design of a shore approach. It is avoided by considering the susceptibility of the soil to liquefaction, based on environmental conditions, and choosing submerge weight of the pipeline and/or burial depth accordingly.

Slope instability

Seabed slopes are caused by sediment discharge from river estuaries. Sediment builds up in layers and slowly moves away from the continental shelf. During this process, these sediments can be subjected to submarine landslide. Landslides can be triggered by:

- Over slope
- Seismic activity
- Storm surges
- Gas release
- Natural gas hydrate decomposition
- Change of tide level
- Seepage
- Volcanic activity
- Glacier activities

They tend to be relatively infrequent, with timescales in the scale of 1000s of years, but devastating to pipelines when they do happen. For example, many areas in the Mississippi delta area have these conditions and failures have occurred. It is possible to determine the frequency by dating organic material in the various layers from bore holes. Slopes can be determined using modern multi-beam seabed profiling equipment. Using these techniques, it may be possible to avoid areas that present a high risk. It is possible to analyse the effect of such an event using stress analysis. In many cases the pipeline will be unable to survive, so the event is treated as an emergency, with the necessary contingencies put in place to be implemented if it does happen.

Active fault rupture

Seabed geological faults are detected using shallow seismic survey techniques. This is often done during the initial pipeline route survey. A geologist should then be able to determine the likely movement and probability of its occurrence. The implications of fault movements (direction and displacement of one or many) can be assessed using stress analysis. In general, not burying the pipeline, or burying it in loose material, significantly reduces the loading.

Advice and a methodology for assessment are provided in Reference [11].

9.2.5 Third party

Shipping

Anchor damage to a pipeline can occur due to the anchor or its chain:

- Being dropped directly onto the pipeline
- Being dragged across the pipeline or,
- Hooking the pipeline

A hooking event with a large vessel can result in the pipeline being lifted and dragged across the seabed. In extreme cases, it can result in a full-bore rupture of the pipeline. Typically, pipelines are not constructed in areas where ships anchor, but ships do encounter pipelines during emergency situations where they move out of shipping lanes (to avoid other ships) or where they are dragging their anchors in a storm.

Shipping can also interact directly with the pipeline in shallow water when the ship is grounded or sinking, but there are a few examples of this. All cases can result in significant damage, including loss of weight coating, crushing, denting, gouging, and even full-bore rupture where larger vessels are involved.

Advice and methods of assessment can be found in Reference [12].

Fishing

There are a wide range of different techniques for fishing, including: lines, trawling, and traps. Most of these have no impact on pipelines however, some fishing techniques involve dragging fishing equipment across the seabed. Seabed trawling equipment can include a beam or trawl doors. These affect a pipeline through the initial impact, crossing of the beam, door and/or wire, and/or hooking. The interaction has the potential to damage the weight coating, produce a dent and/or gouge, and/or to displace the pipeline. Typically, small- diameter pipelines without a concrete coat are most susceptible to damage from fishing equipment. Concrete coating, trenching, sufficient wall thickness, and burial are effective protection methods.

Methods of assessment include the risk of interaction determining the risk of interaction and evaluating the prospective consequences. These can be found in Reference [13].

9.2.6 Dropped objects

A pipeline's susceptibility to dropped objects only occurs where it is routed close to an offshore platform (whether floating or fixed), or an area where construction and/or drilling is taking place. The source of dropped objects can be a platform crane loading or unloading from a support vessel, or from the movement of construction equipment. Typically, areas where dropped objects are likely are avoided, but in congested areas with a moving platform such as with a Floating Production Storage and Offloading (FPSO), this might not be possible.

Dropped objects typically include: containers, scaffold poles, drill tubing, drill collars, anchors, and blowout preventers. If the pipeline is routed in an area where dropped objects are likely, then a risk assessment is conducted, taking account of the type of objects, crane capacity and radius, water depth, and the susceptibility of the pipe to damage from the objects under consideration.

The type of potential dropped objects are taken from the inventory for provisioning supplies and drilling. Mitigation can be provided by mechanical protection such as the use of concrete covers (dog kennels), rock dump, concrete mattresses, or changing the pipeline route, or by the restriction of certain construct activities. For example, to only allow transportation of anchors across pipelines if they are affixed to the deck of the anchor handler, similarly, not allowing blow out preventers to be moved above pipelines.

A methodology for assessing the implications of dropped objects can be found in Reference [11].

9.2.7 Ice

Some pipelines may be under water in cold climates, and if ice might develop or drift, a designer must consider:

- Ice forces on the pipeline system
- Impacts from drifting ice
- Sea-bed scouring
- Ice problems during construction and installation, etc.

Ice presents a high potential for damaging pipelines, but only occurs in a few regions where pipelines are required. These including Northern Alaska, the Canadian Arctic, the Russian Arctic, and the Northern Caspian Sea.

Icebergs

An iceberg is a large section of freshwater ice that originated from a glacier or ice shelf and floats freely in open water. Icebergs can be transported a considerable distance from the place where they were originally formed. The iceberg has a sail above water and keel below water. The keel extends for a considerable distance below the waterline and can ground, scouring the seabed and damaging any pipeline in its path. The principal method of protection of pipelines from icebergs is to bury them deeper than the likely iceberg keel depth, use risk assessment to calculate a reasonable depth, and have contingency procedures in place if this is exceeded, or place the pipeline in areas where there is shallower water between the pipeline and the open sea/source of icebergs.

A design methodology is provided in References [14] and [15].

Pressure ridges

A pressure ridge develops in floating ice because of a compressive stress caused within the plane of the ice. A pressure ridge is like an iceberg in that it has a downward pointing keel which can interact with the seabed. A pressure ridge can scour the soil and damage any pipeline in its path. Again, the principal method of protection of pipelines is to bury them deeper than the likely iceberg keel depth or use risk assessment to calculate a reasonable depth and have contingency procedures in place if this is exceeded.

Strudel scour

Strudel scour occurs in the Arctic every spring when the fresh-water rivers melt and flow out over the sea ice cover of the arctic sea. This fresh water flowing out over the ice drains through imperfections that penetrate the ice. It does so with sufficient high velocity to scour holes in the seafloor sediments below. Strudel scour holes have been observed to have depths of 4-6 m and diameters of 10-30 m. Full strudel scour does pose a threat to a pipeline in that it can cause a large span, but only if directly over the pipeline.

9.2.8 Debris

A seabed can contain debris such as old shipwrecks. Typically, pipelines are not laid in areas where there is substantial debris or the debris is cleared, however, in some situations the debris can be mobile on the seabed and can be moved by current into a position where it impacts the pipeline.

Seabed debris can include:

- Shipwrecks
- Abandoned fishing equipment
- Construction remnants
- Abandoned seabed equipment (including pipelines and cables) and,
- Intentionally dumped material (including ordnance and refuse from other sea users).

Ordinance encountering a pipeline is treated as a hazard and is disposed of using specialist contractors. Other debris in contact with the pipeline can increase hydrodynamic loading and lead to instability. In addition to this, metallic debris can draw extra current from the cathodic protection system on the pipeline and waste its anodes.

Interaction with debris at a pipeline at a river crossing is a special case and can occur when rivers in full flood sweep trees and large boulders downstream. These can be very destructive to pipelines and riverbanks, so crossings that are likely to experience this type of flooding would need substantial protection using gabions and/or concrete.

9.3 Management of threats

9.3.1 Routing

Pipeline routing can have a significant impact on a pipeline susceptibility to damage. A pipeline can be routed to avoid excessive currents, waves, seabed slopes, shipping lanes, shallow water, fishing areas, etc. Reference [18] gives guidance on routing pipelines, Table 1.

Table 9-1. Routing considerations from ISO 13623

Consideration	Pipelines on land:	Offshore pipelines:
Safety	a full hazard assessment will have been conducted.	
Environmentally sensitive areas	areas of outstanding beauty; areas of archaeological importance; areas of designated landscape; areas of conservation interest; natural resources such as water catchment areas, reservoirs and forestry; aquifers and potable water supplies.	areas of special scientific interest; areas of natural conservation importance; areas of marine archaeological importance; marine parks.
Facilities	pipelines; underground and above ground services; tunnels.	pipelines; cables; subsea structures and wellheads; coastal protection works.
Third party activities	land use; mineral workings; mining; military zones.	shipping lanes; anchoring; recreation; fishing; exploration and production; dredging and dumping; military exercises; offloading at platforms; boat landings.
Environmental conditions	1. Geotechnical conditions: uneven topography, outcrops and depressions; instability like faults and fissuring; soft and waterlogged ground; soil corrosivity; rock and hard ground; flood plains; earthquake areas; muskeg* and permafrost areas; areas of land slippage, subsidence and differential settlement; infill land and waste disposal sites including those contaminated by radioactivity or disease. 2. Hydrographic conditions.	1. Geotechnical conditions: uneven topography, outcrops and depressions; earthquake zones; high slope gradients; seabed instability; soft sediment and sediment transport; presence of near-surface gas; coastal erosion; beach movement; high near-bottom currents. 2. Hydrographic conditions.
Construction and operation	access; working width; utilities; availability and disposal of test water; crossings; logistics.	maximum approachable water depth; minimum feasible laying radius; station-keeping vessels; platforms and subsea wellhead approach; tie-ins; shore approach and landfall installation; crossings; logistics.

9.3.2 Monitoring

Submarine pipelines are designed to minimise intervention because of the high costs however, not all threats can be managed in this way. Some threat management can benefit from monitoring and action based on the results. Monitoring techniques include on-line wall thickness measurement, strain and leak detection (numerous methods).

9.3.3 Third party

Awareness

Generally, pipelines are marked on admiralty charts, and offshore platforms have exclusion zones. Shore approaches and river crossings have clearly visible signs indicating the presence of pipelines.

Surveillance

Some port authorities monitor vessel movements and would be able to warn a vessel if it looks like it is going to anchor near pipelines.

Also, some critical pipelines operating in an area subject to heavy shipping traffic are constantly monitored by dedicated vessels.

9.3.4 Inspection

Submarine pipelines are designed to minimize intervention because of the high costs however, not all threats can be managed in this way. Examples are damage from third-parties, exposure due to scour, and instability due to extreme environmental conditions consequently, hydrographic surveys are required of any proposed route before the final selection. This survey covers:

- Seabed geology and features
- Bathymetry
- Environmental and oceanographic data
- Wellheads, wrecks and debris
- Marine growth
- Landfall
- Presence of ordnance or other hazardous items, etc.

Threat management requires inspection and action based on the results of inspections, as there are many obstacles to a pipeline's route on the seabed. There are several different survey techniques which are combined to provide an overall inspection campaign, for both the construction and operation of a pipeline. These are discussed individually below.

General imaging surveys (GI)

This type of survey uses towed or hull-mounted side scan sonar or multi-beam echo-sounders to determine the position of the pipeline and the surrounding seabed. This provides wide seabed coverage, is fast, and relatively cheap.

A GI survey provides an overview of the pipe and surrounding seabed, identifies extent of burial, exposure and spanning including height, start and finish positions, indicates condition of the trench, identifies presence of any debris or signs of third-party interaction local to the pipeline, identifies areas of degradation such as coatings or anode wastage, identifies areas of pipe movement, and identifies features which require additional investigation using General Visual Inspection (GVI).

General visual inspections (GVI)

General Visual Inspections (GVI) use a workhorse remotely operated vehicle (ROV), fitted with cameras and additional sensors.

Corrosion protection (CP) survey, where the potential of anodes is measured, are often undertaken as part of a GVI survey.

Close visual inspections (CVI)

This local inspection is undertaken by a diver or ROV and looks at specific features such as coating loss or other damage. These are typically used on risers to check for bolt loss or other signs of damage. In the shallower sections or a pipeline route this can be considerably impeded by marine growth.

Landfall surveys

Landfalls, particularly in the inter-tidal region, must be surveyed using smaller vessels, because of the relatively shallow water.

QR Code 9-5. Scan here for more information on General Imaging Surveys.

Riser surveys

The above water sections of risers can be surveyed using the 'rope access technique' (RAT). This is where ropes and harness are used to safely lower a person into position on a riser where they can work. This can include general visual inspection or wall thickness measurements, but the technique can also be used for coating repairs and bolt tightening. Some contractors are now offering drones to undertake the inspection aspects of this work.

9.3.5 Spans

Vortex suppression

Any device which modifies the form of the cylinder's surface without being situated predominantly in the wake of the cylinder will interfere with the boundary layer separation, the initiation and formation of the vortices, and cause the VIV amplitude to be reduced. Such devices however will increase drag and decrease lift. They are also expensive to install. Typically, they are only used if VIV is a significant problem. They will not offer any protection from static over stress i.e., bending.

Support

Rock dumping. Precision rock dumping using a fall pipe from a ship can be used to support spans by creating a central berm/support without the need for covering the whole span, with its associated cost implications.

Sandbags. Sandbags are used in conjunction with other forms of protection, e.g. mattresses. They are only effective if they are stable against the effects of wave action, scour, and settlement. Sandbags are susceptible to being displaced by environmental loads, so may require steel stakes to hold them in place. In practice, sandbags often shift due to waves and currents, and are not generally regarded as a satisfactory long-term solution.

Grout bags. Grout bags are cheap and easy to manufacture, flexible and can be placed using an ROV. They are made of fabric, placed in position, and pumped full of grout. Grout bags are typically high mass structures, so can cause further seabed disturbance in the form of scour and settlement. They are ideal for rocky seabeds where undermining by scouring or settlement is unlikely.

Mechanical support. Mechanical supports prop pipelines and are applied using a clamp. They have been used in 50 m to 700 m water. They are expensive to procure and install, so a cost benefit analysis is required to compare with other methods. They tend to be more cost effective for larger span heights (greater than 1.5m).

9.3.6 Protection

Pipe-in-pipe

High-pressure and high-temperature oilfields are often associated with waxy crude oil. This typically results in a requirement for the production flow lines to have very high levels of insulation, which is only practically obtained using a pipe-in-pipe system. A pipe-in-pipe system combines the strength of steel pipe with the insulation properties of an ordinarily mechanically weak insulating medium. In addition to their thermal performance, pipe-in-pipe systems have much higher bending resistance which helps with buckling. The outer carrier also tends to operate at temperatures closer to ambient and therefore reduces end expansion. The higher outer diameter of a pipe-in-pipe system also offers greater resistance to movement from soil cover.

Concrete coating

Concrete coating of pipelines has been found to be very beneficial in protecting pipelines, because it is mechanically strong, impact resistant, and increases the submerged weight of the pipeline which helps to resist displacement due to fishing equipment, debris impact (river crossings), or environmental loads. Many of the threats associated with submarine pipelines can be managed by appropriately specifying concrete coating. This includes selecting the correct rebar, concrete filler and specifying appropriate quality control during the manufacturing process.

Platform mounted fenders

Vessels often must work near offshore platforms and can impact jacket members and risers. Wherever possible risers are designed to be inboard of jacket members, close to the waterline to avoid interaction with shipping. Where this is not possible, perhaps due to the design of the platform such as in a monopod, or due to congestion, shipping fenders can be provided.

Rock

For offshore pipelines, there are two principal methods of applying rock cover:

- Side dump or,
- Fall tube/fall pipe

The fall pipe technique is typically used for precision rock lay. It is graded in a controlled manner for depths from 20 m to 850 m. It provides:

- Stability
- Protection (from snagging and dropped objects) and, to a certain extent
- Thermal insulation

Side dumping, relative to fall pipe, is lower in precision and used in shallower depths (up to 85 m). Consideration also needs to be given to the risk of damaging the pipeline coating if not concrete coasted. Side dumping is much cheaper than using a fall pipe. Good rock dump design is required to ensure stability and prevent degradation of the berm under wave and current action. Typically, rock berms are design as a trapezoid. Figure 9-11. The main design requirements include bulk stability in preventing stones from dislodging from the berm. Prevention of sinking filtration, using a filter layer, is also necessary if the sand is fine or soft.

Gabions

A gabion is a cage, cylinder, or box filled with rocks, concrete, or sometimes sand and soil. They are commonly used for civil engineering projects. There are cost effective and can have a very high density/installed weight and are therefore very effective at providing protection to pipeline in wadi crossings and/or at riverbanks.

Mattresses

Mattresses for use as subsea protection are typically constructed of concrete held together with wires or nylon rope. They are constructed using interconnected cells to provide flexibility. They are not necessarily a span rectification option on their own but can be placed under and over a free span.

Stability of the mattress itself must be addressed, as wave action can lift the mattress and flip it over, and in some instances, it can be carried away completely.

They are often used in conjunction with grout or sandbags to provide protection and scour resistance. Mattresses can be fitted with fronds to promote sediment deposition. This obviously, only works in a situation where sediment is mobile.

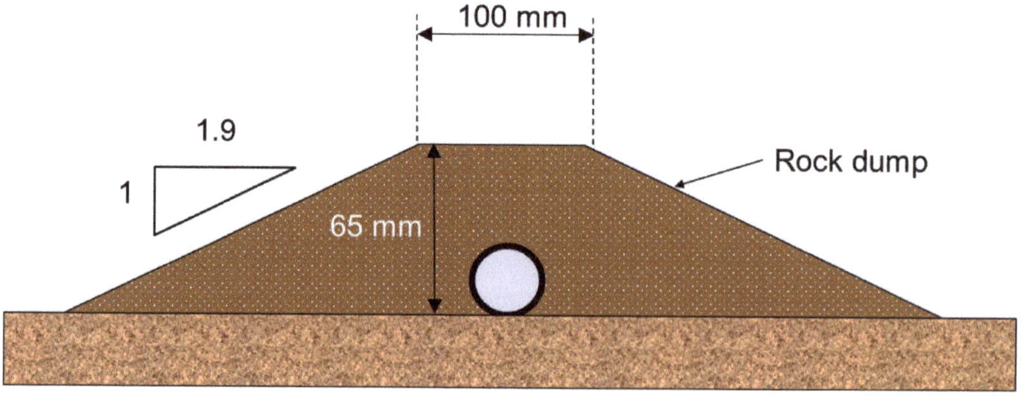

Figure 9-11. Rock dump profile.

Burial/lowering of pipeline

Plough. A plough is pulled over the top of the pipeline and the pipeline falls into the ploughed trench. There is a risk of impact and damage, but ploughing can be fast (dependent on soils). Its effectiveness is dependent on the seabed soils and requires expert advice on trenching. Potentially high stresses on the pipeline will require engineering analysis.

Mechanical cutter. This is typically used for hard soils or rock. Mechanical cutters can produce a neat trench profile. They are mechanically complex and relatively slow.

Jet trencher. A jet trencher uses low volume, high pressure water injection at either side of the pipeline to 'fluidise' the surface of the seabed: the pipeline drops into the fluidised trench. There is a potential risk when used for free span rectification, due to the weight (7-8 tonnes) of the jet sled. There is also a possible problem with fluidisation of backfill material and floatation.

Mass flow excavation. Mass flow excavation uses a high volume, low pressure column of water to remodel soft seabed soils. The water and machine have no contact with pipeline, and therefore present a lower risk. This method is limited to certain soils, but is fast, rapidly mobilised, potentially cheaper than dredging, requires a small deck footprint, and has no physical contact with sea floor or pipe.

Dredging. Dredging (either cutter suction, bucket excavator or suction) is often used for very deep trenches associated with shore approaches, pre-sweeping of megaripples, or lowering associated with crossing a shipping channel.

Self-burial. Where the seabed is highly mobile it is possible for pipelines to bury themselves. This starts by scouring on either side of the pipeline to an extent where the pipeline is only supported by a narrow section: this collapses, lowering the pipeline. This process can be accentuated by fitting fins on top of the pipeline. If fins are fitted, the pipeline will continue to self-bury until the fin is below seabed level.

Advice and an assessment methodology are provided in Reference [16].

Trenchless installation

For river crossings and shore crossings, conventional methods of protection using lowering may be technically feasible, but not environmentally acceptable due to the amount of sediment discharged into the water column, or disruption of other users. Under these circumstances trenchless installation methods may be beneficial including horizontal directional drilling (HDD), pipe jacking, or micro tunnelling.

9.3.7 Marine growth

It is possible, particularly offshore in shallower water where sunlight is more intense, for substantial marine growth to accumulate on pipelines. This significantly adds to weight and hydrodynamic loading. It might be necessary to clean it off and this is certainly required before any inspection can be undertaken. With respect to marine risers, monel cladding or copper impregnated neoprene can be used to prevent marine growth (monel and copper are naturally resistant to marine growth). Alternatively, there is a system which uses a rotating vane energised by passing current.

9.3.8 Isolation

Some threats cannot be eliminated such as icebergs, or submarine landslides. In cases where the consequences of failure are high, isolation mechanisms can be used for example:

- Block valves either side of a river crossing to protect the river from pollution
- A sub-sea isolation valve (SSIV) placed in a pipeline some distance from the platform to protect the platform personnel from the consequences of a failure and the pipeline inventory and,
- Beach block valves that are used to protect the coastal environment from the inventory of the pipeline in the event of a failure.

Valves can be motor driven (actuated), manual, automatic, or check valves (non-return).

9.4 Summary

Threats to submarine pipelines are well-understood and managing them is a question of understanding the operating conditions, the environment and designing accordingly.

Some aspects cannot be completely managed through design, such as ice impact or seabed slope failure, and threats that are not eliminated by design must be monitored through inspection and maintained and responded to in an emergency as required.

There are many publications to assist in managing submarine pipeline threats (for example, References [2] to [16]), and standards (for example, References [17] to [20]) that help with their overall integrity management.

9.5 References

[1] Anon., 'Specification for line pipe', API 5L. 46th edition, American Petroleum Institute. 2018.

[2] Anon., 'Global Buckling of Submarine Pipelines Structural Design Due to High Temperature/High Pressure', Recommended Practice DNV-RP-F110, October 2007.

[3] A.C. Palmer et al, 'Design of Submarine Pipelines against Upheaval Buckling', OTC 6335, 22nd Annual OTC in Houston, Texas, May 7-10, 1990.

[4] D Bruton et al, 'Lessons Learned from Observing Walking of Pipelines with Lateral Buckles, Including New Driving Mechanisms and Updated Analysis Models'. OTC-20750-MS. Offshore Technology Conference. 3-6 May, Houston, Texas, USA . 2010.

[5] Anon., 'On-Bottom Stability Design of Submarine, Pipelines', Recommended Practice, DNV-RP-F109. October 2010.

[6] Anon., 'Free Spanning Pipelines', Recommended Practice, DNV-RP-F105. February 2006

[7] Anon., 'Guidance Notes on Geotechnical Investigations for Marine Pipelines', OSIG Rev 3, Prepared by the Pipeline Working Group of the Offshore Soil Investigation Forum, Society for Underwater Technology, 2004.

[8] Anon., 'Shore Protection Manual', Department of the Army, Waterways Experimental Station, Corps of Engineers, 1984.

[9] Anon., 'Submarine Pipeline Systems', DNV-OS-F101, October 2013.

[10] M Fitzgerald et al, 'Presweeping and Trenching for the Installation of the 40-in. Diameter Zeepipe Gasline on the Dutch Continental Shelf of the North Sea'. OTC-7018-MS. Offshore Technology Conference. 4-7 May, Houston, Texas. 1992.

[11] M Trimintziou et al, 'Designing Offshore Pipelines Facing the Geohazard of Active Seismic Faults', International Journal of Mechanical, Aerospace, Industrial, Mechatronic and Manufacturing Engineering. 9. 1094. 2015.

[12] Anon., 'Design Against Accidental Loads', Recommended Practice DNV-RP-C204. October 2010.

[13] Anon., 'Interference Between Trawl Gear and Pipelines', Recommended Practice DNV-RP-F111October 2010.

[14] G. Davies, M. Marley, K. Mork, 'Limit State Design Methodology for Offshore Pipelines against Ice Gouging - Industry Guidelines from the ICEPIPE JIP', OTC-22037-MS, OTC Arctic Technology Conference, 7-9 February, Houston, Texas, USA, 2011

[15] B Eisler, 'Shore Crossing Design Considerations & Solutions for Arctic Subsea Pipelines', OTC-27453-MS. Offshore Technology Conference. 24-26 October, 2016, St. John's, Newfoundland and Labrador, Canada. 2016.

[16] R Bruschi et al, 'Self-burial of large diameter pipelines: the Nord Stream case', Offshore Pipeline Technology Conference. Amsterdam. 2016.

[17] J R Morison et al, 'The Forces Exerted by Surface Waves on Piles', Journal of Petroleum Technology, Vol. 189, pp. 149-154. 1950.

[18] Anon., 'Petroleum and natural gas industries - Pipeline transportation systems', ISO 13623:2017. 2017.

[19] Anon., 'Managing System Integrity of Gas Pipelines', ASME B31.8S-2014. American Society of Mechanical Engineers. New York, USA. 2014.

[20] Anon., 'Managing System Integrity for Hazardous Liquid Pipelines', American Petroleum Institute. API Recommended Practice 1160. Second Edition. September, 2013.

[21] Anon., 'Pipeline systems Part 4: Steel pipelines on land and subsea pipelines – Code of practice for integrity management', BSI PD 8010-4:2012. British Standards Institution. UK. 2012.

[22] Anon., 'Integrity Management of Submarine Pipeline Systems'. Recommended Practice. DNV-RP-F116. Det Norske Veritas. Norway. October 2017.

CHAPTER 10

Management of geohazards

Dr Andy Young

10.1 Introduction

Geohazards are hazards caused by geological or environmental conditions that have the potential to cause damage to a pipeline or the route corridor. These hazards include landslides and earthquakes.

There is a need to identify and mitigate geohazards at design or manage them during the service life of the pipeline.

10.1.1 Geohazards and new pipelines

For new pipelines, the identification of geohazards generally requires discipline specialists in the terrain, geology and specific hazard types to evaluate the proposed pipeline route corridor. There are many geohazards, including:

- Landslides
- Rockfalls
- Avalanches
- Land subsidence
- Sinkhole formation
- Seismic (earthquake)
- Volcanic
- River and flooding (flowing water) processes (e.g., flooding, or riverbank erosion)
- Permafrost and geothermal effects (such as frost heave or thaw settlement)
- Surface and backfill erosion

Mitigation at the pipeline design stage invariably means avoiding the hazard altogether by adjusting the pipeline route; however, there are circumstances where this is not possible due to other constraints on the route corridor. These situations will require mitigation involving stabilization of the hazard, or measures to reduce the loading effects on the pipeline.

10.1.2 Geohazards and operating pipelines

During operation, the identification of new, or previously unknown, geohazards will be governed by the geohazard management plan that has been implemented. Where no plan exists, the likelihood of pipeline damage or failure is higher. Mitigation options in service are normally limited to hazard stabilization or load reduction.

Re-routing around a developing hazard is only normally considered where the loading cannot be effectively controlled and failure is predicted to occur.

10.1.3 Objective of chapter

This chapter explains the general nature of geohazards, how these present a threat to the integrity of pipelines, and how the pipeline can be protected. The world-wide variation in terrain, geology and climate is enormous, and this leads to a diverse range and complexity of geohazard types and resultant behaviours; for example, Table 10-1.

Table 10-1. Illustration of range and variability of factors which influence the stability. of slopes [after Cooke and Doornkamp[1]]

Category	Factor	Factor	Category
Slope	Slope height	Soil strength	Soil
	Slope gradient	Soil mineralogy	
	Slope shape	Soil plasticity	
Surface & groundwater	Drainage density	Seismicity	Environment
	River gradient	'Solifluction' (flow of saturated soil down a slope)	
	Slope undercutting	Relic landslides	
	Seepage	Deep weathering.	Precipitation
	Ponded water	Rainfall intensity	
	Pore pressure	Rainfall duration	
Rock	Rock joint density	Snow melt	
	Joint opening	Excavation	Human activity
	Joint orientation	Drainage changes	
	Joint gouge	Slope loading	
	Interbedding	Reservoirs	
	Extent of weathering	Deforestation	
	Compressive strength		

Only a brief overview of the main geohazards is covered in the next section, and the emphasis of this Chapter is on the process of managing geohazard threats and maintaining the pipeline integrity.

Experienced pipeline operators and owners generally recognize that the management of geohazard threats for pipelines requires specialist input, because of the severe financial safety and environmental consequences of failure:

- At construction, a team of specialists is invariably required in order to identify and evaluate the range of hazards; and,
- During operation, specialists in particular geohazards may be needed as problems develop.

The Active Earth....

The structure of the earth has three main elements:

1. A dense metallic core that is 3500 km thick.
2. The mantle which semi-solid rock and is 2850 km thick.
3. The outer crust of rock which varies in thickness from 5 km to 100 km thick.

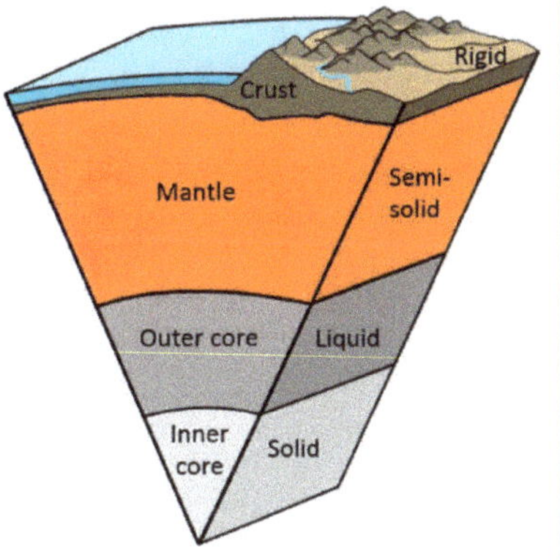

The surface of the earth experiences many different processes that cause change. Tides rise and fall, storms gather and abate, rivers flood and recede, temperatures increase and decrease, ground erodes, topples, falls, slides, flows, and subsides.

Forces deeper in the earth's crust cause mountains to be lifted up, volcanoes to form, and earthquakes to occur.

Consequently, the ground surface undergoes continual change, some is slow and imperceptible to the human eye, whilst others are rapid, widespread, and dramatic.

10.2 Types of geohazard

10.2.1 General

A geohazard threat to a pipeline arises from one or more of the following origins:

- Geological, which relates to the structure of the earth;
- Geomorphological, which is based on landforms and landscapes; and,
- Geotechnical, which describes the behaviour of the earth's materials.

Ground movement leads to relative displacement between stable and unstable blocks of land, and this generates forces in buried pipelines that cross the affected zone, which ultimately may exceed the pipeline's load capacity.

Exposure of pipelines in rivers subjects them to hydrodynamic loading from the current flow and leaves them vulnerable to impact loading from floating debris. In some cases this can lead to structural overload, but in this environment, failure is more common from excessive vibration of the pipeline span causing fatigue fracture.

Many publications deal with the identification and nature of geohazards,[2,3,4,5,6] but the typical hazards that are common to most environments are:

- Landslides;
- Fluvial hazards of river scour and erosion;
- Natural subsidence;
- Mining subsidence;
- Geological fault movement;
- Liquefaction;
- Lateral spreading;
- Erosion.

'Fluvial hazards' are also described as 'hydrotechnical hazards', and these are covered in References 7,8,9.

Geological Complexity

Geological materials are much more complex than common structural materials such as steel, concrete or timber.

Soils are made up of particles where strength and stiffness depend on the confining pressure (typically its depth). In addition:

1. When soils deform they can contract or dilate and change property.
2. When soils undergo shearing, the particles may re-align affecting the shear strength.
3. Water in soil has a major influence on its strength.

QR Code 10-1. Scan here for more on fluvial hazards

10.2.2 Landslide hazards

Landslide hazards are probably the single most widely-recognized geohazard in the pipeline industry, mainly because of the strong link between slope movement and pipeline damage. It involves the down slope movement of soil or rock due to gravity. They are commonly associated with hilly or mountainous terrain; however, they can also develop in shallow slopes with gradients of less than 10%. Lateral spreading is a specific form of landsliding associated with earthquake events, and this may occur on slope gradients as low as 3%.

The factors that influence landsliding are listed in Table 10-2. Two key factors are:

1. The gradient of the slope and how this changes over time (for example, due to river erosion at the base); and,
2. Changes of moisture content or groundwater level in the slope (changes can occur due to climate conditions, typically periods of intense or prolonged rainfall, or rapid transient loads during earthquake shaking).

Table 10-2. Main factors that control the stability of soil slopes. [After Hearn[10]]

Conditioning Factors	Triggering Factors
Slope angle	Prolonged heavy rainfall
Slope height	Intense rainfall
Soil depth	Toe erosion
Adverse relict structures (previous failure surfaces)	Earthquake ground motion
Distinct boundaries (e.g. rock head)	Deforestation or other land use change increasing surface run-off
Soil composition and strength	Excavations
Presence of weak layers	Placement of fill
Groundwater seepage	Changes in drainage

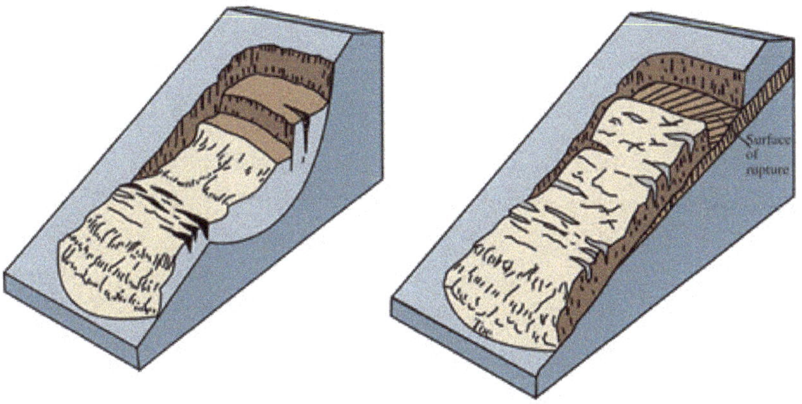

Figure 10-1. Sketches of rotational and translational slides. Credit: U.S. Geological Survey Department of the Interior/USGS U.S. Geological Survey.

Figure 10-2. A pipeline exposed within a rotational landslide.

There are many different types, shapes, and sizes of landslides[11]. Two common types of landslide that can be hazardous to pipelines are rotational and translational slides, where the depth of sliding is below the pipeline Figure 10-1 and Figure 10-2.

Some key issues to consider for slopes that are crossed by pipelines include:

- Is historical landsliding present on the pipeline route?
- Could landsliding develop in the future?
- Will the landsliding affect the pipeline?
- What mitigation can be applied to protect the pipeline?

QR Code 10-2. Scan here for more on landslides.

10.2.3 River crossings

River crossings are active hazard environments particularly during flood events after significant rainfall. This can lead to scour of the riverbed, or erosion of the riverbanks. In some circumstances, the river will abruptly change course within wide, sediment-filled channels, a process known as 'avulsion'.

Data from pipeline right of way (ROW) inspections in North America suggest that river hazards are approximately 5 times more common than geotechnical (primarily slope instability) hazards[12].

The primary threat to pipelines from exposure in rivers, Figure 10-3, is failure due to vortex-induced vibration. This process occurs when the pipeline starts to oscillate in the water current due to the shedding of vortices from the pipe surface, Figure 10-4. The primary factors that influence the onset and magnitude of vibration are:

- The exposed length;
- The velocity of the current;
- The pipeline diameter and mass; and,
- The soil properties at each end of the span.

Wide rivers (for example, exceeding 10 m in width) and estuaries tend to be the main environments where this threat occurs; however, small diameter pipelines may be at risk in smaller river crossings.

Any exposure of the pipeline leaves it vulnerable to impact loading from water borne debris or a mobile riverbed load.

The standard design approach for pipelines at river crossings is to prevent exposure by burying the pipe sufficiently deep so that it lies below the maximum depth of bed scour that is expected to occur. This can be problematic for the following reasons:

1. Calculation of the scour depth requires details on the maximum current velocity during flood events, and the sizing of the particles that form the riverbed soils;
2. Installation by open cut becomes expensive and challenging as the required burial depth across the river channel increases.

Amongst other factors such as bed gradient, prediction of the current, prediction of the current velocity depends on the size of the flood event and this is normally based on the largest expected event within a specific time interval, called the 'return period'. This can be based on measured flood hydrographs for gauged catchments where river flow and water levels are monitored. However, many rivers are not monitored which leads to more empiricism, and consequently less certainty in the predictions.

Pipeline installation by horizontal directional drilling is increasingly common for major strategic pipelines or large crossings to ensure sufficient clearance below the riverbed.

Figure 10-3. Exposure of pipelines at river crossings.

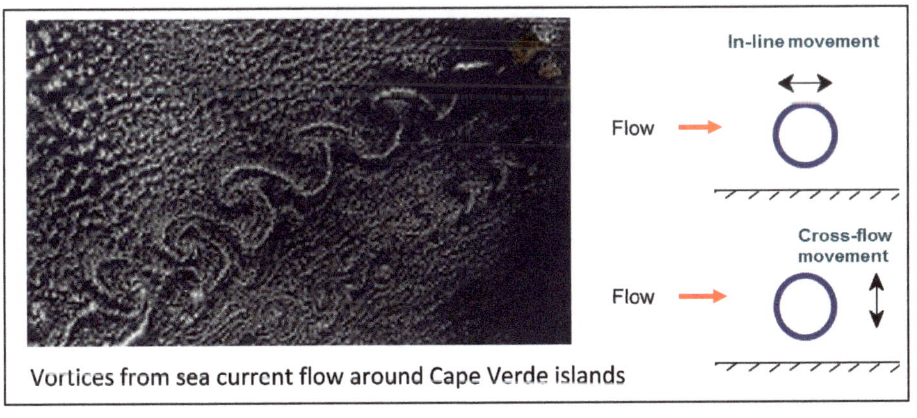

Figure 10-4. Vortices and pipeline movement directions from vortex-induced vibration.

10.2.4 Geological faults

Geological faults are natural, relatively planar breaks in the rocks below the ground surface. Certain parts of the world are tectonically active, which means that there are significant forces in these rocks that can lead to periodic movement on geological faults. This movement is the cause of earthquakes. Pipelines which cross active faults that undergo movement could be subject to large additional loads.

Faults only move periodically because of the gradual build-up of tectonic forces in a region. The movement is not random. The mechanism is well-understood and involves the accumulation of strain at specific locations in the subsurface rock; consequently, faults may appear inactive for considerable periods. It is not possible to predict when a fault that crosses a pipeline route will move because of the complexity of factors that control the build-up of strain and the shear resistance on the fault plane. This includes the influence of the regional structural geology, which may also be complex, see Figure 10-5, as well as the influence of where previous earthquakes have occurred in the region including smaller earthquakes that occur more frequently.

However, the statistical analysis of the occurrence of historical earthquakes means that probabilistic estimates of future earthquake activity can be developed.

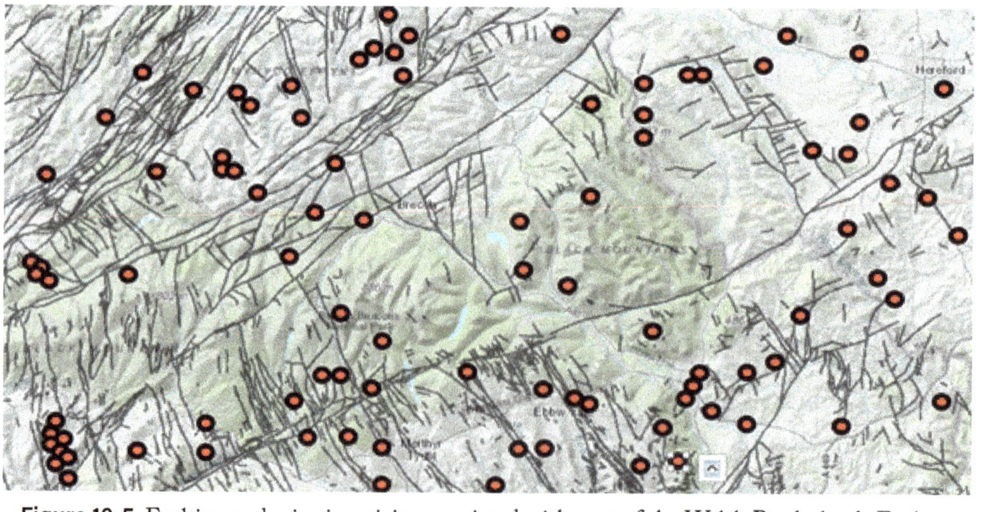

Figure 10-5. Faulting and seismic activity associated with part of the Welsh Borderlands Fault System – one of the more active areas of the UK. Contains British Geological Survey materials ©UKRI [2018]. Base mapping provided by ESRI [Esri, HERE, Garmin, FAO, NOAA, USGS Source: USGS, NGA, NASA, CGIAR, GEBCO,N Robinson, NCEAS,NLS,OS,NMA,Geodatastyrelsen and the GIS User Community | Contains Ordnance Survey data © Crown copyright and database right 2017].

The principal questions that need to be addressed in relation to pipeline integrity include:

- Are faults present at surface on the pipeline route?
- Are the faults classified as active, and therefore could potentially move during the operational life of the pipeline?
- How much will the fault move?
- Can the pipe be designed to withstand the expected movement?

The generally accepted classification of active faulting is the occurrence of at least one movement event during the Holocene geological period, which covers the last 11,000 years. The presence and activity of faulting may be identified from maps of structure geology showing known faults, Figure 10-5, geomorphological features such as sag ponds, Figure 10-6.

The activity and estimates of fault movement can be derived from 'paleoseismic evidence' (evidence from previous earthquakes) or geomorphological features that indicate the magnitude of past events, such as the offset of stream channels, Figure 10-7.

Another common method of estimation uses statistics of historical events, Figure 10-8[13].

Tectonic Plates and Faults

Tectonic forces arise from movement of tectonic plates. There are 15 large plates in the earth's crust that cover the surface of the earth, including all landmass and the oceans.

Underlying these plates is the earth's mantle which is semi-fluidised molten rock, which experiences heat convection currents that make the tectonic plates mobile and push and grind against one another.

Tectonic forces gradually build up against a geological fault until the accumulated force is high enough to overcome the shear resistance of the fault plane. When this occurs, a sudden displacement takes place and the shear energy is released as elastic waves - an earthquake.

However, slow creep movements also take place on some faults. Where this happens, these movements also need to be accounted for in pipeline integrity assessments.

Clearly it is not possible to state with any certainty that a fault that is classified as active, will not move during the lifetime of a pipeline; therefore, the normal approach at design is to take precautions to ensure that the pipeline will withstand the maximum expected movement at the fault crossing.

Details on the identification and characterization of seismic hazards is provided in reference[14]. The design approach for faults is set out in three key references[15,16,17].

In some cases, the presence or activity of the fault is only recognized during the pipeline operation. It is possible to retrofit protection measures, although the array of options are more limited, or will be more expensive.

The ability of a pipeline to withstand the imposed loading is normally established by structural analysis typically using finite element method. This is covered in Section 10.5.5.

Figure 10-6. Sag ponds on the North Anatolian fault in central Turkey.

Figure 10-7. Offset of a stream channel offset on the San Andreas fault, Carrizo Plain, Central California.
Credit: U.S. Geological Survey Department of the Interior/USGS U.S. Geological Survey/photo by Robert E.Wallace.

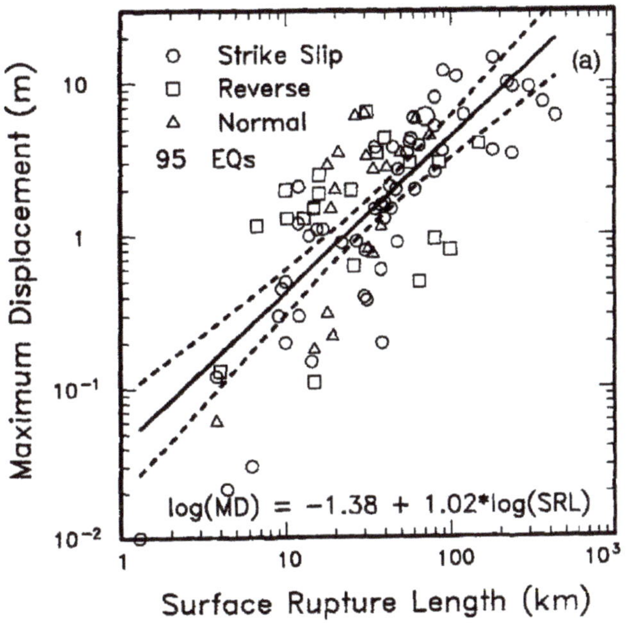

Figure 10-8. Relationship between surface rupture length of fault (strike slip, reverse, normal, and earthquake (EQ)) and the maximum fault displacement. Credit: Wells and Coppersmith 1994.[13]

Earthquake Magnitude and Frequency

A plot of earthquake magnitude on a logarithmic scale against frequency of occurrence shows a linear relationship.

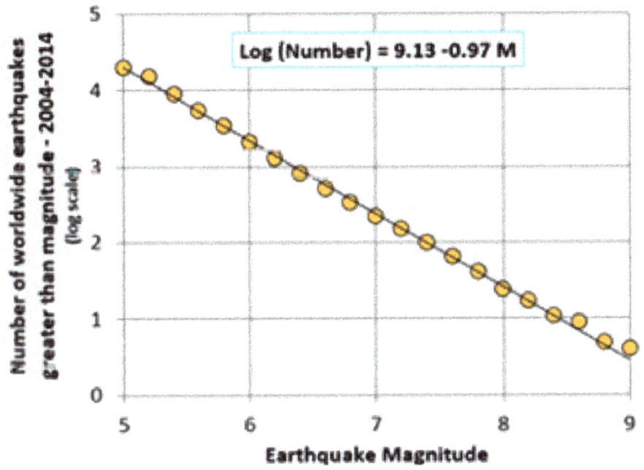

This can be plotted as the cumulative sum of earthquakes exceeding a selected magnitude. The equation describing this line is called the 'Gutenberg Richter' relationship.

This plot is for all earthquakes worldwide over the selected period; however, similar trends are observed in all seismic regions. There is a factor of 10 decrease in the number of earthquakes that exceed a selected magnitude for each integer increase in the earthquake magnitude; for example, the number of earthquakes exceeding magnitude 7 is approximately 10 times less than the number of earthquakes that exceed magnitude 6 over the same time period.

QR Code 10-3. Scan here for more on fault movement.

10.2.5 Liquefaction

Liquefaction occurs in saturated granular soils and describes the temporary change from a solid to a liquid state. This occurs during earthquakes where the seismic shaking causes a transient increase in the pressure of water in the pores of the soil. As a result, there is a reduction in the strength of the soil which ultimately can flow like a fluid.

The reduction in soil strength leads to problems of bearing capacity failure, ground settlement, and ground movement. Pipelines routed through areas where liquefaction takes place may suffer:

- Buoyancy uplift;
- Settlement; or most significantly,
- Dangerously large lateral movements.

Common environments where liquefaction may develop include:

- Flat, low lying areas;
- Flood plains; and,
- Shorelines, estuaries, and deltas.

In these areas, the groundwater is generally within 15 m of the ground surface and this has been found to be typical for areas prone to liquefaction.

Two simple seismic criteria for the development of liquefaction have been defined[18]:

- The earthquake moment magnitude (Mw) exceeds 5.0; and,
- There is a minimum peak ground acceleration (PGA) at the location of interest of 0.1g.

For liquefaction to take place, the soil conditions have to be suitable.

A simple test of whether the soil is susceptible to liquefaction involves the standard penetration test (SPT),[19]. Preliminary guidance on the likelihood of damage based on the SPT blow count, or N value, is provided in Table 10-3[20].

Table 10-3. Relation of liquefaction to SPT blow count.

SPT N value	Damage Potential
0 - 20	High
20 - 30	Intermediate
>30	No significant damage

A more detailed evaluation compares the shear stress developed by the ground motion of the earthquake with the shear resistance of the soil. This is achieved using two parameters: the cyclic stress ratio (CSR); and, the cyclic resistance ratio (CRR)[20].

The potential for liquefaction to occur across geographic areas is described in liquefaction susceptibility mapping. This is based on the presence of susceptible soil deposits and ground motion levels that would cause liquefaction, Figure 10-9. More guidance is provided in [13].

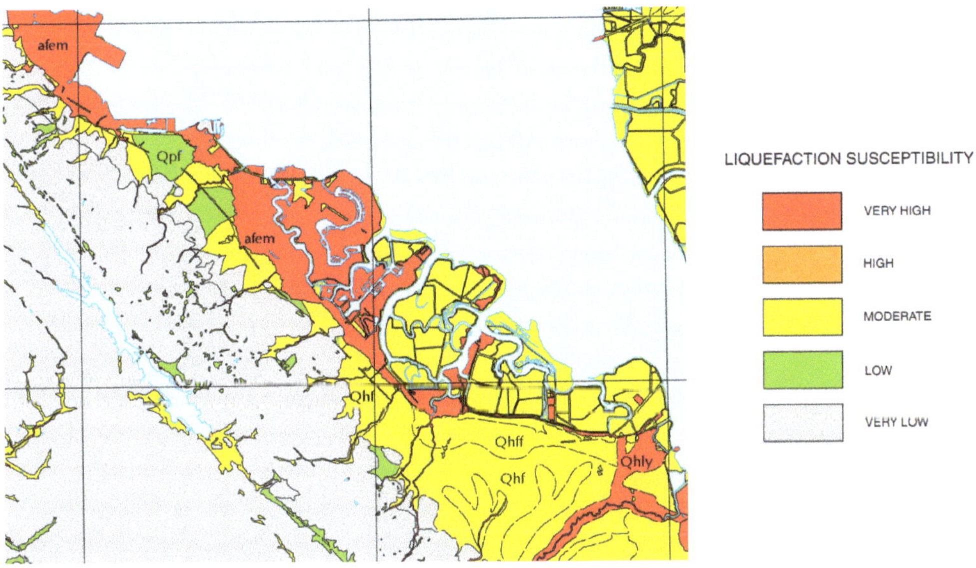

Figure 10-9. Extract of liquefaction susceptibility map of the Central San Francisco Bay Region, California. Credit: U.S. Geological Survey Department of the Interior/USGS U.S. Geological Survey OPEN-FILE REPORT 06-103, 2006.

QR Code 10-4. Scan here for more on liquefaction.

10.2.6 Natural subsidence

Natural subsidence of the ground can occur from a range of causes. A common and well known mechanism of natural subsidence occurs from the solution of limestone. This leads to the development of fissures, gullies, and cave systems which may be filled with soil or water, or remain unfilled and just contain air. These can collapse suddenly creating a sinkhole at the surface. Sinkholes may be up to 100 m in diameter.

Subsidence can also occur in other soluble rocks such as salt (halite) or gypsum. The topography caused by the solution of rocks is called 'karst'.

The development of sinkholes presents a hazard to pipelines because:

(a) Construction activities may trigger a collapse at the surface endangering the project personnel; and,

(b) It may cause a loss of bedding to the pipeline leading to the development of a spanning section, Figure 10-10.

Ground subsidence in soils occurs due to:

- Drainage of peats;
- Hydrocompaction of fine grained soils where fragile clay bonds are broken when saturated (for example, this occurs in loess soils);
- Drainage or loading of clay soils; and,
- Shrinkage of clay soils when desiccated – this primarily affects the highly plastic smectite clays, such as montmorillonite.

Subsidence in soils tends to be less dramatic in scale compared to soluble rocks; however, the development of loads in pipelines can still be significant, especially at the boundaries of the areas of movement if the geological or groundwater conditions lead to high differential movement.

Comprehensive guidance on the nature and investigation of karst terrain is provided in [21].

This includes ground investigation methods which include geophysical surveying techniques, notably seismic surveys, electro-magnetic conductivity, ground penetrating radar, and microgravity. Hazard assessment methods are presented and a number of case histories illustrate the investigation and assessment techniques.

Guidance on natural subsidence hazards is provided in [22], and extensive background detail is provided in [23, 24]. An example of establishing patterns of subsidence based on key controlling geological and hydrogeological factors is provided in [25].

Figure 10-10. A developing sinkhole in soluble rock adjacent to a high pressure pipeline. Credit: Encyclopedia of Pipeline Defects Third Edition, ROSEN Group, Clarion Technical Publishers, Tiratsoo Technical, 2017.

QR Code 10-5. Scan here for more on natural subsidence.

10.2.7 Mining subsidence

During the extraction of minerals by underground mining, a void will be created as rock is removed. Some mining methods such as longwall mining of coal involve the intentional collapse of the roof of the void as mining progresses, Figure 10-11. This occurs contemporaneously with the mining operation. In other methods, such as room (or bord) and pillar extraction or more historical bell pits, the roof remains intact for some period following mining, but may eventually collapse as the rock strength gradually deteriorates, Figure 10-12.

A detailed description of underground mining methods, the nature of ground subsidence, and how subsidence is predicted is provided in reference [26].

The potential effect of mining induced subsidence on pipelines depends on whether the mining is currently active (or proposed), or whether it has taken place historically on the pipeline route.

In many areas of the world, the regulatory body that licenses mineral extraction will require the mine operator to notify all affected landowners and occupiers, including pipeline owners, of the intention to carry out mining likely to cause subsidence. Ideally, early contact is established between the mine operator and the pipeline operator in order to investigate whether the pipeline will be affected by subsidence and how the pipeline can be safeguarded.

For longwall mining, predictions of the subsidence magnitude and distribution generally involve empirical or semi-empirical factors which vary with geographic location[27,28].

Calculations of the pipeline response to ground subsidence involves analysis of a series of ground movement profiles that represent different stages of the extraction process. Predictions are more difficult in areas with complex near surface geology, where geological faults are located and where multi-seam working has previously taken place. The reliability of predictions can be improved by calibration against existing subsidence measurements in the same area.

As indicated in Figure 10-11, longwall mining subsidence results in areas of tension and compression in the ground. Compression loads are particularly damaging and can potentially result in axial local buckling, Figure 10-14.

In areas of historical mining by the room and pillar method, Figure 10-15, the likelihood of mining subsidence taking place needs to be investigated. This involves two key factors:

1. The depth of the mining; and,
2. The geology overlying the mined section, particularly the rock strength.

High-strength rock and increased mine depths frequently result in the choking of the void by arching, bulking of fragmented rock or the development of beam spans in strong rock units as the void attempts to migrate upwards, Figure 10-12.

Typically, a mine depth of 30 m has been used in the UK as a guide to whether subsidence will develop at the surface. Mines deeper than 30 m have been found to rarely develop subsidence problems at the surface, but the actual value will depend on the nature of the mining operation and the geology.

Desk study information covers historical mining plans, regional geology maps, and memoirs, boreholes, aerial and LiDAR imagery, Figure 10-16, and reported subsidence events in the area of interest. This is supplemented with field investigations of geology and subsidence features.

In areas where subsidence is likely and re-routing cannot remove the hazard, mine infilling by grout injection to fill the void is sometimes employed.

Experience of ground movements induced by a wide range of mining operations is reported in [29].

QR Code 10-6. Scan here for more on mining subsidence.

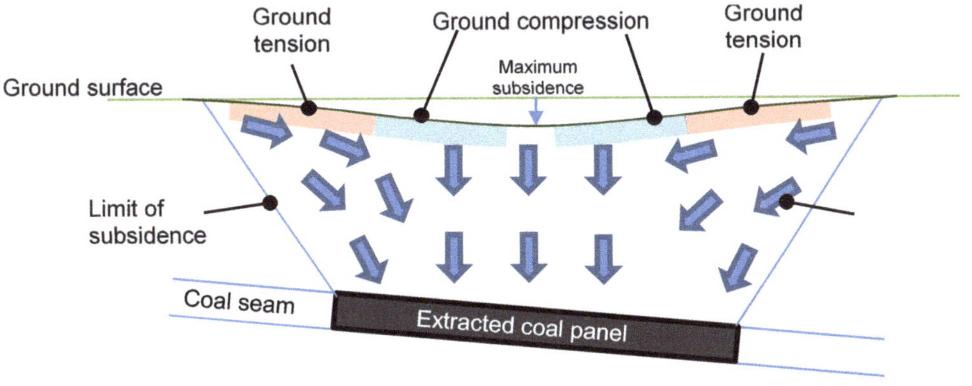

Figure 10-11. Sketch of mechanism of longwall mining subsidence showing areas of ground compression.

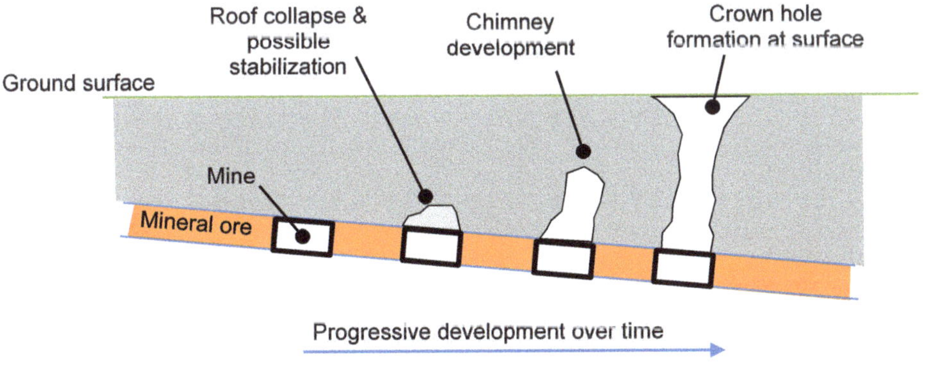

Figure 10-12. Sketch of crown hole formation for historical mining by room and pillar method.

Figure 10-13. Developing crown hole from shallow coal mining adjacent to pipeline right of way in the UK.

Figure 10-14. Axial compressional buckle due to mining subsidence. Credit: Encyclopedia of Pipeline Defects Third Edition, ROSEN Group, Clarion Technical Publishers, Tiratsoo Technical, 2017.

Figure 10-15. Subsidence from coal mining carried out up to the 1920s near Sheridan,Wyoming. Credit: U.S. Geological Survey Department of the Interior/USGS U.S. Geological Survey.

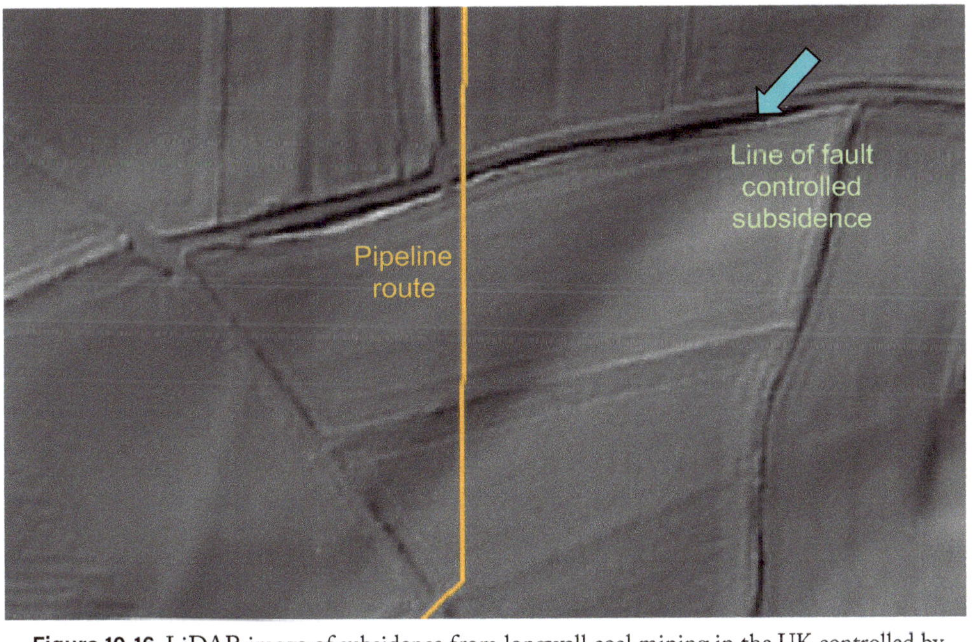

Figure 10-16. LiDAR image of subsidence from longwall coal mining in the UK controlled by structural faulting crossing a pipeline route.

10.2.8 Other geohazards

Other geohazards include lateral spreading and erosion.

QR Code 10-7. Scan here for more on lateral spreading.

QR Code 10-8. Scan here for more on onshore soil erosion.

10.3 Pipeline failure from geohazards

10.3.1 Historical data on failures

The frequency of pipeline failures from geohazards is not easy to establish. In many circumstances, incidents are not widely-publicized; however, an interesting compilation of failure statistics has been gathered by Sweeney[30], Figure 10-17. The data has been plotted from different regions as a frequency per 1000 km of pipeline length per year.

The key points are:

1. The European (general) data involves all terrain and covers many thousands of kilometer years of operation.
2. Most landslides occur in hilly or mountainous terrain; therefore, the frequency of pipeline failures is significantly higher in these areas.

    - The rate in European mountains is 40 times higher than the European general rate.
    - The rate in USA mountains is 16 times higher than the Europe general rate.

3. Data from South America for older pipelines shows a much higher failure frequency: approximately two orders of magnitude higher than the European general rate.

4. The mountains (new) data covers both South American and European experience, where geohazard management has been applied.

Although the sample size is restricted, the study provides a very useful illustration of the influence of terrain and modern construction practice on pipeline failure frequency.

Sweeney[30] has illustrated the benefit of geohazard management. The data from the South American Andes (in Figure 10-17) is shown in Table 10-4 and essentially applies to pipelines where no formal geohazard management has been applied. The high failure rate reflects older, more traditional, practice such as routeing in predominantly straight lines with limited or no specific assessment of hazards (mainly landslides) on the pipeline route.

Table 10-4. Landslide failure frequencies related to geohazard management in mountainous terrain.

Geohazard management	Service (Km Years)	Ruptures	Failure frequency (No/1000 km / year)
None	30,800	65	2.11
Implemented following failures	21,540	12	0.56
Geotechnical input from design	1,378	3	0.17
Full geohazard management	1,214	0	0

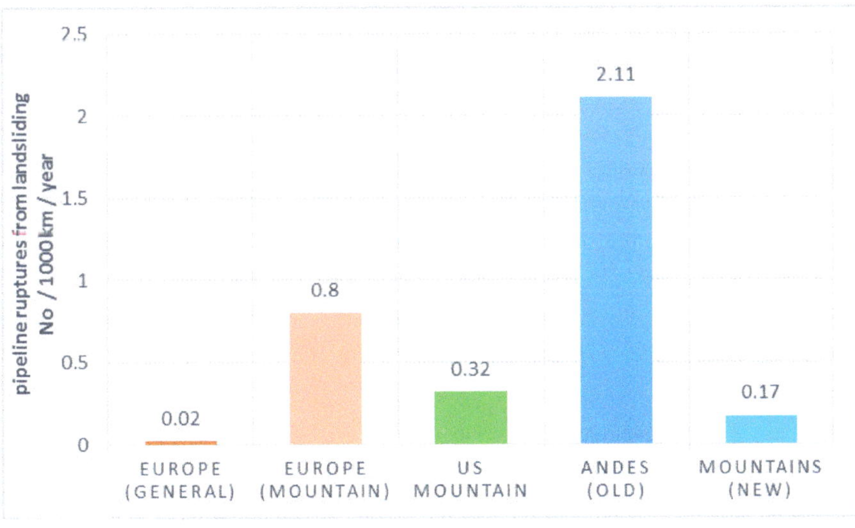

Figure 10-17. Pipeline failure frequencies from landsliding in mainly mountainous terrain (after Sweeney[30]).

A geohazard management program was implemented on some of the pipelines where failures had previously occurred. The reduction in the failure rate on these pipelines as a result of addressing landslide hazards is clearly demonstrated (Table 10-4); however, it is noted that the rate remains high, and reflects the influence of a poorly selected route. Modern practice with geotechnical input from the design stage and also during operation, results in comparatively low failure frequencies. Full geohazard management involves geoteams and geohazard discipline specialists, and this demonstrates significant benefit to date, Table 10-4. These data include some European experience.

10.3.2 Modes of pipeline failure

When geohazards develop on pipelines, action is necessary to understand the level of threat, whether the pipeline has been adversely affected, and what should be done to protect the pipeline from further loading. In some cases, a re-route around the hazard is considered the safest option, and this may be carried out as a matter of urgency. This would typically apply for hazards that develop rapidly, or where interruptions to operation are exceptionally severe. However, in many situations the condition of the pipeline is investigated, and the proximity to the limit state (failure) of the pipeline is established.

A key step in determining the scope of further actions is recognizing the nature of loading that develops in a pipeline as a result of displacement or external loading:

- Partial exposure of a pipe from erosion may not result in earth loads, but could preempt larger scale instability or leave the pipe vulnerable to impact;
- Landslide movement over a short length of pipeline may not cause an appreciable increase in loading for a large diameter pipeline, but could cause failure of a small diameter pipeline;
- Exposure in a river may expose the pipeline to hydrodynamic loads that could result in rapid failure;
- Buried pipelines that are subjected to increasing axial or moment loads from geohazards will ultimately fail by either tensile fracture or local and axial buckling.

10.3.3 Tensile strain capacity of a pipeline

The tensile strain capacity of pipelines is generally governed by the properties of the pipeline's girth welds. This is because welds may contain flaws or imperfections:

- These flaws may lie within the acceptance limits of the design standard but could also be larger than permitted values, due to inadequate inspection or sizing methods;

- The flaws in a weld may grow in size due to stress cycling from changes in operational loads;
- Girth welds may have lower fracture toughness than the pipeline material; and,
- In some circumstances, particularly for high strength steels, the weld may contain weak or softened regions.

10.3.4 Compressive strain capacity of a pipeline

The compressive strain capacity of pipelines is normally defined as the strain at which the maximum moment develops under bending loads, or at which the maximum axial force develops under axial loads. The point of pipeline failure under compressive loads is not clearly defined because failure is normally associated with the loss of containment. Strains beyond the maximum load capacity will result in the onset of local buckling of the pipe wall, and not the loss of product. The loss of containment only occurs when the deformation of the pipe wall reaches severe levels where tensile fracture develops in large bulges.

10.4 Geohazard management – design

10.4.1 Geohazard management

A new ISO standard on geohazard management[31], sets out requirements to identify, evaluate and manage geohazards. The responsibility of the plan to control hazards lies with the operator, and applies from the design phase and throughout the operational life. It is intended to apply retrospectively to existing pipelines. An operator who does not consider a management plan is necessary must provide documentary evidence.

The three main elements of the geohazard management plan are:

1. Identification;
2. Evaluation; and,
3. Mitigation.

An additional, or fourth, element involves the long-term management of the hazard, although this can be included in mitigation. This covers the continued monitoring of the hazard, and implementation of mitigation activities as required. In some cases, periodic re-evaluation of the hazard is required.

An illustration of the application of the geohazard plan through the various stages of the pipeline life is illustrated in Figure 10-18.

10.4.2 Recognising hazards

The variation in hazards is large and the causal factors are wide-ranging. This is illustrated for landslides in Table 10-2.

It is not easy to recognize and understand all the factors that are currently contributing to an active hazard or that may develop under certain conditions at some point in the future. This requires broad expertise in many disciplines which is the basis of the team responsible for managing these hazards – the 'geoteam'. In addition, there is a requirement for discipline specialists to be involved; for example, in landslide risk assessment,[32]. This will particularly apply to:

- Pipelines routed across complex terrains or geological environments;
- At locations of significant vulnerability to geohazard features; and,
- For major strategic pipeline projects.

It is not common practice to undertake detailed evaluations at the location of individual geohazards located on the proposed route of a pipeline. The normal approach is to avoid these locations, but some constraints are unavoidable; for example, when crossing major geological faults. In these circumstances a detailed study is necessary to reduce the likelihood of future problems.

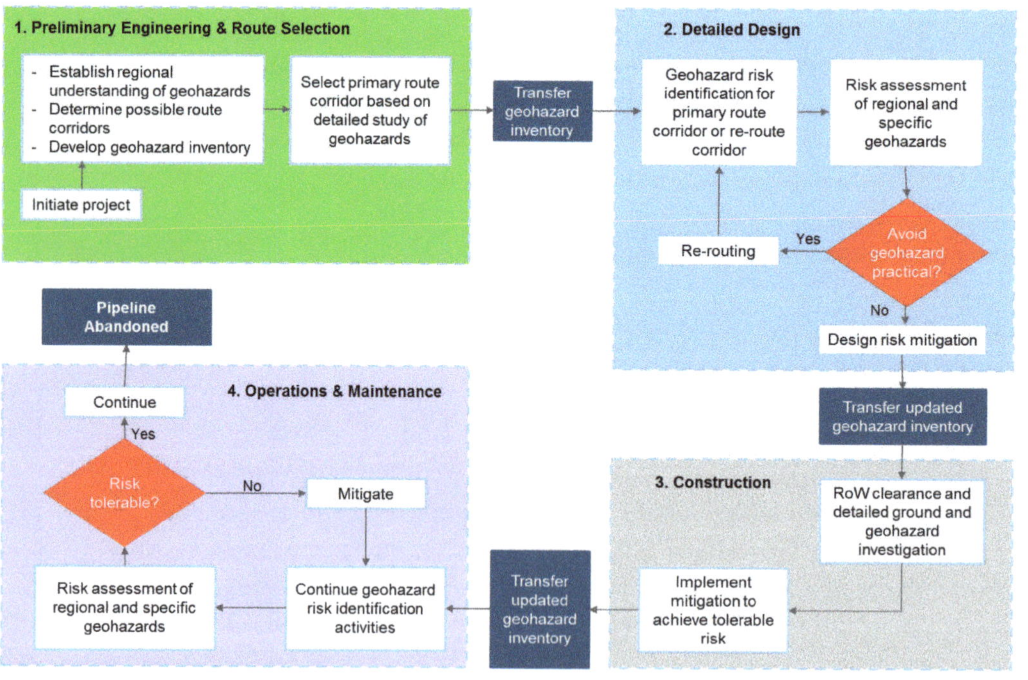

Figure 10-18. Principal elements of the geohazard management plan.

Ascertaining the significance and requirements of individual locations or sections of the pipeline route is a key skill and requires experience. An over-cautious approach may result in unnecessary avoidance of perceived obstacles or hazards with a resultant increase in construction costs. Conversely, and probably the more usual case borne out by experience of operational incidents, is the failure to recognize current or potential future hazards. This can lead to significant operational costs for mitigation works, or worse, a failure incident with costs that can be orders of magnitude higher.

Successful pipeline construction, particularly for larger, strategic pipelines in hilly or mountainous terrain, invariably involves specialists in routeing and in the identification and assessment of landslide, erosion, and river hazards.

Equally, the assessment of fault movement hazards is a complex and specialist activity that requires experienced practitioners.

10.4.3 Assessment process

The key to recognizing geohazards and selecting the optimal route to minimize their impact on a new pipeline is to evaluate the terrain.

Conventional guidance on pipeline routeing [e.g. 33] tends to be based on experience in developed or managed rural land with minimal consideration of geohazards. However, valuable guidance on terrain evaluation methods is provided by a set of conference papers from 2004[34], and experience from North America[35,36].

The process requires experienced specialists and involves the following aspects[37]:

1. Recognize, avoid, or minimize the risks at the earliest possible stage of the project;
2. Identify the issues that may add the most significant costs to construction;
3. Minimize the impact of construction to the environment, particularly ground stability, and erosion; and,
4. Prevent the need for expensive remedial measures during operation.

The integration of terrain evaluation with the development of the different stages of pipeline design and construction is explained in Table 10-5.

A decision chart for the selection of route corridor from alternatives in the feasibility stage is shown in Figure 10-19.

The evaluation of the geohazard is based on qualitative, semi-quantitative, and quantitative risk assessment methods. Selection of the appropriate assessment method will depend on the level of quantification necessary to determine whether the risk is tolerable and the level of detailed information available. The tolerability of the risk is typically set by regulations, the operator, industry standards, or best practice.

A summary of the three main risk assessment methods, with benefits and drawbacks, is summarized in Table 10-6. Factors that should be considered in the selection of the method are:

- What is the purpose of the evaluation, and how significant is the decision that is based on the outcome?
- What is the hazard, how big is the study area, and over what time period?
- What resources are available?
- What data is available?
- What consequences may arise from any adverse effects?
- What are the uncertainties and assumptions?

Qualitative methods may be used early in the design process to compare the basic exposure to geohazards on alternative routes. More detailed semi-quantitative methods may be necessary where routeing constraints are high and mitigation costs expensive.

The use of risk assessment methods to aid the selection of route corridors is demonstrated by Lee and Charman[38]. The application of risk assessment methods in the assessment of landslides is covered in detail by Lee and Jones[32].

The influence of the geo-team is most significant at the reconnaissance and feasibility stage. Early involvement of experienced specialists during the selection and refinement of the route corridor has a major benefit on the reduction in the risk.

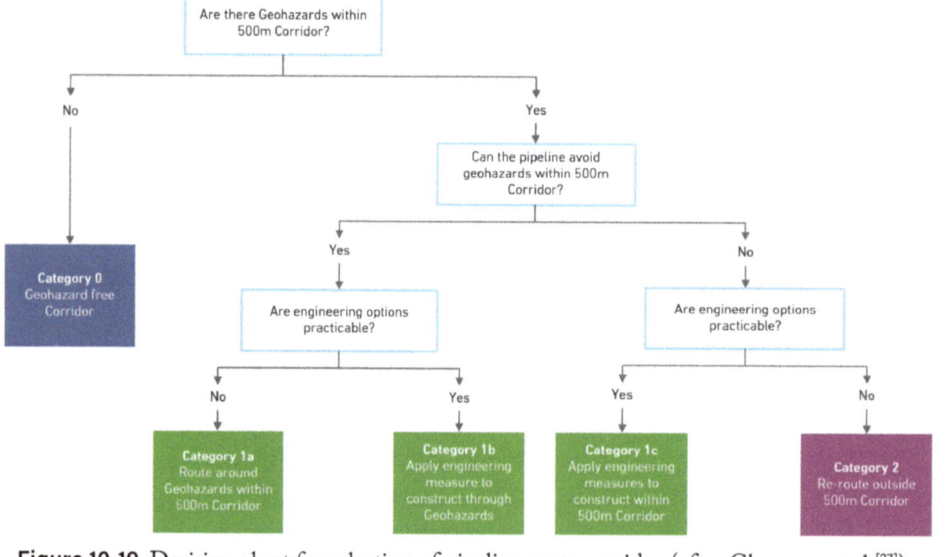

Figure 10-19. Decision chart for selection of pipeline route corridor (after Charman et al.[37]).

Table 10-5. Terrain evaluation in the main stages of pipeline design and construction (after Charman, et al.[37]).

Stage	Objective	Main Content	Resource
Reconnaissance	Identify potential hazards and impact on selected routeing corridors.	Desk study of mapping and remote sensing. Develop initial terrain models.	Geo-team of specialists
Feasibility	Selection of initial pipeline corridor (10 km wide)	Site visits and refine terrain models. Prepare scope of future investigations.	Geo-team of specialists
Preliminary design (front end engineering design ('FEED')	Route narrowing to 500 m	Detailed mapping of specific areas. Refine scope of detailed investigations	Individual specialists for specific areas and hazard types.
Detailed design	Route definition (100 m)	Ground investigations including boreholes, trial pits, in-situ tests, laboratory tests.	Geotechnical supervision, with specialist input as required.

Table 10-6. Comparison of the three main risk assessment methods.

Method	Scope	Advantages	Disadvantages
Qualitative	Hazard probability and consequences expressed in qualitative terms. Risk normally assigned to classes in matrix format.	Based on observation, experience and inference. Enables a basic comparison of different levels of risk.	The method is subjective. Expert judgment is required to execute assessments and reduce inconsistency between comparisons.
Semi-quantitative	Numerical estimates of the relative significance of the key parameters that control the geohazard mechanism and assigning weighting factors. Threat and consequence classes are similar to a qualitative matrix, but with numerical ranges.	More robust and less subjective than qualitative methods. Systematic classification of key factors for simple and rapid evaluation of the geohazard. Provides screening and rioritization of risk. Can be adapted to suit GIS data collection and processing methods.	Typically applies to specific regions and not easily transferrable across diverse geographic or geological environments. Expert judgement is required to ensure consistency between assessments.
Quantitative	Engineering and statistical models are used to quantify the likelihood and consequence of an event.	Perceived as objective and facilitating rigorous comparison of relative risk. Robust approach to reduce uncertainty and increase confidence in assessment results for decision making.	Very demanding in terms of data definition and analysis. The significance or weakness of the assumptions in the mechanism or data may be unclear or concealed within the analysis. Some hazard types cannot be effectively quantified using existing technology/ science given the timellnes and constraints of pipeline projects.

10.4.4 The inventory

This is a listing of all geohazards that affect a pipeline right-of-way either directly or could potentially develop in the future. The data is compiled in standard templates from aerial photography, remote sensing, field observations, or historical records. The records are readily stored and maintained in GIS systems to enable evaluation of geographic position and extent in relation to the pipeline route.

The inventory is updated as new information is gathered during site evaluations during design, construction or in-service.

An example of information that should be recorded is provided in Rizkalla[2]. The use of inventories is demonstrated in reference[39]. Inventories are also used in landslide susceptibility mapping[40], Section 10.4.6.

10.4.5 LiDAR and aerial imagery

Remote sensing is a key part of terrain evaluation. LiDAR (Light Detection and Ranging) provides high density point mapping of the elevation of the ground surface and can penetrate through tree cover and surface vegetation. This provides excellent information on the presence or activity of surface processes such as landslides and subsidence that aids routeing decisions, Figure 10-20.

Air Photo Interpretation (API) is also an important source of data in terrain evaluation[41] for observing the shape of the ground and predicting the soil type and behaviour. This includes slope angle and shape, presence of rock, river and stream characteristics, gullying, soil erosion, landsliding, and areas of poor drainage.

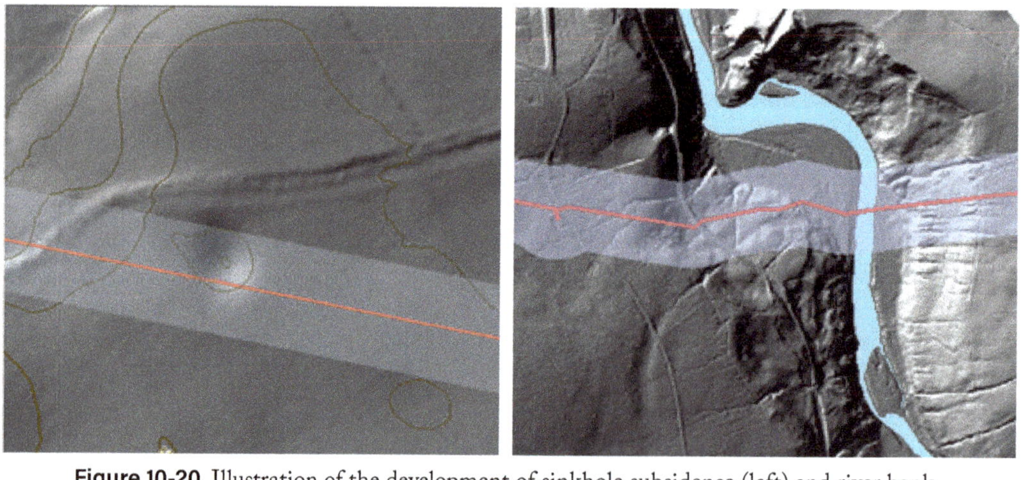

Figure 10-20. Illustration of the development of sinkhole subsidence (left) and river bank instability adjacent to pipeline routes from LiDAR imagery.

Successful interpretation of aerial imagery depends on the skill and experience of the analyst but it is not infallible and there are limitations on detecting what may prove to be important features; consequently, ground checking is a necessary part of the interpretation,[42]. This was demonstrated in a study that showed 50% of landslide features subsequently mapped by field investigations were not identified in the aerial imagery[43].

Active ground movement hazards can be identified from changes between successive LiDAR surveys[44,45]

10.4.6 Susceptibility mapping

Existing geohazards are identified by minimising their influence on the terrain or through the evaluation of the subsurface geology. Features may be subtle or largely masked by land use activities so that the skill of experienced practitioners is required to interpret the hazards, Section 10.4.2.

It is more difficult to identify where ground movement may develop for the first time during the operational life of the pipeline in apparently stable ground. This can be investigated and defined in a process of susceptibility mapping that identifies the potential factors that could lead to ground movement. In most circumstances, first time failures can be rapid, meaning that there is little opportunity to carry out intervention or mitigation works before the pipeline is overloaded. Consequently it is important to understand the level of hazard and implement actions to minimize or manage the threat to the pipeline.

Liquefaction and subsidence susceptibility mapping are briefly covered in Sections 10.2.5 and 10.2.7.

Landslide susceptibility mapping involves:

1. Examining the distribution of existing, or past, landslides to identify the potential location of future slope instability;
2. Identifying and mapping the geographic extent of the controlling factors that have caused existing or historical landslides; and,
3. Evaluating key controlling factors that cause landslides in isolation such as soil /rock type or slope angle.

The two methods commonly used are[32]:

1. *Direct mapping*
   This links the presence of landslides to the geological and geomorphological setting based on field observations. The data sources include:

   a. Landslide inventories;

b. Evaluation of the hazard site geomorphology and extension to neighboring areas; and,

c. Weighted mapping based on identification of the factors that control landsliding in distinct geological, topographic or climatic terrains, Figure 10-21.

2. *Indirect mapping*

This involves the statistical analysis of a large number of parameters that are considered to control landsliding in selected regions. The resultant algorithms enable the prediction of landslide likelihood across the region, based on the distribution of the input parameters.

The proliferation of GIS tools has facilitated the development of susceptibility mapping methods. Despite this, it is not possible to short cut the inputs and these must be reliably defined by geotechnical engineers and geomorphologists who are familiar with, and understand, the process.

Where susceptibility mapping can be linked to landslide frequencies, these can be combined with pipeline structural resistance factors to derive estimates of pipeline failure frequencies. This process is illustrated in UK pipeline standards BS 8010 part 4 and IGEM/ TD/2, but these documents apply to UK conditions only.

Using this method, it is possible to subdivide the pipeline length into segments of uniform failure risk (or probability of failure). A hazard or risk profile along the pipeline is created, that enables inspection, evaluation, or mitigation activities to be concentrated at the locations of highest risk, Figure 10-22.

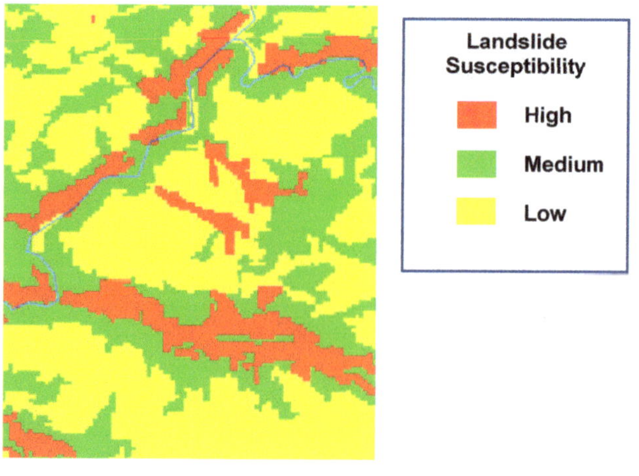

Figure 10-21. An example of landslide susceptibility using direct mapping methods in hilly terrain.

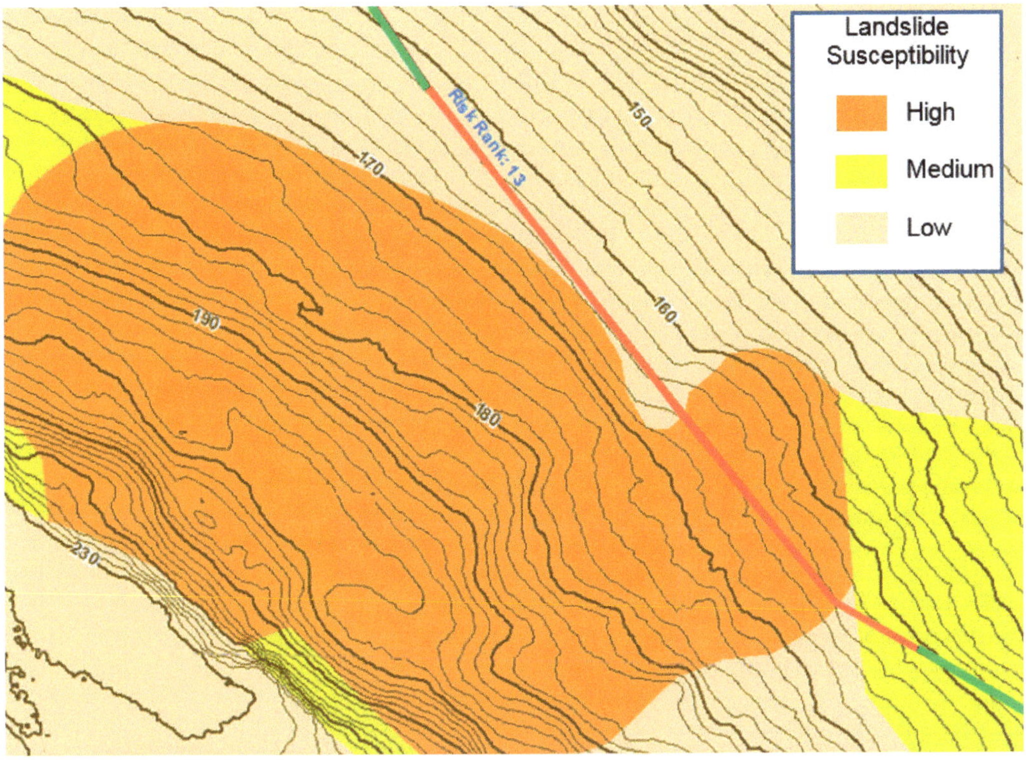

Figure 10-22. Illustration of landslide susceptibility on a pipeline route showing the segmented risk profile and ranked position for the selected network.

10.4.7 Deficiencies and consequences

Pipeline failures due to geohazards can be costly. The loss of product, loss of through-put, clean-up, and compensation all contribute. In addition, the terrain may be remote and difficult to access, slowing down the repair process, and increasing the interruption of supply costs.

There is a direct conflict between the objectives during the design and construction stages and the legacy left for operation. The contractual and financial pressure to complete the pipeline's construction and commissioning means that construction schedules can be severely compressed.

There were four ruptures in the first 30 months of operation on the Camisea pipeline in Peru during 2004 to 2007[46]. Three key issues were identified:

1. Landslides were not recognized as the dominant risk, and these should have influenced the selection of the final route;
2. The short project timescale with financial penalties meant there was insufficient time to undertake an appropriate level of evaluation of the geohazards; and,

3. The pipeline design did not incorporate measures to increase the strain capacity in areas prone to landsliding.

Nyman et al.[47] describe the pitfalls of not scheduling geohazard identification and characterization activities sufficiently early in the project cycle. Geological investigations should commence in the preliminary engineering stage (FEED).

Sweeney[48] presents a pragmatic assessment of the costs and benefits of geohazard investigations weighed against the costs of a failure. Using historical data of failure frequencies for mountainous terrain prone to landsliding, and typical costs of failures, it was calculated that a discounted loss of between \$300,000 and \$1.1 million may apply for each kilometer of a 200 km long pipeline based on one failure over a service life of 25 years. This indicates the level of investment in geohazard management that would accrue financial benefits provided it avoided one failure during the pipeline service life. The experience in Table 10-4 demonstrates the value in geohazard management programs in reducing and avoiding failures.

Pipeline construction involves significant earth moving, storage and disposal activities. The potential for adverse conditions to develop as a direct consequence of construction activities is high in terrain that is prone to instability. Examples of landsliding linked to pipeline construction activities are provided in [46, 49]. The creation of a conduit in the pipeline trench porting water to slopes and river banks and promoting instability is also a common problem.

The time pressures to get the pipeline in the ground means that hazards that are encountered during construction are not always addressed adequately or mitigated in full.

This is commonly paraphrased during construction as "Let's leave it for now, Operations can sort it out later" (after Sweeney [46]).

The experience presented in Table 10-4 suggests these are false economics.

10.5 Geohazard management – in-service

Geohazards such as landslides, or problems at river crossings, may develop during service. The likelihood of these is minimized if a robust design process has been followed; however, despite rigorous and detailed evaluation of the pipeline right-of-way at design, extreme or unexpected events leading to ground movement loading may occur. In some cases the feature is subtle and not identified during the routeing phase.

The content of a geohazard management plan based on an ISO standard has been outlined in Section 10.4.1. The three main elements of the process are described as:

1. Hazard identification;
2. Hazard evaluation; and,
3. Hazard control.

The standard requires that the management plan is also applied during the operational life of the pipeline.

A key difference between geohazards encountered during operation compared to design stage is the potential immediacy of the threat to the pipeline. Although avoidance may be a comparatively straightforward and low cost decision at design, diversion is a major decision with high cost implications during operation. Invariably it requires detailed justification, preferably before damage is sustained and the decision becomes inevitable.

The evaluation of the hazard and its current and future effect on the integrity of the pipeline is a fundamental element of the decision process. Assessment of the structural integrity of the pipeline as the pipeline becomes exposed to loading is a powerful method of demonstrating the need to control the hazard through monitoring or mitigation activities.

10.5.1 A geohazard management process during operation

An example geohazard management process that applies for operational pipelines is presented in Figure 10-23. This involves the three primary elements of hazard identification, evaluation, and control.

Once the presence of a hazard has been identified and characterised, the effect on the pipeline must be established. This primarily involves:

- Structural calculations; and,
- Available measurement data from field monitoring and inertial measurement units (IMU) inspection runs.

The objective is to define whether the pipeline is currently operating within recognised performance limits, and at what point it will exceed the acceptable limits as loads develop in the future.

This provides a decision framework to define and schedule actions on appropriate mitigation measures to protect the pipeline from the hazard.

Some hazards act so rapidly that once initiated, it is rarely possible to execute assessment work activities, and emergency protection or stabilisation measures are necessary. Geohazard susceptibility studies are central to successfully identifying and addressing these risks, Section 10.4.6.

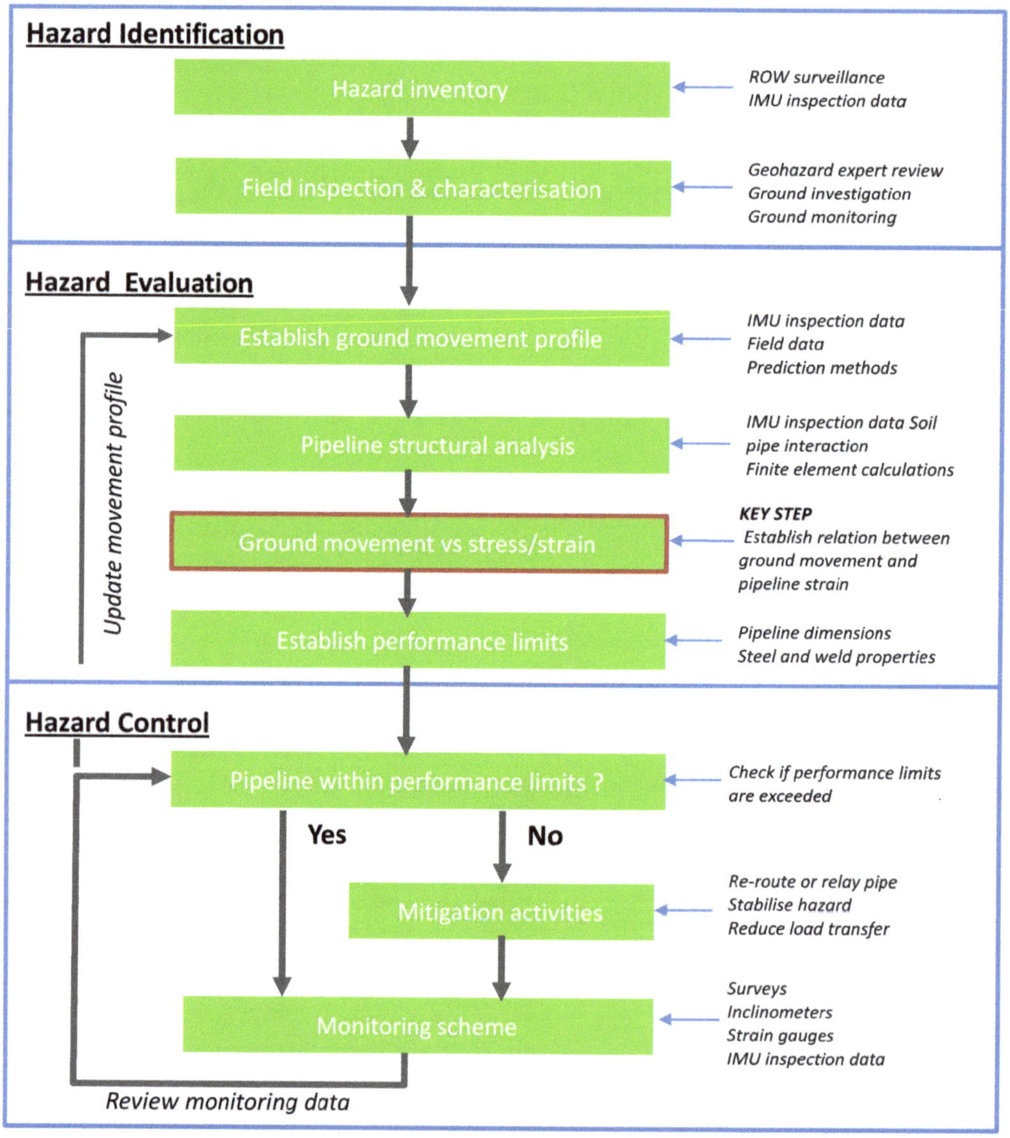

Figure 10-23. Example geohazard management process for pipelines during operation.

10.5.2 Geohazard detection

Detection during service typically relies on either the right-of-way surveillance, Figure 10-24, to identify signs of developing instability or measurement of displacement of the pipeline summarized trajectory by in-line inspection using tools fitted with inertial measurement units (IMU).

Circumstances which may warrant periodic inspection of the pipeline right-of-way by geohazard experts include:

1. An inventory of known hazards exists following construction, and these need periodic evaluation to determine any change in the threat level;
2. The right of way is in challenging terrain and hazards have previously developed during operation, so a detailed inspection is warranted to identify issues at an early stage;
3. The pipeline has had a history of failures, and requires a robust inspection regime to regain control of the hazards, and reduce the failure frequency.

Field visits by geohazard specialists normally apply for major strategic pipelines.

It is possible to develop a geohazard management plan based on in-line inspection services. The advantage of this is that mapping the pipeline provides factual information on the presence and nature of loading affecting the pipeline. The process is summarized in Figure 10-23.

Integration of structural analysis of the pipeline performance with the in-line inspection measurements forms the basis of the hazard evaluation section where the effect of the geohazard on the pipeline is established. As part of this process, the performance limits of the pipeline need to be identified, so that decisions on the scope and timing of intervention activities can be planned.

Figure 10-24. Identification of translational landslide on pipeline right of way, and surveillance by traditional methods.

Control of the hazard involves monitoring and mitigation activities to ensure that the pipeline remains within the acceptable performance limits. Monitoring may involve direct measurement of the loads in the pipeline using strain gauges, or indirect measurements of ground movement. The measurement of bending strain and pipeline movement can also be directly established from in-line inspection.

10.5.3 Geohazard detection using IMU

In-line inspection IMU tools have been used for many years to identify displacement of the pipeline that may be associated with geohazards[50].

These tools work by recording centreline angle and distance information along the pipeline route. This information provides a profile of curvature. Curvature associated with fitted or formed bends in the pipeline is readily identified, so that any remaining curvature represents bending developed during construction and installation, or from external loading sources such as geohazards.

Typical sources of construction curvature include:

- Elevation changes where the pipeline follows the ground topography;
- Horizontal misalignment at tie-in locations;
- Undersized or oversized bends; and,
- An uneven trench bed.

A key advantage of IMU inspection is that it provides direct evidence of loading effects on pipelines. In addition, it is possible to interpret the nature of the loading that has caused displacement of the pipe, and the measurement data can be used to optimise structural modelling of the pipeline.

An example of external loading is shown in Figure 10-25. This occurred when a major road was constructed over a pipeline in soft ground that resulted in ground settlement and displacement of the pipeline.

By running two inspections, changes in pipeline position can be detected. This provides a pipe movement assessment, and is a clear indication of the loading on a pipeline.

An example of pipe movement is shown in Figure 10-26. This shows movement associated with a landslide. There is a significant horizontal component of movement and the vertical movement is downslope. Horizontal out-of-straightness has developed. Review of aerial imagery shows that the pipeline crosses an active mudslide at this location.

Determination of pipe movement requires the comparison of two separate inspections. Where only one inspection is available, it is still possible to identify the presence of ground movement loading. This is illustrated for landslide loading in Figure 10-27 where horizontal changes in bending strain and out-of-straightness occur.

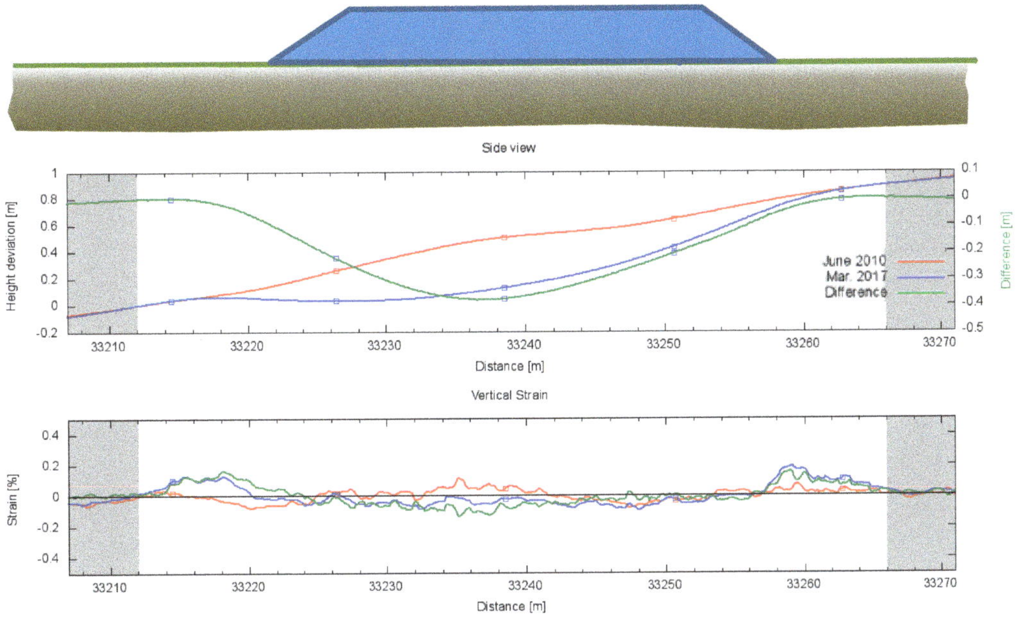

Figure 10-25. Illustration of pipe settlement from construction of new road crossing.

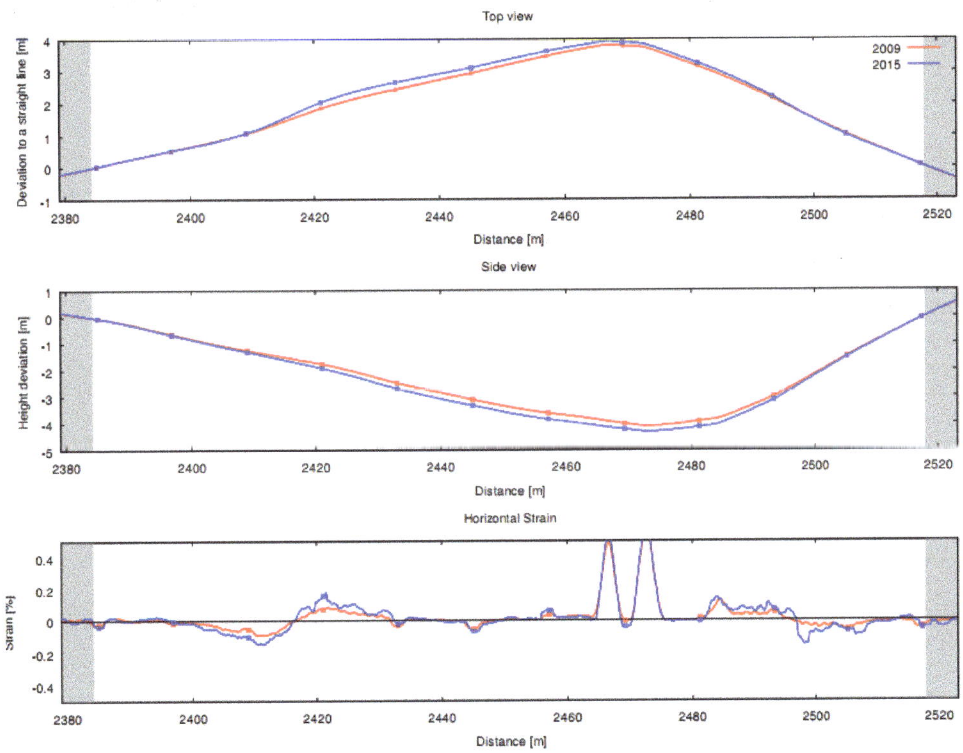

Figure 10-26. Detection of landslide movement from two consecutive IMU inspections.

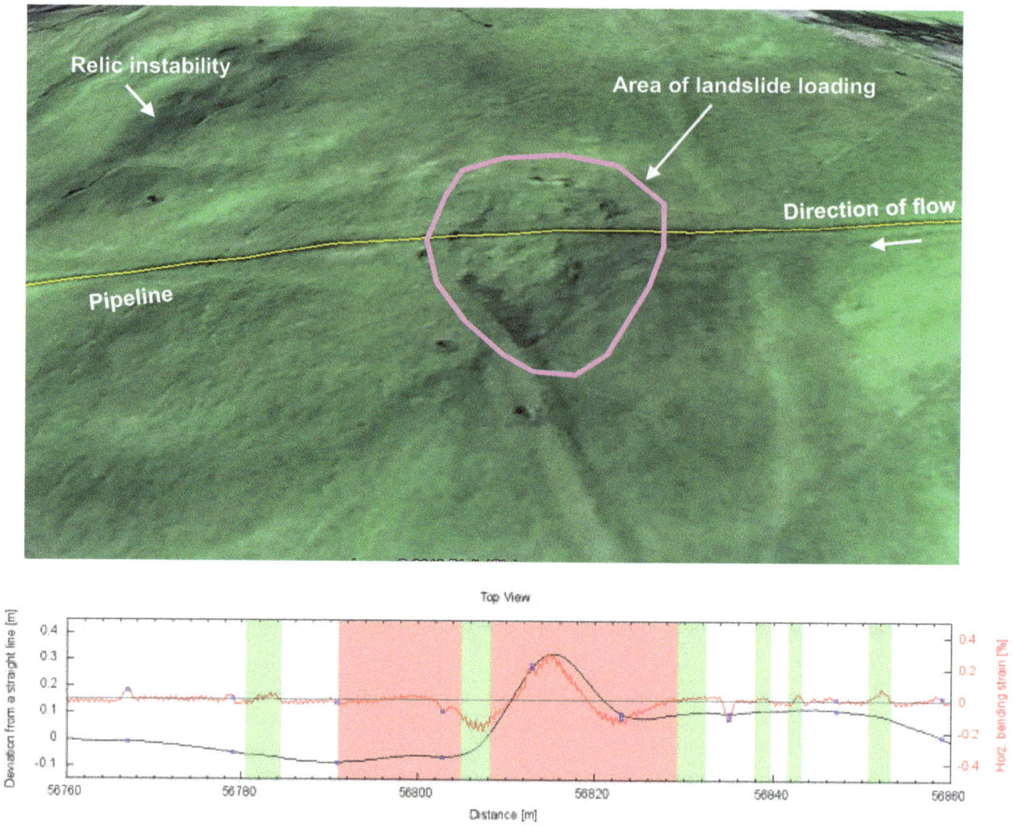

Figure 10-27. Detection of landslide movement from a single IMU inspection.

Measurement of the pipe cross-section geometry using an in line inspection 'caliper' tool (which can detect and size local deformation such as buckles) is an important additional source of data for evaluation of external loading problems. Compressional forces can lead to wrinkling or buckling of the pipe, and ovalisation can develop from moment loads or high soil pressures acting on the pipe.

An interesting case study illustrates the use of both IMU and caliper data on a landslide in British Columbia, Canada[51].

10.5.4 Other technologies

Interferometric Synthetic Aperture Radar (InSAR) involves satellite transmission of microwaves and measurement of the reflected signals from the earth's surface. Ground displacements can be interpreted from changes in repeat measurements with accuracies to centimetres or millimetres. The strengths and weaknesses of the method are illustrated in a case study of a pipeline in landslide[52].

Fiber optics are increasingly used to monitor geohazard events. See QR Code 10-9 for more information on pipeline monitoring.

QR Code 10-9. Scan here for more information on pipeline monitoring.

10.5.5 Evaluation of hazard

When geohazards are identified on the pipeline right-of-way, it is necessary to establish whether the pipeline has been adversely affected. An evaluation of pipeline integrity can be carried out on the basis of direct monitoring from strain gauges, or from using IMU data; however, these methods do not provide a complete picture of the pipe condition:

- Strain gauges are normally only applied once a need has been identified, so that loading may already have developed. Strain gauges only measure changes in pipeline load. It is good practice to determine the initial pipe loads at proposed strain gauge locations by performing stress measurements; for example, using the centre hole method[53].
- Strain gauges record loads at a point on the pipeline. A complete interpretation of moment and axial force loads on a pipe section requires three or more gauges to be fitted around the pipe circumference. The recommended locations are on straight pipe outside the influence of the girth weld or longitudinal seam weld. Gauges can be fitted on bends, but the localisation of loads due to geometrical effects will complicate the interpretation of the moment and axial force components. In these situations, it may be - by good fortune - that the highest load location is monitored due to the steep strain gradients and influence of fabrication effects[54].
- The IMU tool provides bending strain data only. The axial force is not measured. The axial force can be an important component, particularly if the ground movement has a significant component along the pipeline axis. Even where the loading is transverse to the pipeline, axial components become significant at higher displacements.

The most comprehensive assessment of the pipeline condition from ground movement loads is normally achieved by structural modelling because:

1. It incorporates the axial force component;
2. The behaviour of the complete pipeline geometry, including formed bends and elastic bends, is included;
3. Operational loads and the variation of these can be explicitly accounted for;

4. It enables prediction of the future condition of the pipeline as loading develops; and,
5. The influence of proposed mitigation works can be tested.

Establishing how the geohazard load develops in the pipeline with continued movement is a critical input for decisions over the scope and timing of monitoring and mitigation strategies. It can provide the justification for significant cost decisions such as re-routeing. This is highlighted as a key step in the geohazard management process during operation, Figure 10-23. An illustration of the link between ground movement and strain development in a pipeline as a result of landslide movement is shown in Figure 10-28.

An example of how mitigation works can be evaluated using structural modelling is shown in Figure 10-29.

In most cases detailed structural analysis is required to adequately reflect the geometrical and material complexity of the pipeline and geohazard loading.

Assessment of ground movement loads is normally conducted in terms of pipeline strain because the loading in the pipeline is primarily controlled by ground displacement. Ground displacement imposes a specific shape and therefore the strain distribution on the pipeline. The pipe shape and related strains are essentially unchanging unless further displacements take place. This enables the plastic reserve of the pipeline to be utilised.

Beam elements are generally adequate to determine the pipeline stress or strain state. There may be circumstances where a ring model is necessary to define the circumferential loading in the pipeline behaviour particularly where calculations for the cross-sectional deformation are required.

The four main elements of structural analysis involve:

1. Pipeline geometry;
2. Soil-pipe interaction;
3. Ground movement;
4. Pipeline performance limits.

The pipeline geometry can be defined from construction records or from mapping data using an in-line inspection IMU tool which provides a detailed post-construction profile.

Ground movement may be defined from distributed point measurements such as inclinometers, or survey monuments.

A key element of the ground movement is where the transition occurs between stable and moving ground[55], Figure 10-30. Where measurements are not sufficiently detailed, it is necessary to make conservative assumptions regarding the distribution of ground movements and invariably this involves a more abrupt transition.

A continuous profile of ground movement can be interpreted from the pipeline movement recorded by the IMU tool[56]. This requires an understanding of pressure-displacement behaviour of the soils that define the soil-pipe interaction and the nature of the ground movement. An accurate ground movement profile is achieved when the change in pipeline strain between the two IMU inspections that define pipe movement matches the strain predicted by the structural modelling Figure 10-31.

Once the structural model has been validated by the comparison of predicted and measured strains, it is considered to be a reliable representation of site conditions. The validated model provides confidence in the predictions of future pipeline integrity as ground movement continues.

QR Code 10-10. Scan here for more on soil-pipe interaction.

QR Code 10-11. Scan here for more on pipeline performance limits.

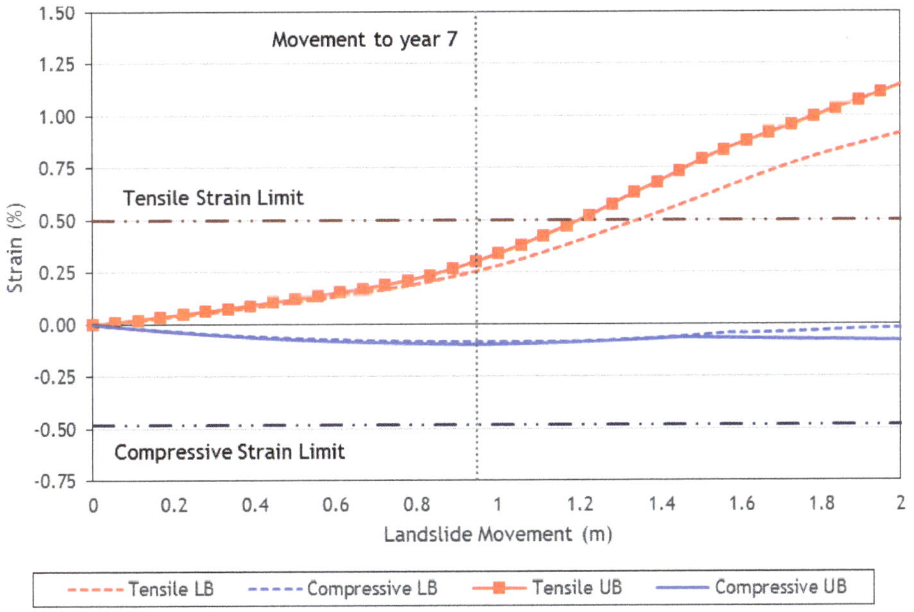

Figure 10-28. Ground movement vs pipeline strain plot.

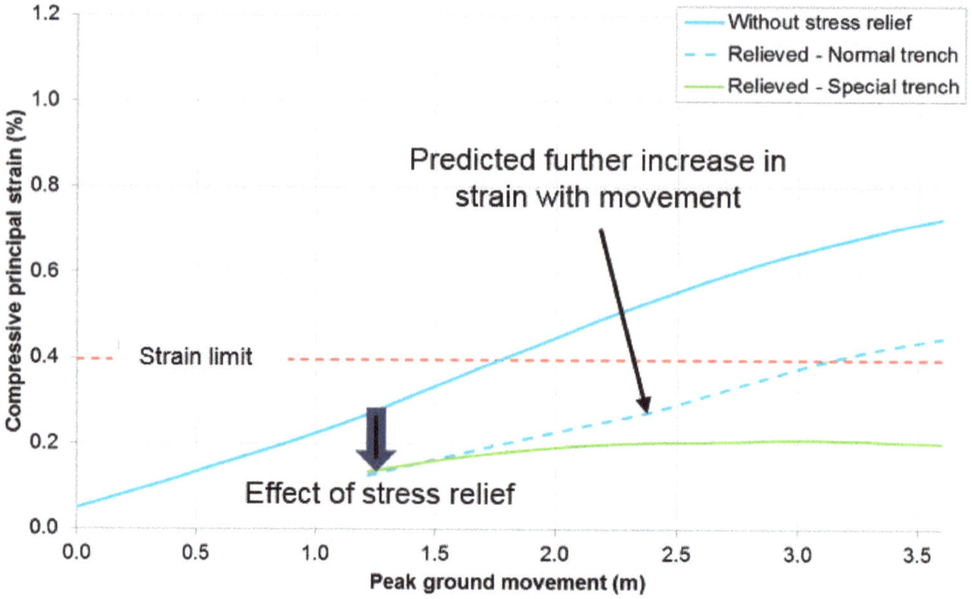

Figure 10-29. Influence of stress relief operation on pipeline strains.

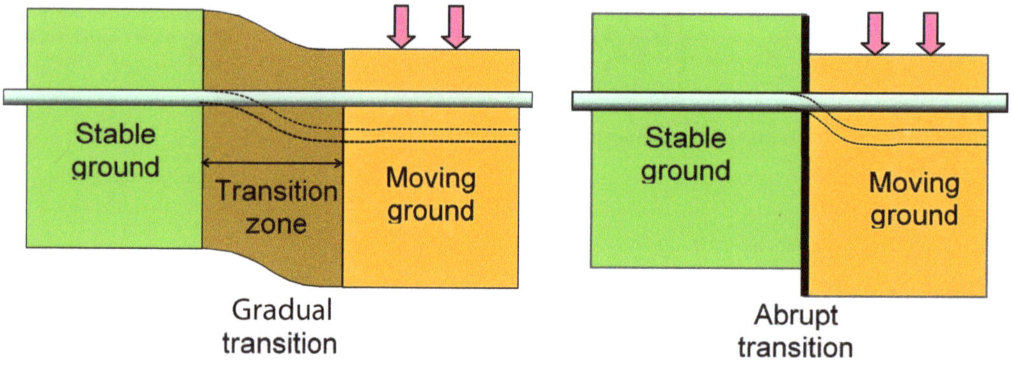

Figure 10-30. Transition zones in ground movement areas.

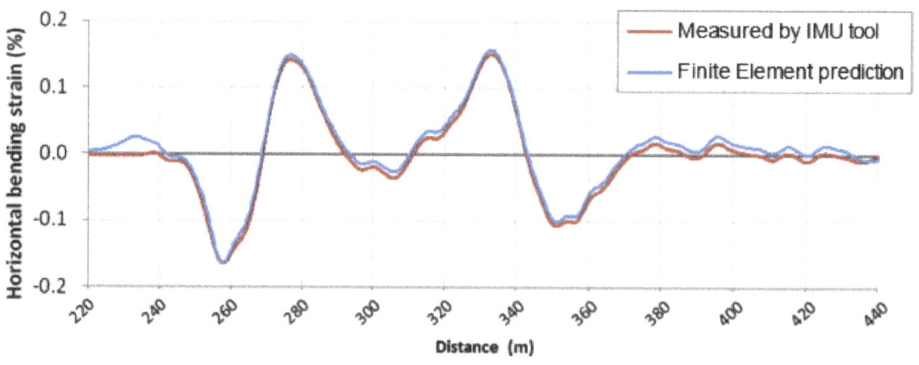

Figure 10-31. Comparison of measured and predicted bending strain change for a landslide area.

10.5.6 Hazard control

The strategy to protect pipelines from adverse geohazard loads is optimised through an understanding of the nature of the hazard, and how the loading on the pipeline develops with future movement. An example of this load development is presented in Figure 10-28.

Monitoring is carried out to confirm the magnitude, rate and profile of load development that is accumulating in the pipeline. This can either be direct monitoring of loads in the pipeline or by monitoring the ground movement and relating that to the load in the pipeline using the relationship such as demonstrated in Figure 10-28.

Many methods of geohazard monitoring have been outlined in Sections 10.5.2 and 10.5.5. The two most common methods of ground monitoring are:

- Slope inclinometers; and,
- Ground monuments or survey markers.

Other methods of monitoring ground movement include InSAR, LiDAR and fiber optics.

The common methods of direct monitoring of pipelines are:

- Strain gauges which can provide a continuous record of changes in strain in the pipeline although measurements are normally on a regular interval for example every 12 hours; and,
- IMU inspection which is periodic with measurements on intervals varying from a few months to several years.

As noted in Section 10.5.5, stress measurements are normally conducted in conjunction with the application of strain gauges. Fiber optics can also be used to monitor pipeline loads.

Monitoring data are used to update the relationship between the geohazard movement and the load in the pipeline. This is illustrated in the geohazard management plan for pipelines during operation, Figure 10-23.

For geohazard mitigation, the likelihood of pipeline failure due to geohazard threats is reduced by:

- Removing the threat: geohazards are normally avoided during the routeing process at design. Where routeing constraints apply (for example due to the proximity to population centres or geographical features such as steep terrain and narrow valleys) it may be possible to directionally drill, or tunnel, below the hazard.

- Reducing the threat: the development of loads on pipelines can be minimised by careful routeing and alignment. This will typically apply at the crossing of active faults where the design may aim to reduce compressional strain. Landslide hazards can be reduced by stabilising slopes through a variety of methods, but typically involves improving drainage, re-profiling the slope, reinforcement of the slope materials, or constructing earth retaining structures. Ridge and spur routeing in unstable terrain also reduces the threat of landslide hazards.
- Increasing the pipeline resistance: the structural resistance of the pipe to withstand loads can be increased by using a thicker wall pipe, or higher strength steel, Figure 10-32. In addition, higher ductility steels result in improved strain capacities, although properties of strength and ductility are normally inversely related.
- Reducing the soil loading on the pipeline: the stiffness and strength of the soil surrounding the pipeline has a dominant influence on the loads induced in the pipe by ground movement. Loading magnitudes can be reduced by designing a softer, weaker soil to surround the pipeline. This is achieved by:

  - Shallow burial;
  - Selecting special fill materials such as loose granular fill; and,
  - Modifying the pipeline trench geometry.

Ground movement loads on a pipeline can be reduced by stress relief, Figure 10-33. This involves trench excavation to remove the soil from around the pipeline so that it can recover any imposed elastic loading. It is important to separate the pipe from the bedding to maximise this recovery. The benefits of stress relief are illustrated in Figure 10-29.

QR Code 10-12. Scan here for more on inclinometers.

Figure 10-32. Replacement with thick wall high strength pipe in area of slope instability.

Figure 10-33. Stress relief operation on a pipeline in a landslide area. Strain gauges have been fitted to monitor the load reduction and subsequent increase in loading.

10.6 References

[1] Cooke, RU. and Doornkamp, JC. Geomorphology in Environmental Management. Second Edition. Clarendon Press, 1990.

[2] Rizkalla M., Read, RS. Pipeline Geohazards: Planning, Design, Construction and Operations. In Pipeline Geo-Environmental Design and Geohazard Management, Second Edition. American Society of Mechanical Engineers. 2019.

[3] Fookes, PG., Lee, EM. and Milligan, G. Geomorphology for Engineers. Whittles Publishing, CRC Press. 2005.

[4] Fookes, PG., Lee, EM. and Griffiths, JS. Engineering Geomorphology, Theory and Practice. Whittles Publishing, CRC Press. 2007.

[5] Guthrie, RH. and Cuervo, VA. Understanding Geohazards, Slopes, Rivers and Coastlines. SNC Lavalin. 2013.

[6] SGS. The Landslide Handbook – A Guide to Understanding Landslides. Circular 1325. U.S. Geological Survey. U.S Department of the Interior. 2008.

[7] Veldman, WM. Trenched and Elevated Crossings. In Pipeline Geohazards, Planning, Design, Construction Operation, ASME, 2019.

[8] Melville, BW. and Coleman, SE. Bridge Scour. Water Resources Publication. 2000.

[9] CIRIA Manual on scour at bridges and other hydraulic structures. Second Edition. C742. Construction Industry Research and Information Association. 2015

[10] Hearn, GJ. Slope Engineering for Mountain Roads. Geological Society Engineering Geology Special Publication No 24. Geological Society of London. 2011.

[11] Cruden, DM., and Varnes, DJ. Landslide Types and Processes. In Landslides – Investigation and Mitigation: Transportation Research Board (eds Turner, AK. and Schuster, RL), Special Report No. 247. National Research Council, National Academy Press.1996.

[12] Baumgard, A., Beaupre, M., and Leir, M. Implementing a quantitative geohazard frequency analysis framework as a component of risk assessment of new pipelines. Proceedings of the 2014 10th International Pipeline Conference. Paper IPC2014-33559. 2014.

[13] Wells, DL. and Coppersmith, KJ. New Empirical Relationships among Magnitude, Rupture Length, Rupture Width, Rupture Area and Surface Displacement. Bulletin of the Seismological Society of America. Vol. 84. No 4. 1994.

[14] Hengesh, JV. and Lettis, WR. Active Tectonic Environments and Seismic Hazards. In Geomorphology for Engineers (edited by Fookes, PG., Lee, EM. and Milligan, G.). Whittles Publishing CRC Press. 2005.

[15] Nyman, DJ., and Bouckovalas, G. Assessment and Mitigation of Seismic Geohazards for Pipelines. In Pipeline Geohazards, Planning, Design, Construction Operation, ASME, 2019.

[16] ASCE. Guidelines for the Seismic Design of Oil and Gas Pipeline Systems. Committee on Gas and Liquid Fuel Lifelines. American Society of Civil Engineers. 1984.

[17] Honegger, DG. and Nyman, DJ. Guidelines for the Seismic Design and Assessment of Natural Gas and Liquid Hydrocarbon Pipelines. PRCI Report No L51927. Pipeline Research Council International. 2004.

[18] Day, RW. Geotechnical Earthquake Engineering Handbook. Mcgraw Hill. 2002.

[19] Clayton, CRI. The Standard Penetration Test (SPT): Methods and Use. CIRIA Report 143. Construction Industry Research and Information Association. 1995.

[20] Seed, HB., Tokimatsu, K, Harder, LF., and Chung, RM. Influence of SPT Procedures in Soil Liquefaction Resistance Evaluations. ASCE J Geotech Eng 111(12), 1425-1445. 1985

[21] Waltham, AC., Bell, FG. and Culshaw, MG. Sinkholes and Subsidence. Karst and Cavernous Rocks in Engineering and Construction. Springer. 2005.

[22] Waltham, AC. Control the drainage: the gospel accorded to sinkholes. The Sixteenth Glossop Lecture. Quarterly Journal of Engineering Geology and Hydrogeology. Volume 49. 2016.

[23] Waltham, AC. Ground subsidence. Blackie. 1989.

[24] Culshaw, MG. and Waltham, AC. Natural and artificial cavities as ground engineering hazards. Quarterly Journal of Engineering Geology and Hydrogeology. Vol 20. 1987.

[25] Edmonds. CN. Karst and mining geohazards with particular reference to the Chalk outcrop, England. Quarterly Journal of Engineering Geology and Hydrogeology. Vol 41. 2008.

[26] Whittaker, BN. and Reddish, DJ. Subsidence Occurrence, Prediction and Control. Developments in Geotechnical Engineering, 56. Elsevier. 1989.

[27] National Coal Board. Subsidence Engineers Handbook. Mining Department. The Board. 1975.

[28] European Commission. Prediction and monitoring of subsidence hazards above coal mines. Research Fund for Coal and Steel. Contract No RFCR- CT-2007-00004. 2012.

[29] Geological Society Engineering Geology Special Publication No 5. Engineering Geology of Underground Movements. Edited by Bell, FG., Culshaw, MG., Cripps, JC. and Lovell, MA. Proceedings of the 23rd Annual Conference of the Engineering Group of the Geological Society. Nottingham. 1987.

[30] Sweeney, M. Terrain and geohazard challenges for remote region onshore pipelines: risk management, Geoteams and Ground Models. The Fifteenth Glossop Lecture. Quarterly Journal of Engineering Geology and Hydrogeology. Vol 50. 2017.

[31] ISO 20074. Petroleum and natural gas industry - Pipeline transportation systems – Geological hazard risk management for onshore pipelines.

[32] Lee, EM. and Jones, DKC. Landslide Risk Assessment. Second Edition. ICE Publishing. Institution of Civil Engineers. 2014.

[33] ASCE. Pipeline Route Selection for Rural and Cross-Country Pipelines. ASCE Manuals and Reports on Engineering Practice No 46. American Society of Civil Engineers. 1998.

[34] International Conference on Terrain and Geohazard Challenges Facing Onshore Oil and Gas Pipelines. Evaluation, Routing, Design Construction, Operation. Edited by Sweeney. M. Institution of Civil Engineers. 2005.

[35] Johnston, K., Waddington, B., Leir, M. and Kenny, C. Re-introducing the benefits of terrain mapping for pipeline routeing and design. Proceedings of the 2016 11th International Pipeline Conference. Paper IPC2016-64285. 2016.

[36] Leir, M., Baumgard, A., and Johnston, K. A staged approach for managing terrain and geohazards on new pipeline projects. Proceedings of the 2016 11th International Pipeline Conference. Paper IPC2016-64175. 2016.

[37] Charman, JH., Fookes, PG., Hengesh, JV., Lee, EM., Pollos-Pirallo, S., Shilston, DT. And Sweeney, M. Terrain ground conditions and geohazards: evaluation and implications for pipelines. International Conference on Terrain and Geohazard Challenges Facing Onshore Oil and Gas Pipelines. Evaluation, Routing, Design Construction, Operation. Edited by Sweeney. M. Institution of Civil Engineers. 2005.

[38] Lee, EM. and Charman, JH, Geohazards and Risk Assessment for Pipeline Route Selection. International Conference on Terrain and Geohazard Challenges Facing Onshore Oil and Gas Pipelines. Evaluation, Routing, Design Construction, Operation. Edited by Sweeney. M. Institution of Civil Engineers. 2005.

[39] Leir, M. and Reed, M. Natural hazard database application – a tool for pipeline decision makers. Proceedings of the 2002 4th International Pipeline Conference. Paper IPC2002-27103. 2002.

[40] PRCI Guidelines for constructing natural gas and liquid hydrocarbon pipelines through areas prone to landslide and subsidence hazards. C-CORE, D.G. Honegger Consulting, SSD, Inc. Pipeline Research Council International. 2009.

[41] Lawrence, C., Byard, R., and Beavan, P. Terrain Evaluation Manual. State of the Art Review 7. Transport Research Laboratory. Department of Transport. 1993.

[42] Fookes, PG., Dale, SG., and Land, JM. Some observations on a comparative aerial photograph interpretation of a landslipped area. Quarterly Journal of Engineering Geology, Vol 24. 1991.

[43] Hart, AB., Griffiths, JS. and Mathers, AE. Some limitations in the interpretation of vertical stereo photographic images for a landslide investigation. Quarterly Journal of Engineering Geology and Hydrogeology, Vol 42, 2009.

[44] Halchak, BE., Bell, J., and Dharmapuri, S. Landslide analysis using multi-temporal LiDAR data. LiDAR Magazine. Volume 7. No. 3. 2017.

[45] Mora, OE., Lenzano, MG., Toth, CK., Grejner-Brzezinska, DA., and Fayne, JV. Landslide change detection based on multi-temporal airborne LiDAR- derived DEMs. Geosciences. Volume 8, Issue 1, Paper 23. 2018.

[46] Lee, EM., Audibert, JME., Hengesh, JV. and Nyman, DJ. Landslide-relatd ruptures of the Camisea pipeline system, Peru. Quarterly Journal of Engineering Geology and Hydrogeology, Vol 42, p251–259. 2009.

[47] Nyman, DJ., Lee, EM., and Audibert, JME. Mitigating geohazards for international pipeline projects: challenges and lessons learned. Proceedings of IPC2008 7th International Pipeline Conference. Paper IPC2008-64405. 2002.

[48] Sweeney, M. Historic risks and modern responses. Terrain and geohazard challenges facing onshore oil and gas pipelines conference, 2004.

[49] Lee, EM., Fookes, PG., and Hart, AB. Landslide issues associated with oil and gas pipelines in mountainous terrain. Quarterly Journal of Engineering Geology and Hydrogeology, Vol 49, p125–131. 2016.

[50] Czyz, JA., and Adams, JR. Computation of pipeline-bending strains based on Geopig measurements. 6th International conference, Pipeline pigging and integrity monitoring, Houston, 1994.

[51] Czyz, JA., and McClarty, E. Prevention of pipeline failures in geotechnically unstable areas by monitoring with inertial and caliper in-line inspection. Proceedings of the Conference Corrosion, 2004.

[52] Guthrie, R., Reid, E., Richmond, J., Ghuman, P., and Cormier, Y. In SAR and the pipeline geohazards toolbox: Instructions for use as of 2018. Proceedings of the 2018 12th International Pipeline Conference. Paper IPC2018-78571. 2018.

[53] ASTM E 837-01, Standard Test Method for Determining Residual Stresses by the Hole drilling Strain-Gauge Method, 2001.

[54] Oldfield, T., Turner, T., and Young, A. Influence of Cold Formed Bending on Strain Based Design Buckling Limits. Proceedings of the 2016 11th International Pipeline Conference. Paper IPC2016-64342.2016.

[55] Laing, M., and Young, A. Recognising the influence of landslides transition zones in the assessment of pipeline integrity. Proceedings of the ASME 2017 International Pipeline Geotechnical Conference, Paper IPG2017-2512. 2017.

[56] Lockey, A., and Young, A. Predicting pipeline performance in geohazard areas using ILI mapping techniques. Proceedings of the 2012 International Pipeline Conference. Paper IPC2012-90496. 2012.

CHAPTER 11

Material and construction defects

Dr Brian Leis

11.1 Background and related history

Pipeline systems involve linear segments that connect the point of origin for the product to discrete facilities that process and drive the product toward delivery or storage. This chapter considers the linear segments that comprise pipes (a thin-walled cylinder or 'can'), which can be seamless (SMLS) or be closed axially by a longitudinal (long) seam that runs straight or helically, that are then joined end- to-end using girth welds. Many steps are involved in pipe-making, from preparation of the edges for welding through forming the can and making the long seam if used. The forming step for 24-inch (610 mm) diameter helical-seam pipe is presented as Figure 11-1. Further details for some types of pipe production can be found at QR Code 11-1 on the next page.

Many steps are also involved in constructing a pipeline. Figure 11-2 illustrates progress for the pipeline illustrated at which point the adjacent strings of pipe have been 'lowered-in' and the tie-in weld between them has been completed, with this section now ready to be bedded and padded to avoid the presence of rocks in the native backfill. Further details illustrating the steps for typical cross-country pipeline construction can be found at QR Code 11-2 on the next page.

Material and construction (M&C) defects are, as the words imply, defects that threaten pipeline integrity that trace to the materials that comprise the pipe and its construction into a pipeline system. The work done to prove the utility of hydrotesting in establishing the safe operating pressure of pipeline systems circa 1968[1] indicated that more than one-half of the defects failing the initial pressure test for new pipeline construction then involved features specific to pipe making or pipeline construction. A similar percentage of in-service failures likewise were traced to M&C defects.

Figure 11-1. Pyramid forming of helical pipe.

QR Code 11-1. Scan here for the steps involved in pipe making.

Figure 11-2. Tie-in made: ready for bedding.

QR Code 11-2. Scan here for details on pipeline construction.

> **Some definitions...**
>
> API 1163[2] defines a 'defect' as a physically examined anomaly with dimensions or characteristics that exceed acceptable limits that in its current use suffices to cause failure. An 'anomaly' is a deviation from the norm that remains unexamined. Similar terminology has been adopted by the Pipeline Operators Forum[3]. More generally, a 'feature' is a physical object detected by inspection technology.

But, while problematic materials and construction procedures were abandoned in the late 1960s, and others have greatly improved, M&C defects remain a significant threat to pipeline safety. This is evident through inspection of Table 1 in the Chapter herein titled Mechanical Damage, which lists the causes of casualties and fatalities for transmission pipelines in the US for the period from 1994 through 2012. It is apparent from that tabulation that M&C defects remain a major cause of casualties and fatalities – particularly for natural-gas pipelines. In part this reflects the fact that the benefits of improved higher-strength steels have been offset by issues in construction: for example, girth-weld alignment issues are magnified due to the thinner wall afforded by the use of higher-strength steel.

It follows that the material and construction threats addressed in this chapter remain an important focus for all involved with pipelines – from design and specification for new construction on through integrity management for existing systems. Following this background section, and an introduction specific to M&C defects, examples of a range of such defects will be presented and discussed relative to early construction, as well as for pipelines being designed and built today.

Given their significance as just outlined it is not surprising that understanding and characterizing M&C defects has been a focus of pipeline research and technology development (R&TD) since the early days of such work. The evolution of stronger steels coupled with increased demand for hydrocarbons and an aging infrastructure have kept these aspects in the crosshairs of the industry since. Important developments began in the late 1950s with that work being done under the auspices of the American Gas Association (AGA)[4], in what was dubbed AGA Project NG-18, which eventually morphed into the Pipeline Research Committee (PRC) of the A.G.A. Such work motivated the first AGA Symposium on Line Pipe Research (LPR), which was held in 1961, being followed since by eight others, all of which considered pipeline defects.[5]

As time passed work on the effects of pipeline defects also was initiated in Europe under the auspices of the European Pipeline Research Group (EPRG), which was organized in 1972. Collaborative seminars began between the PRC and the EPRG in 1975, with much of the early work in Europe being done at the British Gas Corporation (BG) at their Engineering Research Station (ERS)[6]. Over time it became clear that the R&TD program in Europe had scope and depth that paralleled that under the auspices of the A.G.A., such that the LPR Symposia gave way in 1996 to the ongoing joint seminars between the PRC and the EPRG. In the years since the transition to the PRC-EPRG Seminar, which since has been renamed the Joint Technical Meeting (JTM), defects of one type or another have remained a focal point at each meeting. As their causes were understood the work has transitioned to managing defect integrity via detection and sizing, and to methods to quantify their severity. QR Code 11-3 provides further details on this history and links to two of the organizations that were involved.

QR Code 11-3. Scan here for details concerning history and background regarding M&C defects.

11.2 Introduction

Based on the literature associated with the PRC-EPRG meetings noted above, the published literature, and file data, this chapter illustrates material and construction defects. Such defects can fail as the pipeline enters service, or they can grow to failure over time by mechanisms like corrosion, environmental assisted cracking (EAC), and fatigue. Their failure also can result from a random threat such as mechanical damage or outside forces acting on the defect. While these aspects can contribute to failure at material and construction defect sites, their role is dealt with more generally in other Chapters devoted to such topics. Accordingly, the focus here is the possible failure of defects specific to the line pipe and its construction into a pipeline system via plastic collapse or fracture. For present purposes material defects reflect the steel the line pipe is made from, as well as pipe-making including the longitudinal (long) seam if one is used. Some of the defects in the plate or skelp[1] noted in standards include features like blisters, hard spots, inclusions, laminations, laps, plug scores, scabs, slivers, and slugs[7,8][2]. While some among this list are no longer used for pipe manufactured to modern standards such as API 5L (e.g., scabs), most of the others remain plausible, even for mills producing PSL 2 pipe. The key difference for most modern production is that various types of quality control (QC) are in place during production to limit such concerns, with quality assurance (QA) following production to largely weed out what the QC did not address.

The practical significance of some features on this list, such as blisters or hard spots, is as dependent on their eventual service, as much as it is on the production process for the steel and its hot rolling. The key difference again is the use of QC during production and QA subsequently to detect what QC missed to the extent it is consequential.

---

1 Skelp is strip steel that is thinner than plate, which is received at the pipe mill as coiled stock. For related steel terminology see https://www3.epa.gov/ttn/chief/old/ap42/ch12/s051/reference/ref_03c02s04_2008.pdf

2 Reference 7 titled Standard on Imperfection Terminology defines the imperfections in steel and line pipe under the auspices of the American Petroleum Institute (API). Reference 8 currently titled Specification for Line Pipe and denoted API 5L considers some features that are not addressed in Reference 7,and specifies acceptable limits.

Collapse vs Fracture Control...

History indicates that fracture and plastic collapse are the two most prevalent mechanisms of pipeline failure. If the steel's toughness (fracture resistance) suffices then initially sharp defects blunt, with the limit pressure then controlled by the ultimate strength of the steel.Otherwise the initially sharp defects grow through-wall under fracture control. Further details follow later in this Chapter. QR Code 11-4 provides additional discussion.

QR Code 11-4. Scan here to read a Primer on Collapse-and Fracture-Controlled Failure.

It follows that most of what is discussed in this Chapter remains relevant for modern pipeline systems, which is evident in the still significant statistical presence of M&C defects as a cause of failure, as detailed above. Key in this context is the observation that the US incident statistics do not indicate older pipelines are more prone to incidents than is modern construction[9] – a trend that tends to hold across the categories reported. The reality in that context is that mechanized processes are still susceptible to upsets in the mills and on the spreads, so care must be taken to manage their potential presence.

With defects and anomalies as they have been defined, the above-noted 'usual list of suspects' also should include aspects that limit the steel's fitness for service regarding attributes like strength and weldability. While various forms of QC are in place during steel production, and QA is done to address what QC did not, the reality is that some cross-country pipelines recently have been built with steel that failed to meet the rudimentary strength requirements for the Grade specified[10][3]. Issues also have emerged for weldability.

Material defects for pipe also exist in connection with:

- plate or skelp edge preparation and pipe forming,
- completing the long seam if one is used,
- cutting pipe to length,
- cold expansion, and end-bevelling.

Such aspects will be considered in regard to modern seams made using double-submerged arc welding (DSAW) and electric resistance welding (ERW) specific to

3 As the scope here subsequently considers the sources and causes of defects it is noted that the cause of such pipe being recently built into several long-distance pipelines was among other factors the use of inadequate pipe acceptance requirements, discussion of which follows later (see References [10] and [11] for details).

preparation, edge-alignment, and process upsets. Because 'vintage'[4] pipelines continue to operate in several countries, some material defects unique to such line pipe also will be considered specific to low-frequency ERW seams, single-side submerged arc welds (SSAW), and lap welds.

Construction defects are features introduced in any manner as the pipe moves from the mill until it begins service. These kinds of defects can be introduced in connection with shipping and handling, staging and stringing, girth and tie-in welds (alignment, process upsets and control), fittings (involving valves, transitions, bends (related to fabrications and settlement / secondary loads)), road and river crossings (support/settlement at casings, flood scouring, directional drill / pull issues), and construction damage (coating and bedding/padding issues). While unique issues can occur in contrasting urban joint by joint construction versus the usual cross-country practices, these aspects are not considered.

Reference [6] identifies 20 types of defects related to the pipe steel/body, and another 23 (with some redundancy) associated with the long seam. This scope expands depending on the era and method of construction. Because the intent is to be comprehensive in balance with the other chapters, it follows that it is not possible to deal with all plausible features.

This chapter begins considering the sources and causes of defects. It then examines their possible failure via collapse or fracture to develop the means to address their integrity management based on detection and sizing and severity assessment. But because inspection, assessment, and integrity management are addressed in other chapters, this discussion of defects is specific to the nature of the defects, and their failure by collapse or fracture.

11.3 Sources and causes of material and construction defects

As the sources of many such defects emerge in the components of a pipeline system and the effects of associated secondary loads, it is appropriate to present some background. For those interested in these aspects, this information can be accessed through QR Code 11-5, which presents key aspects of pipeline systems and outlines possible secondary loads that could cause material and construction defects to fail.

QR Code 11-5. Scan here for a Components and Loads Primer.

4 The term vintage as used here and elsewhere (e.g., [9]) reflects construction from ~1940 to ~1970.

11.3.1 Material defects in the pipe body

This subsection opens with a review of the physical defects found in the plate or skelp. Images and discussion follow specific to laminations, blisters, hard spots, laps, plug scores, and slugs – as such features are known to the author to have caused failures either in pre-service hydrotests or delayed failures in service. All such features are known to have caused incidents in vintage systems. Such features either no longer occur or do so with lesser frequency and severity in modern construction. The subsection closes with discussion of pipe expansion and the major rehabilitation programs that were undertaken to put new construction back into service following issues that developed due to the inadvertent use of low-strength pipe.

Laminations. A lamination is a planar discontinuity in the plate or skelp that tends to form mid-wall and lie parallel with the surface, but also could be inclined to the pipe's surface, or be off-center in the wall. Such discontinuities form within bands of segregation rich inclusions or other impurities during solidification or hot rolling. While they are more prevalent in early dirty ingot production, continuous cast (concast) production is not immune to their formation.

As laminations are strong ultrasonic reflectors their presence in modern production should typically be identified by mill QC. Likewise, they should be identified by inspection focused along the edges of plate and skelp being prepared for welding. Any laminations missed by mill inspections that might pose a safety concern should be exposed in pre-service hydrotests. They could however lead to in-service concerns in construction prior to the routine or mandated use of high-pressure hydrotesting.

Figure 11-3 makes clear that expectations do not always play out in reality. The failure through the lamination shown in cross-section in Figure 11-3 occurred in steel and pipe produced in the late 1960s by an established integrated US mill. This in-service failure occurred at the maximum allowable operating pressure (MAOP) about one year after a pre-service hydrotest that targeted a pressure corresponding to the specified minimum yield stress (SMYS). The subject pipe steel easily satisfied all requirements of the then current API Specification 5LX.

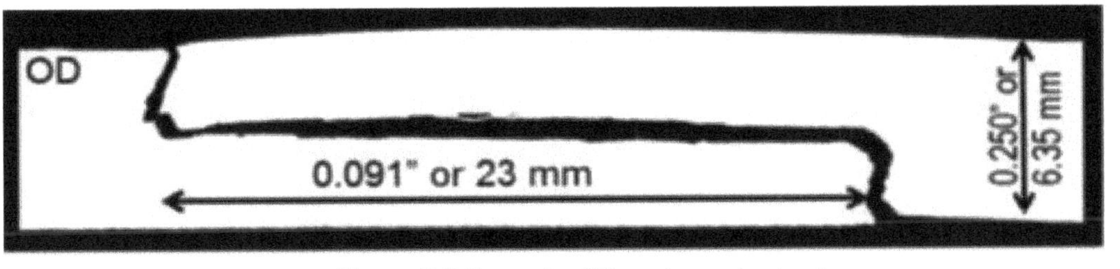

Figure 11-3. In-service failure along a lamination

Laminations that lie parallel to the pipe's surface do not reduce the cross-section and so in concept are not expected to fail. Failure occurred in this instance due to inclusions that as oriented here were distributed at the right end of the lamination. The weakness associated with these stringers grew stably into a crack-like feature during the high-pressure test, turning the path of the lamination toward the inner surface of the pipe. The resulting crack-like features continued to grow via stress-activated creep leading to delayed failure, apparently first for the inner ligament and subsequently for the exterior ligament, leading to failure about a year after the pressure test.

While laminations that lie parallel to the pipe surface can fail under unusual circumstances, they usually are not a problem because they do not cause a reduction in the pipe's cross-sectional area. Some laminations, however, can cause structural problems as they can attract hydrogen and so might lead to cracking and/or form blisters if they are in sour service or are in a hydrogen-charging environment. The inclusions associated with laminations can be problematic intersecting weldments: such circumstances are considered in the next subsection. Multiple laminations through the wall thickness as well as laminations that are inclined to the surface can form leak paths or cause ruptures. QR Code 11-6 illustrates how and why such features affect a reduction in wall thickness and can lead to rupture.

QR Code 11-6. Scan here for a discussion about Inclined Laminations.

Blisters. Blisters develop as pipe-body defects from features present in the plate or skelp as the pipe enters service. They can develop from laminations or inclusions, or any other hydrogen trap in the steel. Atomic hydrogen (H+) available on the inside pipe wall due to related corrosion caused by the transported product, or on the outside wall as a reaction product, is adsorbed and diffuses to sites where it can recombine to form molecular hydrogen, and thus becomes trapped. As the amount of trapped molecular hydrogen (H_2) increases, the volume of the cavity grows to accommodate it via stepwise planar delamination. This continues until delamination is resisted by sound steel, after which the blister eventually splits. Failure of the blister if large enough promotes failure of the remaining wall by collapse or fracture depending on the local pipe properties, either as a leak or a rupture.

Figure 11-4 presents images of cross-sections cut through the domes of axially adjacent blisters located in a pipe joint that failed in high- pressure hydrotesting through another larger blister.

This late 1960s construction was in service transporting natural gas that also contained hydrogen sulphide, water vapor, and carbon dioxide. The subject pipe steel easily satisfied all requirements of the then current API Specification 5LX. It is apparent from this figure that these blisters formed in heavy bands of segregated material determined to be stringers of nonmetallic inclusions. The blister that failed broke through the interior ligament and thereafter through the remaining ligament, much like the process evident in Figure 11-3. Blisters are easily detected using ultrasonic technology (UT) in-line inspection (ILI) tools, and so should not pose an integrity threat through use of timely inspections. Limiting external access to atomic hydrogen can be affected by managing the cathodic protection (CP) system, whereas internal access can be affected by better quality control of the transported product and or mitigation as needed.

Hard spots. As the name implies, a hard spot involves a significant local increase in the hardness of the steel that from their microstructure forms by localized quenching of uniformly hot steel. Hard spots are stronger than the surrounding pipe, and so do not yield to the same extent when the plate or skelp is formed into pipe, creating local flat spots. As such they are subject to a bending stress that is tensile on the outside diameter (OD) of the pipe, in addition to the usual membrane tension. Hard spots are prone to brittle fracture through the quenched (hardened) microstructure, which can be aggravated by the presence of adsorbed atomic hydrogen making them prone to failure by mechanisms like hydrogen-stress cracking. Issues with hard spots first became evident in the 1950s and have recurred since associated with mill-process upsets, as has recently been evident for steels produced using on-line accelerated cooling.

Figure 11-5 shows the fracture surface from one such failure, which in this case resulted in a leak in a pipeline that was constructed in the early 1950s.

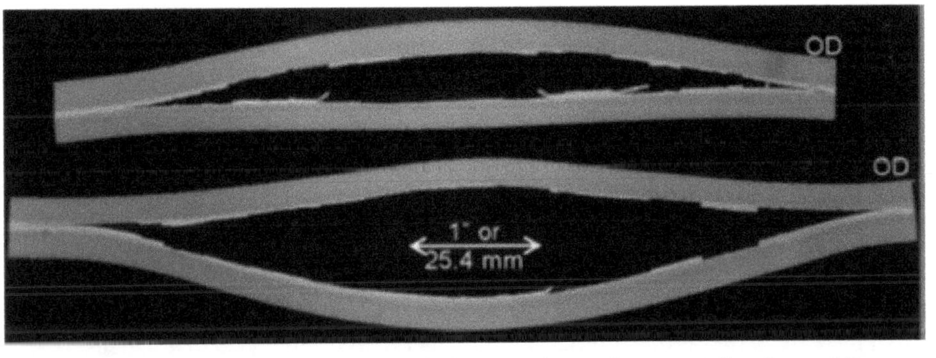

Figure 11-4. Cross-sections through the centerlines of intact axially adjacent blisters.

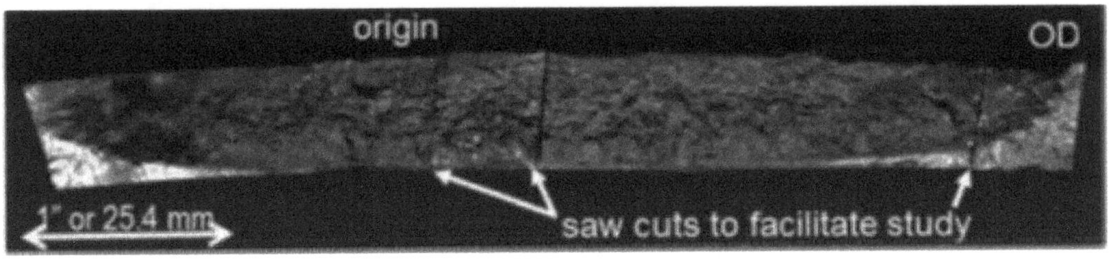

Figure 11-5. Fracture surface through a hard spot.

This failure occurred decades after the line went into service. The hardness on the OD surface in this case ranged from well above 500 Knoop (500 kg load), and decreased sharply to about 300 Knoop throughout much of the wall – whereas the background hardness was about 215 Knoop. While a single origin that was brittle in character is noted in the figure, several other thumbnail features are evident along the OD. As anticipated given the gradient in toughness, the brittle character of this cracking transitioned to ductile response through much of the pipe wall. The leak path represented in Figure 11-5 developed through-wall from a crack whose OD length was about 5 inches (127 mm).

This pipeline operated in Location Class 4 (i.e., a densely populated area adjacent to the pipeline's right- of-way (RoW)). This line typically transported sweet gas at a pressure corresponding to its Class 4 service. It was pre-service pressure tested to in excess of the current mandated requirements for Class 4. This line experienced major swings in peak gas temperature along with modest swings in pressure. It was reported that the CP nominally ran at near optimal conditions and that the coating condition was poor in places. Unfortunately there was no data supporting the claimed gas quality, and upsets in CP do occur, such that both could be factors driving a hydrogen mechanism. Consistent with that, the cracking from the OD surface was brittle and the line had survived a lower-pressure hydrotest consistent with its service. In conflict with that, the presence of a degraded fracture surface and the transition to ductile propagation, suggest the possibility of a progressive failure process. Unfortunately, conclusive evidence to prove or discard either of these scenarios could not be deduced from the evidence available.

Generally hard spots are not a primary threat. However, they can be activated in various ways and can increase the susceptibility to other threats, particularly in sour service as has occurred recently for accelerated-cooled steels. As such, they should be managed in conjunction with related condition assessments, which can be done for example using an ILI tool that sensed the decreased magnetic permeability of the steel within the hard spot.

Laps and plug scores. An example of a lap is shown in Figure 11-6a whereas pits and plug scores are shown in Figure 11-6b on the inside-diameter (ID) of a backlit seamless (SMLS) pipe. Such features were not uncommon in early pipe steel, and in SMLS pipes, respectively. Laps could be a carryover feature from the ingot for early line pipe, or could develop during hot-rolling and so be evident in modern production. Plug scores are specific to SMLS production as they are a consequence of debris that collects on the mandrel as it was withdrawn from the pipe. Such features reflect the SMLS production process, which while it has been modernized over time remains the same in concept today as it did decades ago. Accordingly, plug scores could occur in modern production. Like laps, plug scores are much less likely and less severe today as compared to what might be found in vintage or earlier construction.

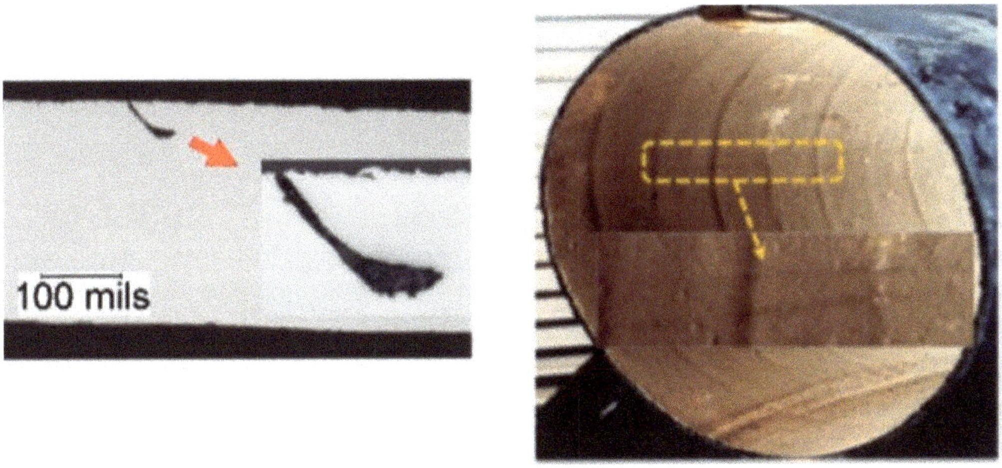

Figure 11-6. Images typical of a lap and a SMLS ID pitting and plug scores:

(a – left) lap (100 mils = 2.54 mm);

(b – right) pits and plug scores in backlit SMLS pipe.

The relatively shallow lap shown in Figure 11-6a survived a hydrotest to well in excess of 100% SMYS, and apparently opened somewhat due to that pressure. This feature was found by coincidence in pipe produced in the early-1950s in hydrotesting done to study ERW scam failures. Like the very shallow plug scores, the pits evident in Figure 11-6b reflect debris collected on the mandrel, which for this case reflects early 1930s production.

Because laps and plug scores are axially aligned features, which could be sharp, they posed a threat to either leak or rupture. These types of defects were quickly recognized as integrity threats, with changes made to the production process decades ago to limit their occurrence and severity. The upside today for operators with vintage systems is that the worst of such features quickly led to failure, or were exposed in subsequent hydrostatic tests done to re-establish the serviceability of the pipeline. Any remnant features, if large enough to pose concern, should be detectable via ILI done as part of pipeline revalidation or other integrity management activity.

Slugs. The term slug refers to foreign material that has inadvertently become part of the pipe wall. In seamed pipe they reflect material that was dropped onto or picked up as the slab was handled or while the plate or skelp was hot rolled. In SMLS pipe they represent material forced into the OD during forming or sizing, or into the ID by the mandrel during forming. Slugs have been seen as larger pieces of foreign material or as a compacted aggregates. While slugs like plug scores reflect the pipe-making processes that in concept have changed little over time, modifications to such processes have been made to minimize the frequency and severity posed by such features.

While the slug illustrated in Figure 11-7 is not the largest known to the author, much large features were infrequent even for vintage production, with relative large features known to be found into the 1970s. Regardless of their size, slugs reduce the local wall thickness and could introduce crack-like re-entrant corners, and so pose a threat to leak or cause a rupture. The slug illustrated was found in a smaller-diameter SMLS lateral during pre-service hydrotesting, with the leak detected at 64% of SMYS. Slugs as large as or larger than that illustrated tend to pose little practical concern, as such large features could be exposed during the mill pressure test, or should be exposed in pre-service hydrotesting – as was the case for the slug shown. However, smaller features could survive even high-pressure testing, and so could pose an integrity threat depending on the pipeline's service. Such features tend to separate from the pipe wall with increasing pressure such that still adherent slugs should be detectable using UT ILI tools, whereas those that drop out of the wall should be detectable using magnetic flux leakage (MFL) ILI tools.

11.3.2 Breakdowns in QC, QA, and issues with acceptance testing

As alluded to in the Introduction, with the terms *defect* and *anomaly* defined as they have been, the usual list of physical material defects expands for the pipe body to include issues that limit the pipe-steel's fitness for service. While comparable issues are known in regard to weldability, this subsection illustrates issues related to low yield strength as compared to SMYS, through the details and illustrations made public through a Freedom of Information Act (FOIA) request filed with the US Pipeline and Hazardous Materials Safety Administration (PHMSA) in 2010 that included Reference [12].

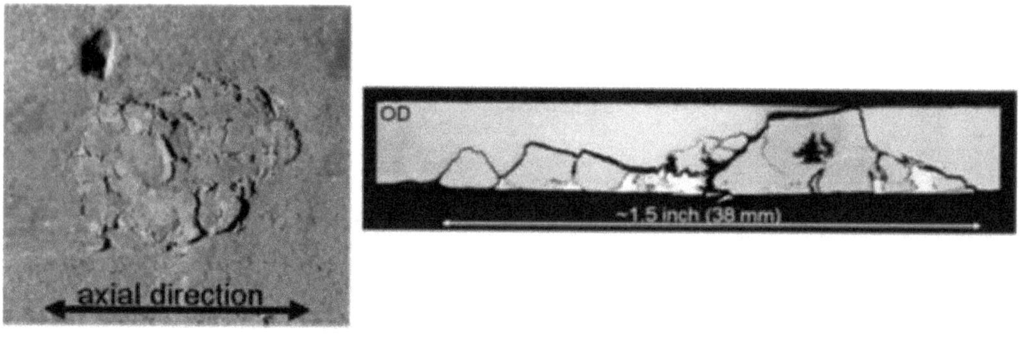

Figure 11-7. Slug comprising a compacted aggregation of foreign material:

(a - left) ID view of the slug;

(b - right) axial cross-section through the slug.

Figure 11-8 contrasts the results of properties trending of the yield stress measured in the mill – labelled in the figure MTR (mill-test report) – and subsequent replicated tensile testing[12]. The tensile tests represent samples cut from pipes that were removed after the pre-service pressure test because they had suffered significant expansion. Related trending also was done based on chemical composition and other parameters measured on the pipe excised following the pre-service test.

Figure 11-8a presents cumulative distributions (CD) for the 0.5% total strain definition of the yield stress for each of the four sources of steel involved in construction from 2006 to 2008 during which interval low strength pipe was a Regulatory focal point in the US. Figure 11-8b shows the corresponding distribution of the MTR data. The results in Figure 11-8a are shown as the composite CD for all steel sources derived from the MTR data, which is the uppermost trend shown. This figure also shows the tensile yield stresses for the joints of pipe that expanded excessively during the pre-service hydrotest, which are represented by the four dashed trends. Also shown is a horizontal dot-dash line, which denotes the value of SMYS for this pipeline. It is readily apparent that the trend for the pre-service hydro-expanded pipes lies from 3 to 8 ksi (21 to 55 MPa) below the trend developed from the MTR results. Deviations from the specification were also evident in chemical composition in this sampling, with the expanded data representing leaner compositions.

These trends can be explained by a breakdown of process QC and QA controls that lead to variable steel quality quantified by its strength and chemical composition. Inadequate pipe acceptance testing requirements that failed to detect a hint of such variability contribute to situation.

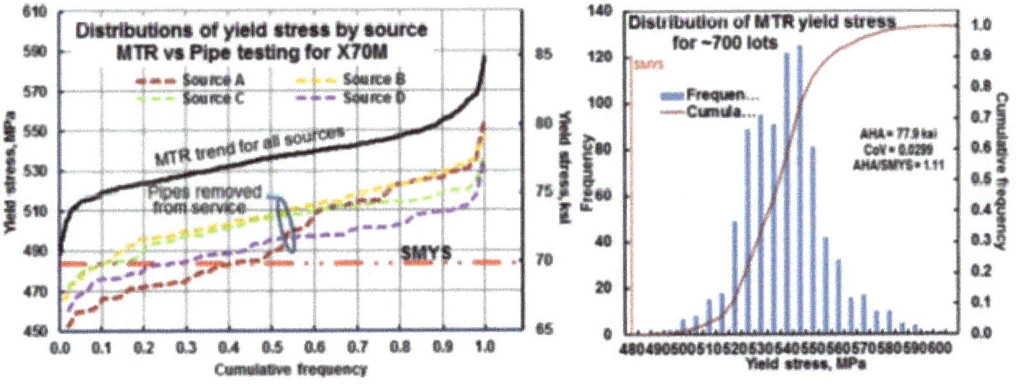

Figure 11-8. Properties characteristic of the low-strength issues in modern US construction:

(a - left) cumulative distributions for each steel source;

(b - right) histogram of the MTR yield stresses.

Rectifying this situation required several large-scale rehabilitation campaigns to get this new construction back into service as designed – whose cost was many orders of magnitude greater than would be the budget needed to identify and resolve the root cause, to avoid the recurrence of this issue. Oversight developed by Plains Justice[10] has noted in this context that "globalization of steel pipe supply chains has made quality control more challenging and increased the need for greater domestic measures to ensure discovery of defective pipe."

11.4 Material defects in or along the longitudinal weld

This section considers defects either in the deposited weld metal or the bondline, or in the pipe body where the microstructure has been altered by making the seam – which includes the upset in an ERW seam and the heat-affected zone (HAZ) of fusion welds. The seam types considered are those in common use over the years in large-diameter hydrocarbon transmission service.

According to Kiefner and Clark[13], early pipe production relied on butt-welding. Circa 1916 Reference 14 indicates that the use of such seams was largely limited to pipes with diameters less than 2.5 inch (approx. 63 mm), whereas hammer-welded and lap-welded pipe were available in larger diameters. Discussion in References 14 and 15 makes it clear that both hammer and lap seams are autogeneous welds (i.e., made without filler metal) that joined the overlapped edges of the skelp, with the extent of that overlap leading to a bonding length much larger than the pipe's wall thickness. The hammer process used skelp with square (slit or sheared) edges[15] whereas the lap process used matched scarf'd edges[14]. Of these, lap-welded pipe was available in the diameters sought for early hydrocarbon pipelines, whereas the hammer-welded pipe was only available in diameters from 24 to 96 inches (610 to 2438 mm), being used primarily as water pipe. Accordingly, this subsection considers long seams made using lap, submerged arc, and electric resistance welds (including direct current (DC), low-frequency alternating current (AC), high-frequency AC variants, and DC flash weld (FW) that was made solely by A.O. Smith (AOS) in the US.)

This subsection opens with illustrations and discussion of defects characteristic of some vintage lap-welded seam welds. As noted in regard to the body defects, vintage features are considered because a significant amount of this mileage remains in service in some countries.

Thereafter, typical long-seam defects produced using modern seam technologies are illustrated and discussed. This section closes with brief consideration of some worst-case images illustrating how not to make long-seam welds – as such can occur in production from new pipe mills that are still well back on the learning curve. For purposes of condition assessment, which is covered fully in a later chapter, the features considered are classified in terms of their acuity ranging from blunt volumetric features, to somewhat sharper notches such as undercut features, to crack-like planar features. Some features will be surface connected, while others will be buried. Typical process upsets leading to such features can result in a lack of fusion (LoF) between the joined edges, or lack of penetration (LoP) into the joined edges. Edge preparation and proximity while being joined are among the keys to a quality seam in this context.

11.4.1 Lap welded pipe

Lap welded seams have failed due to 'burnt' steel, wherein the steel's ductility and strength were degraded by the continued heating of the steel during welding at a temperature too close to solidus, which leads to a film of sulfur-rich liquid on the austenite grain boundaries. Failure through these lap-welded seams also has occurred due to forging the seam at a temperature below that for which the mill scale is a liquid (2370 - 2500°F (1300 - 1370°C)) and so could flow out of the bondline. Joining at lower temperatures thus leaves oxide trapped along the bondline, thereby reducing the bond length, and so also its load-carrying capacity. Figure 11-9 shows a cross-section through a lap weld in pipe made in 1910 by the National Tube Co that illustrates a seam largely free of burnt steel and large trapped oxides. While there is some evidence of trapped oxides, as highlighted in the inset box, and also evidence of an intermetallic stringer running above and parallel to the bondline, this seam remained fit for its service until the line was replacced.

Where both the bondline and the steel are fit for service, failure generally occurs in lap welded and other long-seamed pipes remote to the seam, as was the case in regard to the cross-section shown in Figure 11-10, which was taken local to the origin. The pipe in this figure lacks curvature as it was largely flattened during the course of the failure. This failure occurred in 1929 Spang-Chalfont[5] lap-welded pipe. It initiated in a shallow top-side ploughed gouge that coincidentally occurred next to the seam and ran more or less axially along it for some distance.

The failure occurred when the line was pressure tested at nominally 1.25 times MAOP for this pipeline system. Fractography showed that the failure occurred by a mix of brittle and ductile response, which was consistent with the high-transition temperature of this early open-hearth steel. Metallography showed a polygonal ferrite microstructure with interspersed small islands of pearlite, which is common for such early pipe. Polished cross-sections did not show evidence of burnt or overheated steel, nor did samples etched to reveal such damage.

5 Company records were used to establish the construction date. Evidence of the pattern embossed over the lapped seam, which is unique to each producer, was clear enough to determine the pipe maker. Sometimes termed spellerizing [e.g.,13], this pattern differs from the roll-knobbling process that also was termed spellerizing[16].

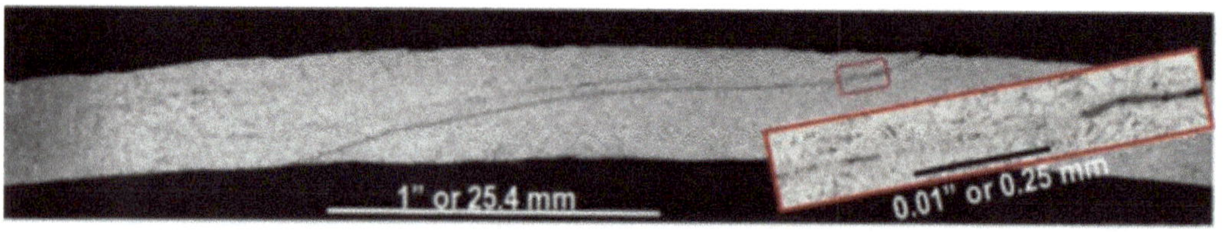

Figure 11-9. Fit-for-service lap weld in pipe removed in a rehabilitation campaign (OD is up).

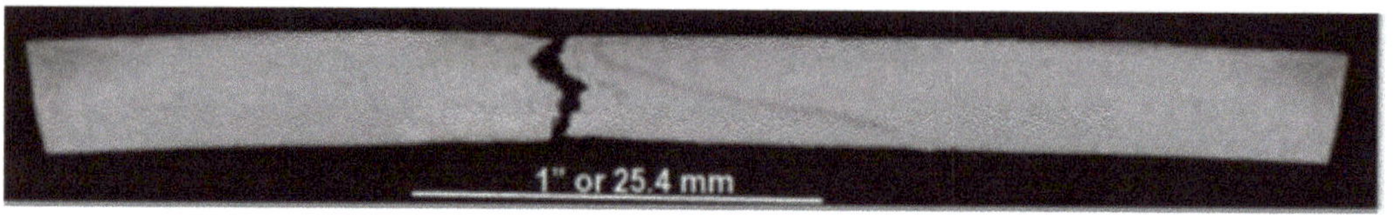

Figure 11-10. Failure adjacent to lap weld associated with mechanical damage (OD is up).

11.4.2 Resistance welded pipe

Clark *et al.*[17] indicate that the shift away from lap-welded pipe began in the mid to late 1920s with the emergence of pipes made by electric resistance welding, flash welding, and other viable seaming processes. Like the lap and hammer welds, both electric resistance and flash welded seams are autogenous forged welds.

The *Handbook of Electric-Weld Tubing*[18] indicates that tubing was first commercially produced using a rotating electrode AC- ERW in 1921 by Steel and Tubes Inc., which was bought in 1928 by what became Republic Steel Corporation (RSC). Based on language in early ERW patents, the frequency for processes using AC sources ranged from 60 Hz up through 900 Hz, all of which relied on wheel (rotating) electrodes[19]. Over the years this low-frequency (LF) electric-resistance process, which progressively joined the abutting surfaces along the length of the seam, was scaled up in size. Diameters suitable for hydrocarbon transmission service were available circa 1929 from RSC [20]. The patent literature indicates that the Direct Current Welding Company and the Youngstown Sheet and Tube Company (YS&T) co-developed a DC variation of the rotating electrode process[e.g., 21, 22]. Circa 1930 YS&T was producing pipe. A few years prior (in 1928[23]) AOS had begun producing large-diameter pipe in 40-foot (12.2 m) lengths using a DC FW process[6]. Unlike the gradual end to end progression of the ERW process, the FW simultaneously coalesced the entire area of abutting surfaces. In the years that followed into the mid-1950s and beyond, many others produced such pipes using similar ERW practices. QR Code 11-7 helps to understand the key differences between such seaming processes.

6 AOS began making line pipe in 1927 producing 30 foot (~9.1 m) long joints of large-diameter pipe made using a shielded-metal-arc weld (SMAW) seam. The efficiency gained with a FW as compared to the SMAW process was such that arc-welded production ceased in 1930, whereas capacity was added for FW pipe production[23]

QR Code 11-7. Scan here for Patent Perspective on Early ERW Seam Processes.

The use of higher frequencies focuses the current flow near the edges being joined, the economic benefits of which were already under study in the 1930s[7]. But it took until the 1950s for the HF ERW process to become practical[25]. Such seams were developed wherein the HF current was introduced by induction[e.g., 26, 27] (HFI) or by contact via shoes[e.g., 28,29] (HFC). Review of the patents for the LF and HF processes indicates that the only significant difference between the HF processes and the historic LF processes lies in the frequency of the current used to heat the abutting faces of the skelp. However, technology has emerged to better affect process control, and to provide online inspection and improved skelp quality – with these effects being most evident for post-1980s production. Aside from these differences, the LF and HF processes both create a forged weld, both require pressure between the abutted edges that are brought to a molten state instantaneously before the abutted faces meet to expel the oxide and create the upset to make the seam. Many of the defects that follow derive from issues in the process and controls related to these aspects, most of which can occur in HF as they did in the LF vintage seams.

Figure 11-11 shows cross-sections through LF ERW and FW seams along with that for a modern HF induction (HFI) (non-contact) seam. These images reflect typical production, as benchmarks for what follows. As they reflect typical production, all three show some deviation from what would be a 'perfect' seam. The LF ERW seam and the HFI seam are slightly misaligned and over-scarf'd on the ID, while the FW shows some asymmetry in the scarf – ID to OD. QR Code 11-8 discusses of the similarities and differences between the modern HFI and HF ERW processes versus the vintage LF ERW processes.

QR Code 11-8. Scan here for Compare -Contrast HF and LF ERW Seams.

7 For example, while the introduction to a patent filed granted in 1937[24] vaguely notes "using an alternating current of the necessary frequency" that patent later makes reference to using a frequency the order of 30 KHz.

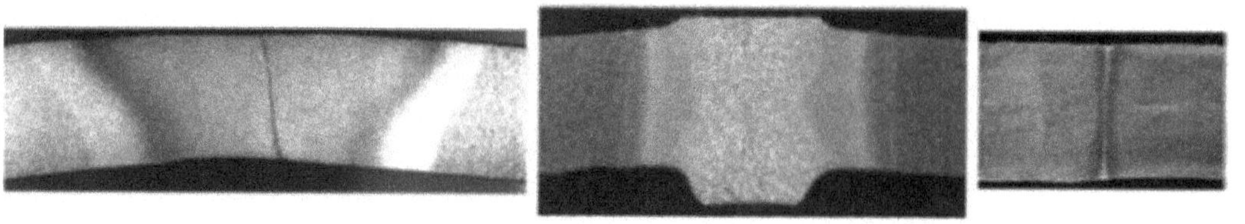

Figure 11-11. Examples of vintage and recent electric-resistance long seams (OD is up):

(a - left) LF ERW;

(b - center) FW;

(c - right) HFI.

These images represent production in mills in either the US, or Canada, in similar thicknesses of skelp (nominally 0.219-inch (5.56 mm) or 0.250-inch (6.35 mm)) and grades (X42 or X52). Each has been proportionately scaled to a common thickness. The left image reflects 1963 production from a mill that was then still producing LF pipe, the middle image is from FW[8] pipe made in 1949, with similar traits evident in 1960s production. Finally, the right image represents modern HFI production (2003), with similar traits evident for this type of HF process as far back as the 1970s.

Differences between the images in Figure 11-11 reflect the combined effects of the heat input and the mechanical upset that were involved in making these seams, and the thermal cycle of its post-weld heat treatment (PWHT), except for the middle image as this seam was not subject to PWHT. Differences in seam quality when the pipe leaves the mill develop due to process control issues in heat input and upset force, subject to other preparatory factors like alignment and edge trim and/or post-weld issues during expansion, mill testing, and other mill functions. Among the features that are or might become defects are: cold welds, stitched welds, hook cracks, weld-area cracks, contact marks/burns, improper PWHT, inclusions, penetrators/pin-holes, offset edges, and excessive/inadequate trim. QR Code 11-8 presents a broader view of ERW/FW defects and related discussion, and/or see References [19], and [30 to 33].

Cold welds, paste welds, stitched welds, and hook cracks. Figures 11-12a and 11-12b illustrate cross-sections associated with in-service ruptures due to a cold weld and a hook crack respectively – two of the more common defects in resistance welds. Figures 11-12c and 11-12d show images along the fracture surfaces of stitched seams that failed in pressure testing.

Figure 11-12c shows stitching that has formed through the wall (TW) whereas for Figure 11-12d the stitching has only developed part way through the wall (PTW).

8 Welding textbooks (e.g., see [30]) indicate that although flash welding heats the faying surfaces by combined resistance and arcing it is classified a resistance welding process, and is termed such herein.

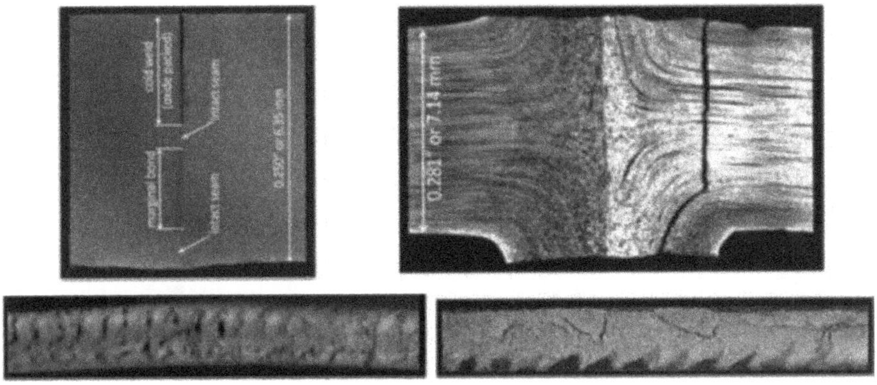

Figure 11-12. Views of a cold weld, a hook crack, and stitched welds:

(a – top, left) OD partial cold weld origin (OD is up);

(b – top, right) ID hook crack origin (OD is up);

(c – bottom, left) TW stitched weld (t = 0.312in or 7.92 mm);

(d – bottom, right) PTW stitched weld (t = 0.250in or 6.35 mm).

Cross-sections through stitched seams would look much like that in Figure 11-12a, with the depth of the crack varying depending on where along the seam the cross-section was made.

The image of the cold weld was made at a section cut though intact pipe slightly removed from the origin, whereas the image of the hook crack was made local to the origin, as were the fracture surfaces for the stitched welds. Cold welds and hook cracks can form in pipes made by the vintage processes as well as in pipes made using the modern HF processes, whereas the stitched weld was unique to intermittent current flow that occurred in the vintage lower frequency AC rolling electrode processes, causing short repetitive variations in the welding heat. The failure pressure for such planar defects depended on the length and depth of the defect, as well as the local properties. The repetitive nature of the process issue that caused stitched welds led to segments of seam that were intact and strong more or less in balance with segments that were locally cold welds. The resulting "effective" seam strength thus ranged up from about one-half of the collapse pressure for the seam, such that most stitched seams either remained stable over time or were exposed by pressure testing the pipeline. In contrast, hook cracks and cold welds reflected process issues of varying duration, such that both leaks and ruptures occurred depending on the circumstances.

The annotated cross-section in Figure 11-12a shows a YS&T DC-ERW seam for which the ID and OD weld-upset have been neatly trimmed. This image shows that at this location the weld was cold and oxide filled through a depth close to one-half of the wall thickness. The remainder of the pipe reflects a partially sound bondline over which portions are intact while the remainder has marginal strength. Cold welds form for one or more of several reasons. For example, the abutting edges might have been adequately heated but the skelp was under-width such that the necessary upset force was not generated. Alternatively, the abutting edges were not adequately heated such that the oxide formed in heating did not flow out in the presence of adequate upsetting force. Another cause is skelp that was not slit or sheared with square edges, such that upsetting brought either the ID or the OD edges into contact with sufficient force to flow out the oxide and forge a weld across part of the thickness, whereas the other edges remained agape – which appears to be the scenario here as sufficient heat and upset were present to form a bondline over part of this seam. Such cracking can initiate either at the surfaces of the pipe, but also has initiated from buried weakness within the thickness. Cold welds in vintage pipe are often filled with a compacted black high-temperature oxide (Fe_3O_4). Due to the presence of the oxide a cold weld is a planar crack-like feature. A similar planar feature has been observed for the HF processes for which a weld develops, but has little strength that has been termed a paste weld[33, 34]. Cold welds tend to be strong reflectors from an UT perspective, which facilitates their detection. However, paste welds develop continuity between the faying surfaces – which means they cannot be detected as readily prior to failure. Paste welds are considered further in a subsequent section.

The image chosen here to illustrate a hook crack in Figure 11-12b is typical of a well-trimmed FW seam that was made in split-skelp involving a quite dirty steel. Dirty steel is key to the formation of hook cracks, as they initiate and grow along films or bands of pancaked/strung-out inclusions or other impurities. These films or bands turn toward the ID or OD pipe surface as the upset forms, creating the characteristic hook shape that underlies the name of this defect. The interface along such inclusions/impurities is weak, which leads to crack initiation where the interface is more or less perpendicular to the hoop stress. Such cracking can initiate from one side of the pipe, either the ID or the OD, but on occasion cracking initiates from both sides, and grows more or less symmetrically depending on the circumstances. As 'dirty steel' underlies such defects they were most problematic until the 1960s when steel cleanliness was recognized as a critical factor for higher toughness at lower transition temperatures. While banded dirty steels have been more an issue in vintage systems, centerline segregation has led to hook cracks forming and failing in modern construction using concast steel. QR Code 11-9 presents additional images and related discussion of hook cracks, including consideration of split-skelp, dirty steel, banded microstructures, and centerline segregation, and their roles in ERW seam failures.

QR Code 11-9. Scan here for Hook Cracking.

Figures 11-12c and 11-12d respectively show a TW stitched weld (t = 0.312" or 7.92 mm) and a PTW stitched weld (t = 0.250" or 6.35 mm). Both show the presence of the high-temperature black oxide evident in Figure 11-10a in areas where the forged weld was incomplete. As noted earlier, because such welds reflect closely spaced repetitive defects with an effective area that is a large fraction of the full wall, the remnant wall is rather strong, and so is not prone to fail in-service, even as a leak. Failure for the situation shown in Figure 11-10c occurred at 94% of SMYS in a field hydrotest. The failure associated with Figure 11-10d opened up along a seam split that occurred at 98% of SMYS in a laboratory burst test.

Penetrators, pin-holes, and inclusions. Figures 11-13a and 11-13b present images of fracture surfaces for pin-holes while Figure 11-13c is a fracture surface showing an inclusion. Figure 11-13a is specific to a HFC seam in 2004 production whereas Figures 11-13b and 11-13c are specific to a 1963 DC ERW vintage seam. The pinhole through the HFC seam was detected along with several others during the pre-service hydrotest. This feature was exposed by saw cutting close to it after which fracture surface was exposed by wedge-loading after chilling in liquid nitrogen (LN_2). The features in the vintage seam were exposed in association with an in-service seam split that was triggered by a cold weld that then propagated through them. While the feature in the HFC seam was an open leak path in the pre-service test, those in the vintage seam were initially oxide-filled. As time passed, in some cases the oxide broke-up as the pipe was pressure cycled creating leak paths. The full wall thickness was shown for the pinhole through the HFC seam whereas the black upper portion of the images shown for the vintage seam was the oxide covered continuation of that cold weld that extended slightly beyond the wall thickness shown in these images. The inclusion was not analysed to establish its origin, but as it was trapped as the bondline solidified it presumably was some high temperature oxide.

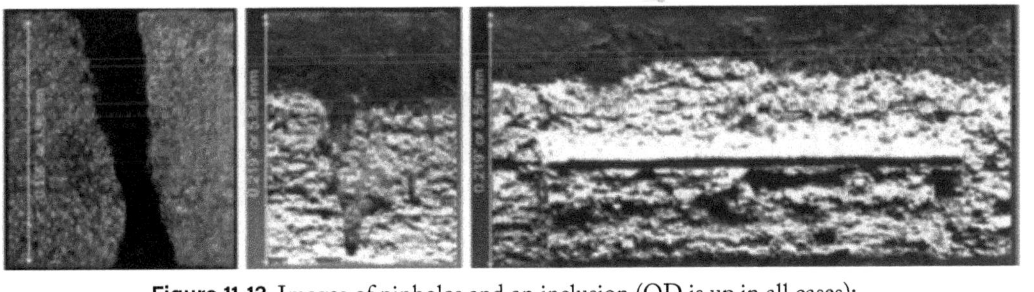

Figure 11-13. Images of pinholes and an inclusion (OD is up in all cases):

(a - left) pinhole: HFC;

(b - center) pinhole: LFERW;

(c - right) inclusion: LFERW.

As defined in API 5T1[7], a pinhole is a short unwelded area in the bondline extending through the pipe thickness so that fluid will leak out through the area very slowly, whereas an inclusion is foreign material or non-metallic particles entrapped within the bondline during solidification[9]. The thinner the pipe-wall the more likely it is for upsets in the seaming process to develop a through-wall leak path. Thus it is not a surprise that the pinholes shown in Figures 11-13a and 11-13b respectively lie in smaller-diameter pipes that were 0.156 inch (3.96 mm) and 0.219 inch (5.56 mm) thick. While open pinholes are leak paths, small inclusions are of little practical concern.

Weld area cracks and paste welds. Figure 11-14a shows the fracture surface for a feature that is defined specifically for FW seams termed a penetrator whereas Figure 11-14b shows the fracture surface for a similar-appearing feature termed a weld-area crack, which is defined in regard to both the LF and the HF ERW seams. Penetrators cannot be leaking, even slowly, or they would be either pinholes or weld area cracks. As defined in API 5T1[7], a penetrator in a FW seam is a localized spot of incomplete fusion (i.e., a weak bondline). It is subtly discriminated from the similar-appearing weld area crack (WAC) in that the penetrator is considered to have some strength whereas the WAC is defined as a stress-induced "separation" of the metal (in the bondline or weld upset), which is axially stable because "without any other influence (it) is insufficient to cause complete rupture of the material."

Note in this context that a WAC is defined for FW production in addition to the penetrator that represents a weak bondline – whereas API 5T1 does not define a low-strength feature analogous to the penetrator for either the LFERW or HFERW processes. Personal experience beginning in the late 1990s makes clear that weak bondlines have formed in HFERW seams, which opened as origins of seam splits during pre-service hydrotests. Fractography indicated that such features were free of oxides and had limited strength, which in subsequent discussion with the producer was termed a "paste weld".

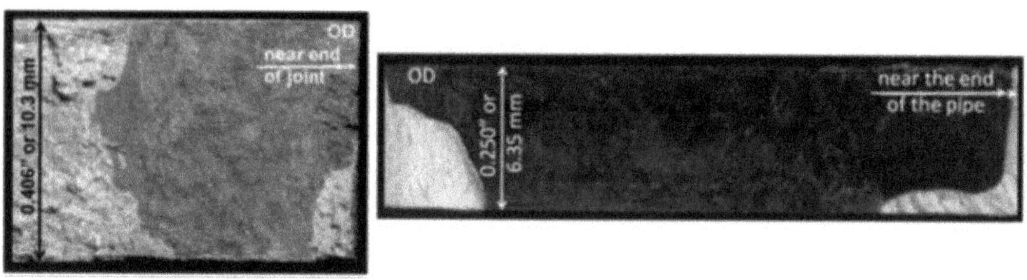

Figure 11-14. Images of a penetrator and a weld-area crack (OD is up):
(a - left) FW penetrator;
(b - right) HF weld-area crack but also occurs for LF ERW.

9 The literature specific to HFERW seams also has referred to features such as this pinhole as penetrators. [35,36]

The open literature then made little mention of such features, whereas they have been more broadly discussed more recently[33, 34]. While not currently defined in API 5T1[7], perhaps they should be.

Consistent with the definitions in API 5T1, the penetrator in 1963 FW production shown in Figure 11-14a did not leak during its pre-service hydrotest or in service, but did leak many years later during a hydroretest at a somewhat higher test pressure. Also consistent with API 5T1, the weld area crack in 2003 HFC production shown in Figure 11-14b was one of several similar leaks that occurred during the pre-service hydrotest for that pipeline. As the images shown in Figure 11-12 involve leaks, these fracture surfaces were exposed by saw cutting close to them after which fracture was opened by wedge-loading after chilling in LN_2. Both features were close to a girth weld, which for the FW pipe implied a process issue toward the ends of such seams, as that process produced pipe a joint at a time. However, for the 2003 HFC pipe the nearby presence of the girth weld is coincidental, as this joint was cut from continuous seam/pipe production.

As defined in API 5T1 the maximum size of a WAC is limited in that the definition requires that such features remain axially stable under pressure loading. As axial stability is determined by the local properties, a WAC could be quite large in very tough steel. The definition of the penetrator likewise fixes the size of this feature as it must not initially leak. As for the WAC, penetrators could be somewhat larger in tougher steel – subject as well to the constraint that it must not leak.

Figure 11-15 shows views of lightly etched cross-sections illustrating process upsets as found in vintage and modern ERW seamed pipes.

All show intact seams: because these sections were cut well away for the origin of the failure being investigated, and more importantly because such process upsets have not typically been causes of in-service failures. The upper pair of images represent smaller diameter pipes whereas the lower pair were cut from larger diameter pipes. These images are sized to illustrate the features and as such the magnifications differ somewhat among these images.

Offset edges. Figure 11-15a, which reflects later 1960s ERW production under API 5LX[10], shows an image that illustrates the effects of offset edges. API 5T1[7] does not define offset skelp or plate edges with regard to ERW seams, but for DSAW pipe defines this feature as the radial offset of (skelp) plate edges in the weld seams. While not considered in API 5T1, Section 8 of the then current API 5LX titled Workmanship addressed offset edges in §8.5b noting there for ERW pipe that the radial offset shall not be greater than 0.060 inch (1.52 mm) – by which the offset shown in this figure is

10 The current scope of API Specification 5L was at one time covered by three such specifications. The 33rd Edition of API 5L (March 1983) was the first to combine these separate although similar specifications. Previously these were denoted 5L for 'Line Pipe', which then dealt with Grades A25, A and B, 5LX that dealt with the 'Higher-Test' X grades of straight-seam pipe, and 5LS that dealt with 'Spiral-Seam' pipe for all grades.

acceptable. Paragraph §8.5f added, in regard to ERW pipe, the requirement that the depth of groove that results from scarfing the internal flash shall not exceed 0.015 inch (0.38 mm) for this pipe size. It is evident from the dimensions overlaid in this figure that this seam marginally exceeds this requirement – although this aspect had no influence on the related failure.

Excessive ID trim. Figure 11-15b is a cross-section through the HFC seam in 2005 production made under API 5L that illustrates excessive ID trim in pipe that was made using split skelp. API 5T1[7] defines excessive trim by reference to the relevant paragraph in API 5L[8], which here is §7.8.7. This requirement has not changed in the decades since it was introduced for the X-Grades in the early 1960s. As above, for this pipe size API limits the depth of groove resulting from scarfing the internal flash to less than 0.015 inch (0.38 mm). And, as above, the dimensions overlaid in this figure indicate that this seam marginally exceeds this requirement – although again this aspect had no influence on the related failure. Note in that context that this pipe contained a thick band of centerline segregation that is evident on the upper right side of the image as the dark band that is out-bent approaching the bondline.

Inadequate ID Trim and arc burns. Figures 11-15c and 11-15d are cross-sections through the seams of 1943 DC welded production by YS&T, which given their production date were made prior to mention of resistance welded pipe in API 5L. Both seams show clear evidence of offset edges as now defined in API 5L. Because the edges are offset, the eccentric loading imposed by the roll-cage to upset the seam causes one heated edge to slide relative to other, which leads to the offset and inclined seams that are clearly evident in these images. Figure 11-15c shows what today would be inadequate trim, although there were no quantitative requirements then concerning either offset edges or seam trim on the OD or the ID. While API 5T1 does not define arc burns in regard to ERW pipe, such features are apparent in Figure 11-15d. Arc burns are localized spots or linear intermittent features where shallow surface melting occurred due to arcing between electrode or ground and the pipe surface. Early releases of API 5T1 do define a related feature termed contact marks, which are distinguished from arc burns as they are intermittent tracks free of surface melting that run adjacent the weld line.

Features such as inadequate trim and arc burns (contact marks) have been largely controlled by changes in the ERW process and its controls. However, offset edges largely trace to issues with the skelp supply that can result from the use of split skelp, or cambered skelp, or wavy skelp edges. While this feature is unlikely to pose in-service concerns it can necessitate repairs during the pre-service hydrotest. The same can be said with respect to process upsets involving the PWHT.

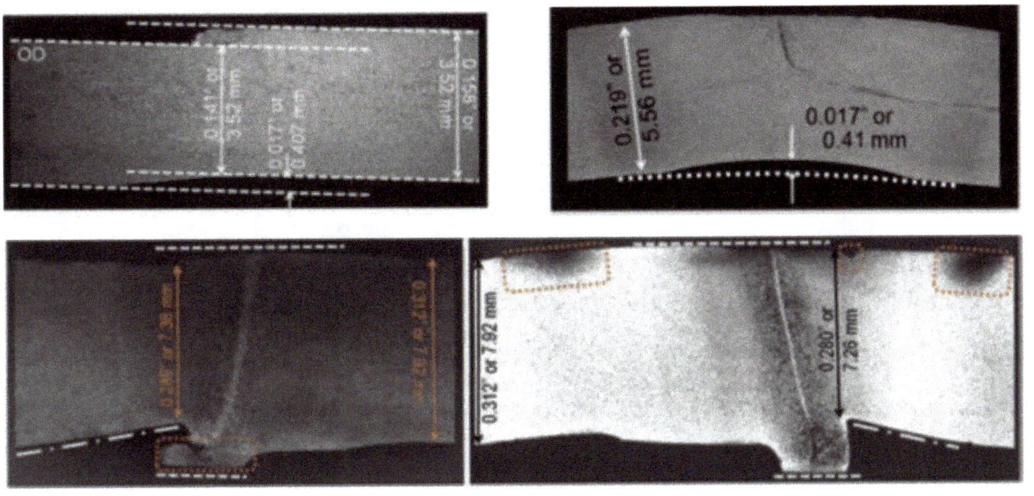

Figure 11-15. Images of seam-process upsets:

(a – top, left) offset edges;

(b – top, right) excessive ID trim;

(c – bottom, left) inadequate ID trim;

(d – bottom, right) arc burns.

11.4.3 Submerged-arc welded pipe

The submerged-arc welding process emerged in a practical context in 1930 as a development of the National Tube Works Co[37]. Consistent with that and pipe-making history as outlined in Reference 38, Clark et al.[17] indicate that submerged-arc welded (SAW) line pipe was first produced about the same time that the first larger-diameter ERW pipe was made. This early SAW process involved a single submerged-arc weld (SSAW) made from the outside of the pipe against a backing shoe that was positioned at the inside surface, which is needed if adequate fill and full penetration are to be achieved. They further indicate that DSAW pipe came into use in the late 1940s, the essential difference being that the ID bead was added to the production sequence, being made with the seam at the 6 o'clock position. QR Code 11-10 provides further details, videos, and related discussion of the DSAW process as used in pipe making.

QR Code 11-10. Scan here for more on DSAW pipe production.

Views typical of SAW seams. As for the vintage versus the modern ERW processes, the vintage SAW process has not fundamentally changed in comparison to the modern SAW process – although the quality of the outcome has improved significantly. Thus, DSAW seams can be prone to similar issues as have occurred for the SSAW process – but are free of the ID imperfections that were not uncommon in SSAW pipe production. As time passed and technology developed, the pipe production sequence has evolved to include process improvements and capitalize on inspection and quality assurance. Figure 11-16 shows etched cross-sections through SAW seams that illustrate typical microstructures for an early SSAW seam and a modern DSAW seam. Cross-sections illustrating and discussing production defects through SSAW seams follow in Figures 11- 17 and 11-18, while defects in DSAW seams are considered in regard to Figures 11- 19 to 11-21.

Figure 11-16a shows an etched section through a SSAW seam, which is typical of such pipe production. Porosity is evident near the ID surface on the lower left side of this seam. The columnar grain patterns in this image develop from the beveled boundaries of the skelp, as well as from the ID backing shoe whose shape is well defined by the profile on the right side of this seam, whereas the weld was slightly underfed on the left side. As evident from the columnar structure, this SSAW seam is a single pass process made in a single-V bevel with what appears to be a relatively broad root face. The thin skelp relative to the size of the bead and related heat input led to a quite large HAZ, which is evident out to roughly one-half of the wall thickness beyond the bevels. Higher magnification views indicated that the transition from the columnar weld deposit into the base metal showed a coarse grain structure, followed by a band of fine grains, and finally by an inter/sub-critically heated structure, as is typical of such seams. This is as expected based on the literature. [e.g., 39, 40]

When mills that were set up to produce SSAW pipe transitioned to DSAW production they initially continued making the OD bead first, after which they rotated the pipe 180° to complete the ID bead. The microstructure of the ID bead would in such cases appear imprinted over that of the OD bead in cross-sections like that shown in Figure 11-16b. In contrast, inspection of Figure 11-16b that reflects a DSAW seam made in a modern X80 Grade pipe indicates the reverse of this sequence, because this image shows the microstructure of the OD bead is imprinted over that of the ID bead. As expected based on the literature[39,40], higher magnification views of such seams also show that the transition from the columnar weld deposit into the base metal first involves a coarse grain structure, which is followed by a band of refined grains, and finally by an inter/sub-critically heated microstructure. In contrast to the SSAW seam, this transition in modern production leads to a narrower HAZ, particularly given that the wall thickness for the DSAW seam is three times that for the SSAW seam.

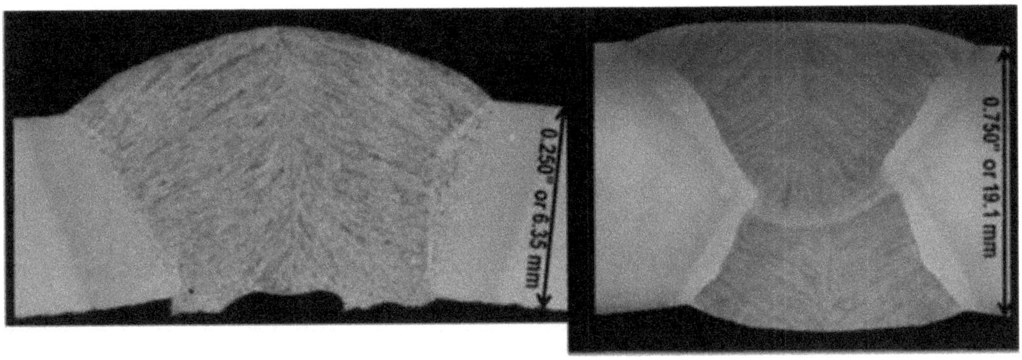

Figure 11-16. Typical etched cross-section views of SAW seams (shown at quite different scales):

(a - left) SSAW in vintage Grade B (notice the void);

(b - right) DSAW in modern X80.

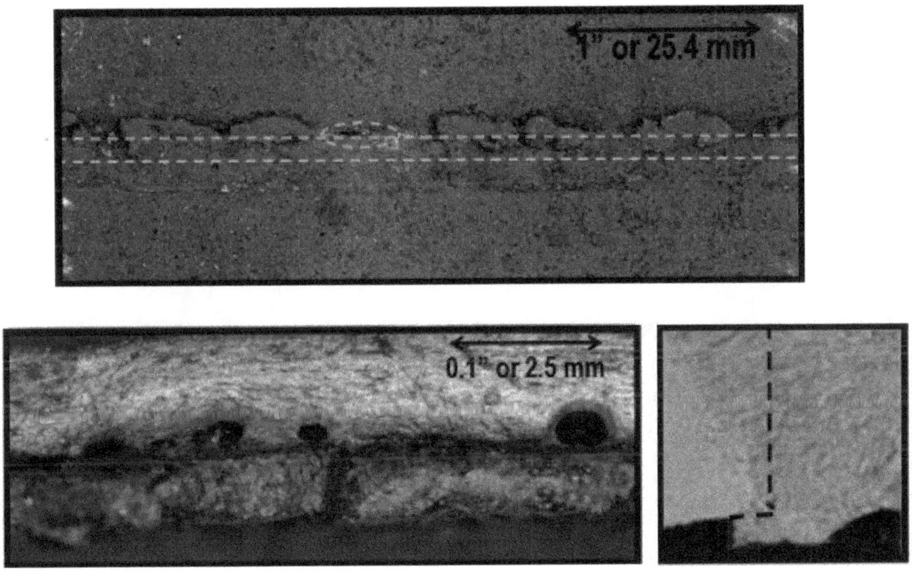

Figure 11-17. Views of a SSAW seam and related fracture surface:

(a - top) ID surface across the seam – dashed lines highlight the track of the backing shoe;

(b - bottom) fracture surface in the weld (lighter coloured band) oriented as in the adjacent section.

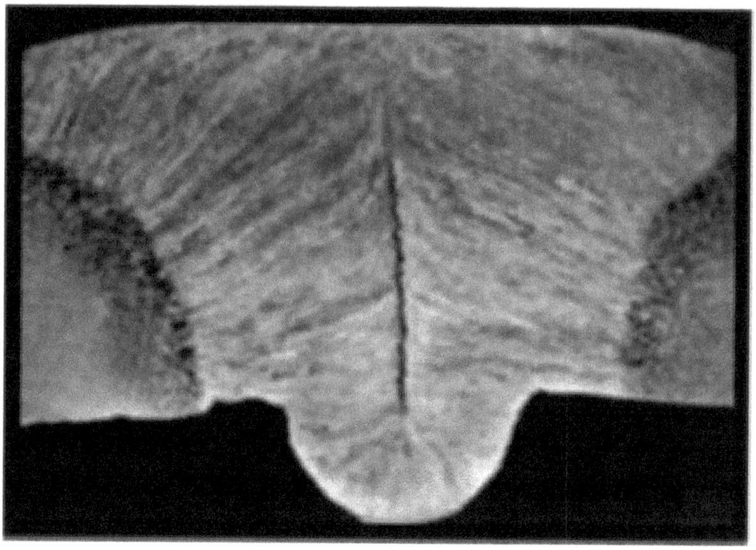

Figure 11-18. View of a typical solidification crack in a SAW longitudinal seam (cropped at top).

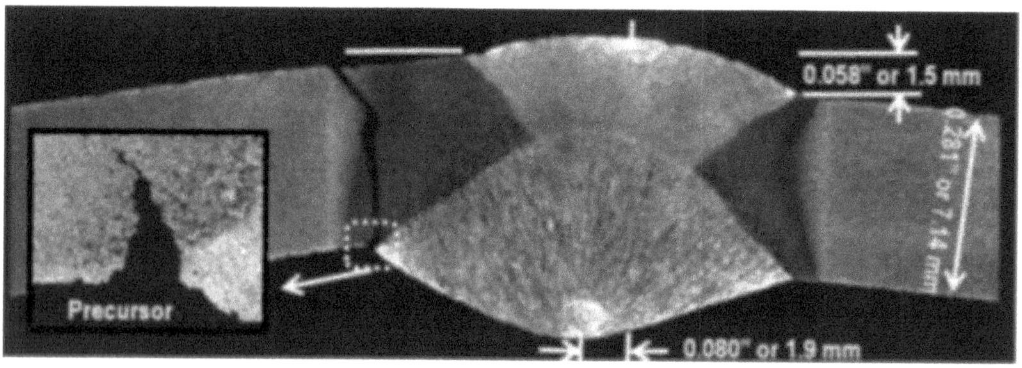

Figure 11-19. Toe cracking due to local bending caused by coupled edge and bead offsets.

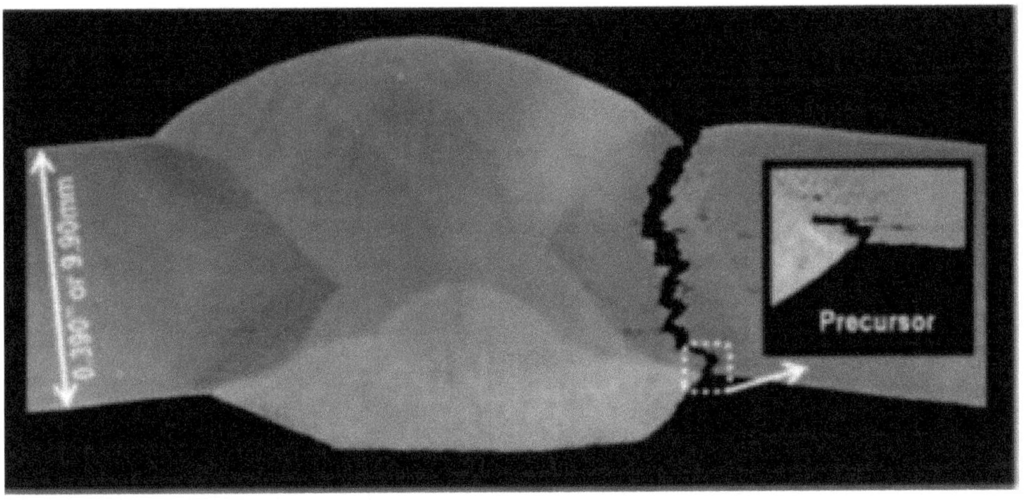

Figure 11-20. Toe cracking due to an inclusion cluster in the fusion zone near the toe.

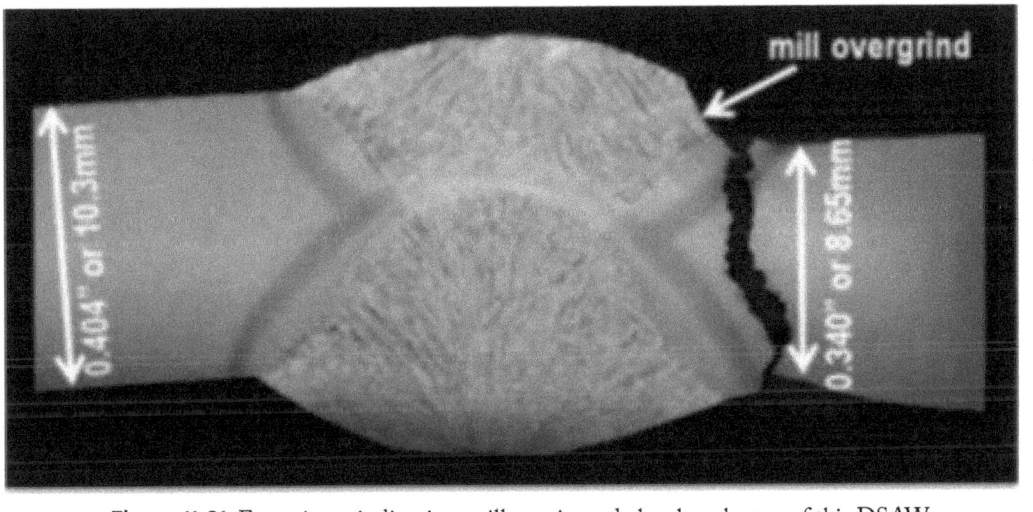

Figure 11-21. Excessive grinding in a mill-repair made local to the toe of this DSAW seam.

Fracture features along a SSAW seam. Figure 11-17a shows a view of the inside surface along the SSAW seam near the cross-section shown in Figure 11-16a. Figure 11-17b shows the fracture features associated with a burst test of the pipe containing this seam with this typical image oriented as shown in the adjacent cross-section. The voids evident in Figure 11-17b are comparable to that apparent in the cross-section shown in Figure 11-16a. In some cases the voids that formed led to surface-connected volumetric features that run axially parallel to the shoe tracks, such as that within the dashed ellipse in Figure 11-17a. Such volumetric features were small for this seam, and so generally not an issue. In contrast, the overlapped weld deposit generally lays unfused to the ID wall. This gives rise to a linear feature that if coupled with continued lack of fusion up the root face of the bevel could lead to through-wall cracking. The features in these images indicate that the SSAW process can be prone to producing a lower-quality seam, which might be expected for a process that was developed in 1930 and abandoned in the late 1940s in the transition to DSAW line pipe production.

As noted in the caption for Figure 11-17b, the dashed lines in that image highlight the track of the backing shoe used to make this weld. The cross-section in Figure 11-16a indicates that overlap occurred onto the ID surface on the left side of this shoe (as oriented in that image), with that overlap apparent as the darker band below the weld deposit and voids evident in Figure 11-17b.

While process improvements and inspection technology have largely eliminated features that have been sources for historic failures in early SAW pipe production, there is a significant amount of pipe in the ground that contains such features, which although stable for decades could be activated by changes in the pipeline's operation or the conditions along the RoW.

Experience has shown that cracking either in or along the weld deposit or adjacent to weld toe poses an integrity concern, as have issues associated with the alignment of the seam edges or the weld beads. API 5T notes nine defects in the context of a DSAW seam many of which are also relevant to SSAW seams: specifically excessive reinforcement (excessive overfill), incomplete fusion, incomplete penetration, out-of-line or off-seam weld beads, offset edges, porosity, slag inclusions, undercut, and weld-area cracks. In addition to these defects, process upsets can lead to hot cracking and toe cracking. Likewise, off-seam welds and offset edges can initiate cracking due to local bending across the beads. Skelp issues involving clusters of inclusions also can promote cracking when present in the vicinity of the weld toe or the fusion boundaries. Among these, this section focuses on aspects that involve or can lead to cracking.

Hot (solidification) cracking. Figure 11-18 is an etched cross-section through a SSAW seam that illustrates hot cracking, a term that is used in regard to solidification cracks like that shown here. The term hot cracking also is used in regard to liquation cracks and cracks resulting from hot-shortness / low ductility at high temperature.

As evident from the above discussion, such cracking is not noted among the features listed in API 5T1[7]. While the view shown reflects cracking in a SSAW apparently made without an internal shoe at the ID surface, the appearance of this cracking lying between the columnar structure developed from both edges of the bevel is similar to the appearance of such cracking within either or both beads of a two-pass DSAW seam.

DSAW toe cracking with edge and bead offsets. Figure 11-19 shows an etched cross-section through 1967 DSAW production. In contrast to the view for the modern X80 production shown in Figure 11-16b, the first pass of this weld was made from the OD as was done for the SSAW process, after which the ID bead was completed. It is largely for this reason that the HAZ of this weld looks like that for the SSAW in Figure 11-16a, rather than that for the X80 pipe shown in Figure 11-16b. It is apparent from the labeling included in this figure that this seam shows offset skelp edges, as well as offset beads, both of which are cited in API 5T1 in reference to the requirements stated in the relevant API 5L specification. The labelling for the section in Figure 11-19 indicates that the edges were offset by 0.058 inch (1.5 mm). While this is just in excess of 20% of the wall thickness and leads to local bending across the seam, it was acceptable under the workmanship requirements in §8 of API 5LX. As stated there the offset was limited to less than 0.0625 inch (1.59 mm) for SAW pipe. It is also apparent from the labelling of Figure 11-19 that the beads are somewhat offset, by the order of 27% of the wall thickness. This too was acceptable according to the 1967 Edition of API 5LX, which then stated that misalignment of the weld seam (i.e., an off-seam weld) shall not be cause for rejection provided that complete penetration and complete fusion have been achieved as indicated by nondestructive examination.

In response to pressure loading the edge and bead offsets evident in Figure 11-19 develop ID tension at the toe of the weld on the left side of the seam as it was oriented in this image. Cracking through the wall of this pipe as shown originated local to where this cross-section was cut. It developed from an origin like that of the precursor feature that is imaged in the inset, which was captured from a cross-section cut from the same joint of pipe that was identified based on the results of penetrant inspection. Inspection of the details evident in the inset indicates that this precursor feature developed from the toe of the weld and grew into the wall by ductile tearing. While there were no limitations then stated in 5LX regarding the size ID reinforcement for DSAW pipes, numerical analysis[41] indicates that it leads to significant stress concentration and so likely was a contributory factor.

Given that the highest pressure experienced by this joint of pipe occurred during the pre-service hydrotest that in 1960 was increased for API 5LX to the current minimum pressure of 90% SMYS, such toe-crack failures were often attributed to the above-noted factors. However, there also were other causes, such as the presence of inclusions and other such body issues in the vicinity of the weld fusion zone.

DSAW toe cracking with fusion-line inclusions. Figure 11-20 presents an etched cross-section through the DSAW seam of 1967 production that illustrates cracking emanating from a crack-like feature formed prior to when the pipe received its internal epoxy coating. This timing for the initiation of the cracking was established by the observation that un-failed but cracked segments along the seam that were opened were found lightly tinged to filled with a red-epoxy coating. These paint-decorated features were very long as compared to their depth, which was typically the order of the precursor feature inset at the right margin of Figure 11-20. This image was captured from a cross-section cut from the same joint of pipe that was identified based on the results of penetrant inspection along either side of the bead.

This pipe was made by the U then O then Expand (UOE) process, with the expansion following the mill hydrotest that API 5LX then required be taken to 90% SMYS for this X60 pipe. To affect the expansion required stresses in excess of the actual yield stress such that the highest hoop stress prior to internal coating occurred during expansion. The compressed columnar structure of the ID bead and its flattened shape suggests that the expander was either not grooved to accept the ID seam or the pipe was indexed incorrectly.

Failures such as that in Figure 11-20 are due to the presence of clusters of intermetallic inclusions that get strung-out during rolling. Crack initiation for such circumstances occurs in response to the most severe hoop loading, which given it was U-O-E pipe occurred during expansion, being aggravated in this case by the local tensile bending stresses induced by contact of the ID bead with the expander and the toe stress concentration due to the reinforcement[41]. Such failures thus tend to occur prior to service (i.e., in the mill test or in the pre-service hydrotest) in cases where the inclusions clusters were severe. However, they also have occurred in service in cases where the cracking that initiated during expansion or in the pressure testing was subcritical and the pipe survived, only to have in-service cycling or another mechanism propagate that precursor crack either to leak or rupture – as occurred for this joint. Clear evidence of initiation prior to service was evident in this case by the paint-decorated long shallow features identified along the ID seam toes.

DSAW failure at overgrinding. Figure 11-21 shows a cross-section through a mill grinding repair that was made local to the toe of a DSAW in 1982 U-O-E X65 production, which was made under what was the last Edition of the API 5LX specification. As usual, the expansion in the U-O-E production sequence followed the mill hydrotest that API 5LX then required be held at 90% SMYS. This failure occurred during the pre-service hydrotest that local to this failure developed 104% SMYS, in a test protocol that this operator then typically targeted in excess of 100%.

As can be seen from the image, the toe-grinding removed a portion of the reinforcement and the weld-toe, but in so doing also locally thinned the nominally 0.404-inch (10.3mm) wall to 0.340-inch (8.65mm). The result was a local wall thickness that was reduced to less than the workmanship allowance in API 5L then set at 12.5% of the nominal wall. The actual yield stress for this joint was 65.5 ksi with a UTS at 82.5 ksi (451 and 568 MPa, respectively), such that this hydrotest failure occurred as expected very close to the pressure predicted based on net-section-collapse analysis. Further discussion of this mode failure follows later in this chapter, with more details included in QR Code 11-4 presented earlier.

The DSAW seams used in making the pipe reflected in Figures 11- 20 and 11-21 were made by the same producer in the same plant but were separated in time by 15 years. As such, these images indicate the evolution in DSAW seam production from the late 1960s (1967 for Figure 11-20) into the early 1980s (1982 for Figure 11-21). It is apparent that during this interval the production sequence did not change, as the OD seam was made first for both pipes, after which the ID seam was completed. It also can be seen that the ID reinforcement for both seams is flattened at the crown, indicating that the expander used for these joints was likely not grooved to accommodate the longitudinal seam, as it is unlikely that two randomly selected pipes would both be mis-indexed. What is clearly different between these images is the patterns developed in the HAZ of these seams. As both show the OD seam was made first, these differences indicate that the heat input and related SAW process changed significantly over the 15-year interval covered by these images. Such changes to either save energy to reduce production costs or to improve quality were typical as the steel and pipe mills worked to develop and/or retain a competitive position.

11.5 Construction defects

This subsection considers construction-related features, presenting images of construction damage and defects, and discussing their causes and significance. Construction defects for present purposes involve features that might be introduced once the pipe making process is completed. This involves the interval from before mill-coating through pipeline certification and acceptance after the pre- service hydrotest and the benchmark ILI for modern pipelines, or the corresponding interval for their vintage counterparts. QR Code 11-2 on Construction Processes provides access to video, images, and related discussion outlining the construction sequence and details typical of cross-country transmission pipelines.

Many steps comprise the construction process as just defined. Some of the steps that can damage the pipe or otherwise lead to defects include shipping and handling issues, girth-welding issues, lowering in damage, tie-in issues, transitions, components, and fittings issues, settlement issues, and issues during crossings and drills, selected examples for which follow.

11.5.1 Handling and shipping issues

Once each lot/heat of finished pipe has been tested, inspected, and accepted consistent with the terms of the pipe order those joints begin their move toward the construction spreads. Initially by horse-drawn wagons and early versions of truck and/or train, modern pipe moves first to the coating mill, and then to the spread, by truck, train, barge, or boat. In contrast, for vintage systems the pipe went uncoated or was coated on the spread, typically getting there by truck, train, or barge.

In the early years strong backs, lifting hooks, etc., were used, which could lead to some damage, whereas today soft slings and magnetic lifts are used with little concern for damage. While today the threat of fatigue crack initiation along long-seams due to ovalization caused by shipping loads is a known concern that is managed by careful support and racking, fatigue crack initiation during shipping did cause a number of in-service leaks and ruptures before standards[42] were developed to manage this issue. In addition to the threat posed by fatigue crack initiation, handling pipe opens to the possibility of contact with stanchions and other metallic contact, such as with the rivet or bolt heads that were used in railroad gondola and flatcar construction. Some representative defects due to handling and shipping are presented in Figure 11-22.

Figure 11-22a shows a lightly etched cross-section through one example of fatigue crack initiation associated with pipe shipping via railcar whereas Figure 11-22b shows an example of a surface scar on the OD of a pipe that occurred during pipe handling in transit to the coating mill. While such shallow shipping fatigue cracks would be small enough to survive the pre-service hydrotest, depending on the pipeline's service and operating environment such cracks could grow in service by SCC or fatigue, and possibly several other mechanisms, and over time eventually threaten the pipeline's integrity.

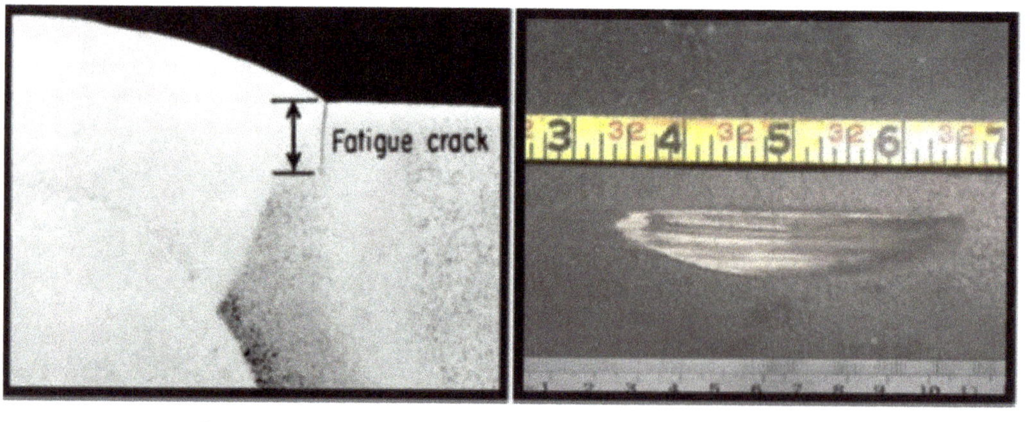

Figure 11-22. Views of some issues with shipping and handling:

(a - left) shipping fatigue OD toe crack (after [43]);

(b - right) surface scuff on line pipe.

In contrast, very shallow gouges that occur prior to modern coating practices should be identified visually and should be dressed out and inspected for evidence of remnant cracking prior to coating, and in that context would not pose an integrity threat. Care should be taken to complete an engineering critical assessment (ECA) when dealing with crack-like or significant metal-loss features.

11.5.2 Typical girth-weld issues

As for any weld, defects can occur along the bond or fusion line, in the deposited metal for fusion welds, and/or in the HAZ. Fusion-line defects can occur along the bevel or between weld passes being due to LoP and/or LoF. HAZ defects in fusion welds tend to reflect underbead cracking and/or environmental effects, while defects within the deposited metal trace to hot-cracking (like that in Figure 11-18), and also environmental factors. Prevalent causes of girth-welding (GW) issues include poor alignment or fit-up (which in turn can develop local secondary stresses), the use of poor welding procedures and unqualified practices (which can lead for example to LoP, LoF, and cracking), and the effects of outside forces (such as soil slips and subsidence).

As many such scenarios have occurred, this section presents a few illustrative examples involving alignment and procedural issues, with outside forces scenarios being discussed in Settlement or Support-Related Issues later in this chapter. Where instructive, images of satisfactory welds will be contrasted to or discussed relative to problematic welds.

GWs and LoP-induced cracking. Figure 11-23 illustrates failure of a GW located immediately adjacent to the outboard end of a cased crossing. Figure 11-23a is a view from the OD whereas Figure 11-23b is the corresponding view from the ID, while Figure 11-23c shows its fracture surface. The OD view of this failure occurs through near the mid-plane of the weld, indicating that the weld-toe was not a factor. The ID view is typical of the combined effects of a LoP defect and tensile loading acting normal to such planar defects. In this instance the tensile axial stresses developed due to differential settlement that in turn developed secondary bending that occurred between the cased and uncased portions of this pipeline.

It is apparent from the ID view in Figure 11-23b that there is little evidence of a root bead over much of the length of the weld shown in this image, which comprises a LoP defect along the ID of this GW. Inspection of the ID surface over some intact portions of this GW confirms this observation, as does the view of a portion of the fracture surface shown in Figure 11-23c. This LoP is almost the full depth of the root-face of the bevel and runs almost the full width shown in this image. Voids (apparently gas pockets) and slag also are evident. This LoP and overall weld quality coupled with the secondary stresses were sufficient to initiate and continue cracking leading to failure of this GW.

The GW featured in Figure 11-23 reflects 1948 construction that involved FW pipe. While at that time fusion welding had been codified for building construction [44], a comparable code for pipeline construction was not published until 1953 as API 1104[45]. Thus, the GWs in this pipeline were made consistent with that company's practices of that era and were subject to the skills of the welder. Company records did not indicate what practices were used, nor whether field girth weld radiography, which then was used selectively by some companies[46], had been used. Inspection of the intact portion of this cracked GW weld showed segments where the root bead had marginally developed full penetration, along with segments with inadequate intermittent penetration, as shown in Figures 11- 24a and 11-24b. Similar inspection of other rings containing GWs cut from this pipeline showed satisfactory full penetration welds, as shown for example in Figure 11-24c.

GWs with fit-up induced cracking. Fit-up issues also can lead to cracking, particularly when combined with other causative factors, as illustrated in Figure 11-25. This figure shows a code-acceptable offset between X70 pipes made of plate reported as sourced from two mills with the GW made using a fast-freeze type 8010-G Shield Arc 70+ electrode. The chemical compositions of the plates led to high carbon equivalent (CE) values that were at or marginally below current API 5L[8] limits for PSL2 pipe, for both the IIW and the critical metal parameter (Pcm) measures of CE[47]. While the pipe was preheated to 150°F (65.6°C) in an effort to minimize this high CE, it did not adequately offset the effects of the high nickel and chromium levels present from the electrode in the weld pool coupled with its fast-freeze design – as martensite was found local to the fusion line, along with slag inclusions. The combined effects of the inclusions and brittle microstructure lead to cracking that propagated into the hot pass (the first pass after the root pass) and beyond. The slight LoP evident at the root pass also could be a contributing factor.

It is apparent from the failures characterized in Figures 11- 24 and 11-25 that failure at GWs often involves the coupled effects of weld quality, alignment, and secondary loads or stresses. This pattern continues to be evident for the GW failures that develop in subsequent phases of pipeline construction.

Lowering-in damage and issues introduced during bedding and/or padding. While GW failures due to bending that develops during the lowering-in phase have occurred, they are much less frequent as compared to damage that can occur to both the coating and the pipe as the pipe-string is lowered-in, particularly into tight rocky machine-ditched trenches. Damage also can develop as the pipe settles onto the supports or 'benches' that are used to keep it off rocky trench floors, or if it subsequently settles onto rocks and/or across rock ledges.

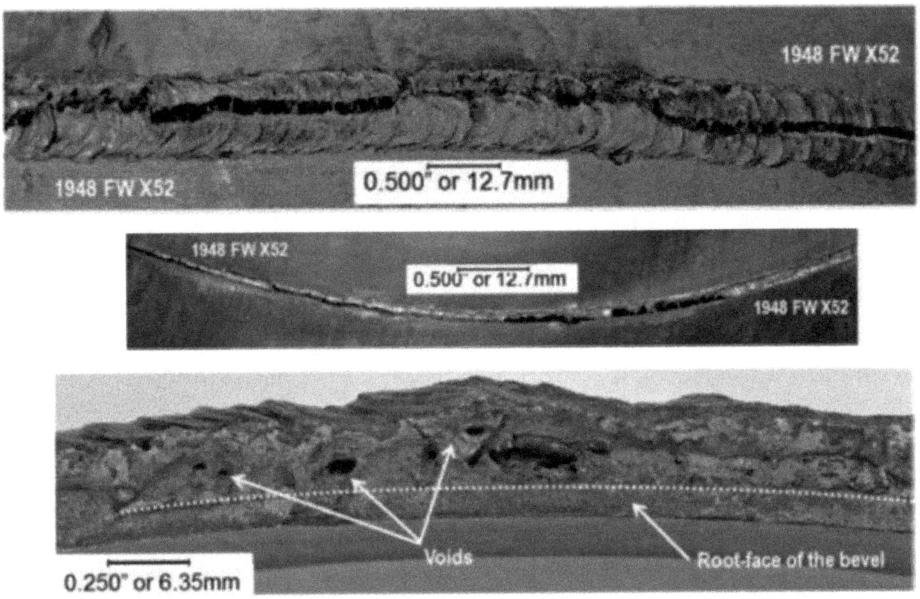

Figure 11-23. Views of GW cracking due to axial tension across the weld:

(a - top) OD view from above the cracking;

(b - center) ID perspective view of the cracking (slightly different scale versus the image above);

(c - bottom) a portion of the fracture surface of this 1948 GW showing LoP at the root-face.

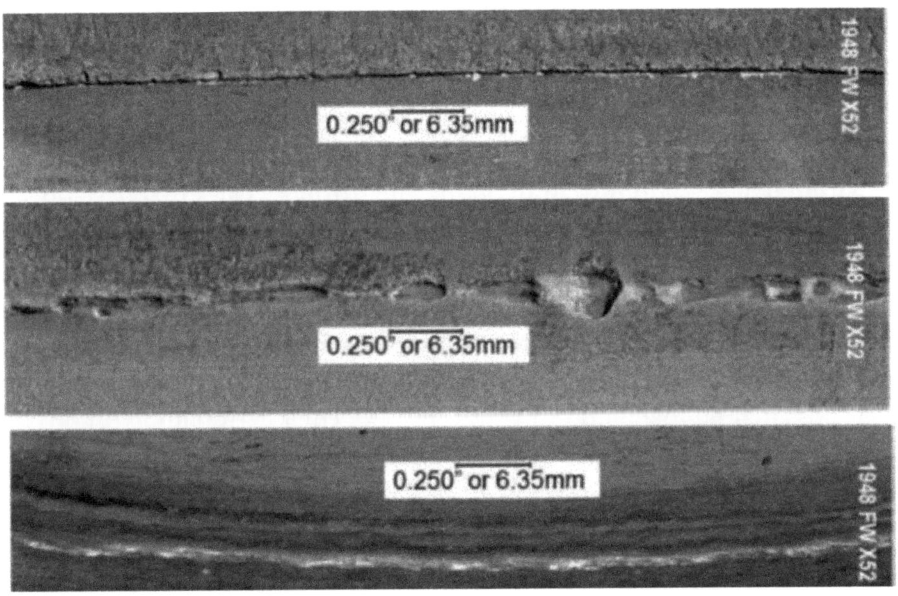

Figure 11-24. Variable states of GW quality:

(a - top) inadequate penetration between the abutting edges (LoP);

(b - center) intermittent and inadequate penetration;

(c - bottom) full penetration but slightly under-filled visually acceptable weld.

Figures 11-26a, 11-26b, and 11-26c illustrate some of the current methods to support the lowered-in pipe above a rocky trench floor, such that bedding material can flow under the pipe and thereby limit contact with any rocks that remained in the ditch. The view in Figure 11-26a illustrates the use of sandbags that have been stacked to maintain clearance above the trench floor for this over-bend. Sandbags have been used in this context for many decades, but for thin-walled large diameter line pipe have been in some cases stiff enough to cause the pipe to deform around them, leading to shallow bottom-side (restrained) dents. In spite of the intent to keep rocks away from the pipe it is apparent from Figure 11-26a that such contact does occur – in this case along the sides of the pipe. Preformed dense foam and sprayed foam pillows (also termed benches) have been developed with a view to better spread the loading and avoid bottom-side dents, such as the pillows shown in Figures 11- 26b and 11-26c. Whereas the sandbags can be too stiff, foam pillows can be too compliant, thereby letting the pipe settle onto the trench floor if they crush too much, as has occurred in Figure 11-26c. As this over-bend was continuing to settle onto the trench floor as it was being lowered-in (see the light pipe imprint evident either side of the pillow), the pipe was raised to insert a preformed dense-foam shim, after which lowering-in continued.

Figure 11-27a further illustrates the threat posed by large rocks in the ditch, whereas Figures 11- 27b and 11-27c illustrate the extent of coating damage that has occurred after the line was "jeeped" to locate and repair coating holidays prior to lowering-in. This damage in this instance reflects scraping along rocks that lay along the upper edge of what in places was a tight and deep trench. While coating and rock damage do not immediately threaten pipeline integrity they can and do pose longer-term issues.

Tie-ins and transition issues involving components and fittings. Issues that develop when making tie-in girth welds or when making girth welds involving transitions are much like that discussed regarding Figures 11- 23 to 11-25. The major difference in the case of tie-in and transition welds is that they are often subjected to secondary stresses due to forces applied to bring the adjacent strings of pipe and/or fittings into alignment. Transition welds are further subject to local stresses that develop due to differences in wall thickness and potential metallurgical issues that can emerge in joining different Grades of steel. Comparable unique issues can also be evident with some vintage construction fabrications. Such can also emerge during rehabilitation or upgrading of older pipelines, which can involve transition GWs where local stresses develop due to differences in wall thickness and Grades of steel. Figure 11-28 illustrates some of the issues that develop in making tie-in girth welds.

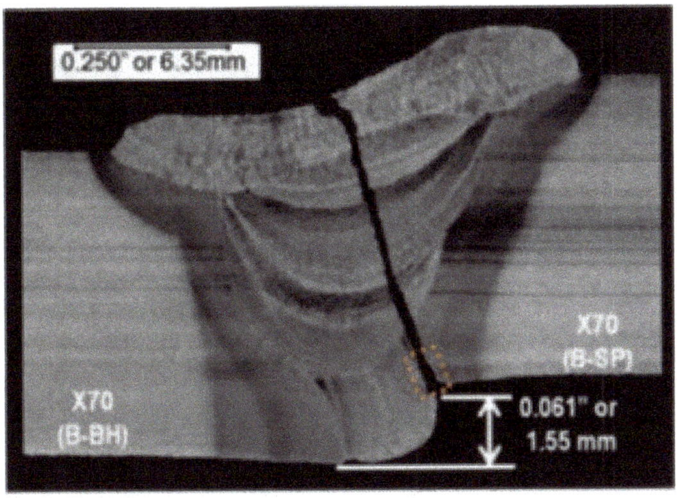

Figure 11-25. Cracking due to the combined effects of microstructure and an offset.

Figure 11-26. Pipeline trench-bottom supports:
(a - left) stacked sandbag support
(b - center) preformed foam pillow
(c - right) sprayed-in foam pillow

Figure 11-27. Some potential issues associated with lowering-in of a pipe string:
(a - left) rocks under pipe
(b - center) coating damage
(c - right) unrepaired coating damage

Figure 11-28. Secondary forces often associated with tie-in welds and at fittings:

(a - left) come-along used in aligning a tie-in weld;

(b - right) secondary loading involved with realigning at fittings.

Figure 11-28a shows a tie-in weld in process on a large diameter 0.429 inch (10.9mm) wall X70 pipe. The arrow in this image highlights the role of a 'come-along' – which here comprises a sling coupled to a side-boom – to force one side of the tie-in into alignment and stabilize their position while completing this GW. Once the weld is completed the external force and sling are removed, equilibrium develops across the seam, creating stresses in addition to those due to pressure and any other secondary loads. Figure 11-28b illustrates a similar scenario except that here once the blind flanges are removed the lateral must be leveraged into alignment to complete a flanged connection to a valve. In turn that valve has been welded to a tee on another large-diameter pipeline that also involved 0.429 inch (10.9mm) wall X70 line pipe. It is evident that the mainline is currently supported by two large permanent concrete pads, whereas the lateral sits on temporary cribbing, as does the valve (although it is not evident in the image). The loading that develops once the blind flanges are removed and the flanged joint is completed depends on the quality of the alignment in the initial fit-up, as well as in the level of skill in trimming the lateral and in welding on the flange. It also depends on the permanent support under both the lateral and the heavy valve.

Alignment is an important potential causative factor in many tie-in welds as it is often difficult to achieve along with a viable root gap. In cases where the tie-in is well aligned and not a factor, welding quality and metallurgical issues still could compromise the GW integrity, just as has been noted in regard to Figures 11- 23 to 11-25.

Figure 11-29 illustrates a GW failure that occurred in transition tie-in weld that was made to join strings of 2003 X70 pipe that was being retrofitted into a 1948 X52 pipeline due to the increasing population around the pipeline, causing a 'class change' whereby nominally thicker pipe had to be used. Long stretches of a cross-country pipeline were urbanizing, creating Class 4 high-consequence areas (HCAs) over those segments. All retrofit GWs were completed and inspected in accordance with the current edition of API 1104 circa 2003. The operator indicated that their welding procedure specification (WPS) utilized shielded metal-arc welding (SMAW) and cellulosic coated electrodes (EXX10) with downhill progression, and added that their WPS was qualified in accordance with Section IX of the ASME B&PV Code[48]. Their WPS required a minimum preheat temperature of 150°F (65.6°C), which presumably reflected concern for high CE / Pcm values for the pipes involved. In spite of this preheating, the operator reported that some moisture was known to exist on the ID pipe surface as many of the welds were being made. The operator also stated that all welds passed an initial radiographic inspection, whereas about 48 hours later it was evident that delayed cracking had occurred, which was centered at about the 6 o'clock position, after which the weld was cut out.

As the labeling in Figure 11-29 indicates, the 1948 construction made use of FW pipe that had a 0.312- inch (7.92mm) thick X52 wall. The retrofitting joined that early construction to 2003 X70 pipe with a 0.408-inch (10.4mm) wall, consistent with the 0.40 design factor for its Class 4 service. As designed, these welds met the requirements of the ASME B31.8[49] Code regarding wall thicknesses and the ID angle for such transitions, as set forth in the Mandatory Appendix I titled End Preparations for Buttwelding. As that Appendix deals specifically with joints having the same SMYS, the combination of wall thickness and Grade involved in these transition tie-ins were selected such that they satisfied the requirements (e.g., [50]) to affect collapse-controlled failure in the line pipe versus possible failure within the weld-deposit.

Inspection of Figure 11-29 indicates that failure initiated at the ID on the FW side of the seam, and that the cracking initiated at a slag inclusion in a weld that had a large root-gap. The eccentric loading in a transition is comparable to the offset at the abutting ends in Figure 11-25, which can be aggravated by ovality. Measurements however indicated this was not a factor. While the worst-case offset at 0.17 inch (3.18 mm) exceeds the API 1104 allowed 0.125 inch (4.32 mm) for GWs made between matched pipes, as noted above this transition design falls within Appendix I Code requirements [49]. That said, further analysis was undertaken to determine if there were systemic issues respecting the WPS and the operator's usual GW practices. This included dissecting the failed GWs through detailed metallography and fractography, supported by hardness traverses and testing to quantify the mechanical properties, and characterizing the chemical compositions of the line-pipe steels involved.

Hardness traverses were made at 0.1-inch (2.5 mm) increments starting in the pipe base metal of the DSAW pipe and ending in the base metal in the FW pipe on one cross-section. These results were not unusual regarding the base metal and the weld deposit. There was however some local hardening evident in the HAZ in the vicinity of the fusion line. The mechanical properties were consistent with the requirements of the Editions of API 5L relevant to the vintage and grades of the pipes involved. The composition of the FW pipe was found to be typical of the semi-killed C-Mn linepipe steels produced in the 1940's, with that for the 2003 X70 pipe likewise typical of low carbon, low sulfur, microalloyed (V+Nb) Grade X70 production. While little can be concluded from these results, regarding the cause(s) for the delayed cracking, the CE levels that derive from their chemical compositions proved instructive[11]. It was found that the IIW CE for the FW steel was 0.49 which, while unspecified in the 1940s, is quite high as compared to the current maximum specified at 0.43. In contrast the Pcm for the X70 was just 0.17 relative to the maximum specified in 2003 at 0.23. It follows that delayed cracking is indicated on the FW side of this GW, but not on the X70 side, as was the case. Delayed cracking is also favored on the FW side due to the presence of the MnS inclusions that serve as traps for monotonic Hydrogen, which recombines locally to drive such cracking.

Figure 11-30 presents views of this transition GW from the ID and OD of the pipe. Figure 11-30a shows OD view with the bottom-button (convergence site of the opposed vertical down passes) centered in the image. Figure 11-30b is the corresponding ID view, although it is not exactly centered about the bottom-button as above. It is apparent from the images shown in Figure 11-30 that the cracking evident in the cross-section view in Figure 11-29 consistently tracked the OD toe as well as ran into and along the ID root pass of the GW. This is consistent with the role of delayed cracking wherein conditions local to the fusion line and in the adjacent weld metal affect this cracking.

The image showing the OD of the pipe indicates that this GW is typical of modern SMAW practices. The appearance of the weld cap pass indicates that the WPS was implemented – at least to the extent evident in this image. In contrast, examination along the weld root at the ID identifies several imperfections among those noted in API 1104 Section 9, as illustrated in Figure 11-30b. These include inadequate penetration (IP), incomplete fusion (IF), and internal undercutting (IU), with some evidence also of what appear to be repaired burn-through sites. While many minor shortcomings are evident in the root passes in areas remote to the cracking, comparable examination in those areas indicates a viable GW, more typical of modern SMAW practices. Grinding marks are evident on the thicker X70 side of this GW, which apparently was done to minimize the internal mismatch and its possible adverse effects on weld integrity. It follows that in addition to the large slag inclusion highlighted by the inset image to Figure 11-29 weld quality also was a factor in this delayed failure. Such inclusions along the root pass and elsewhere along the fusion boundaries create planar defects that have the same net effect as a LoF defect. QR Code 11-4 presents additional images illustrating aspects of lowering-in and during bedding and padding that open to potential integrity issues.

11 The literature shows that CE above specified minimum levels is a strong indicator of susceptibility to delayed (hydrogen-induced) cracking, and that avoiding delayed cracking in steels with an IIW CE of 0.49 requires a higher preheating temperature than was specified in the WPS(e.g., [51]).

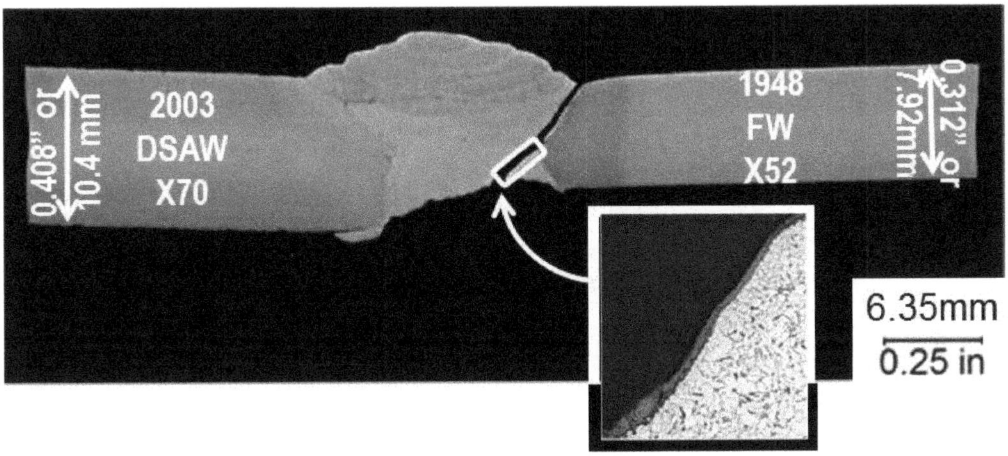

Figure 11-29. ID initiation along a slag inclusion in a failed transition tie-in GW.

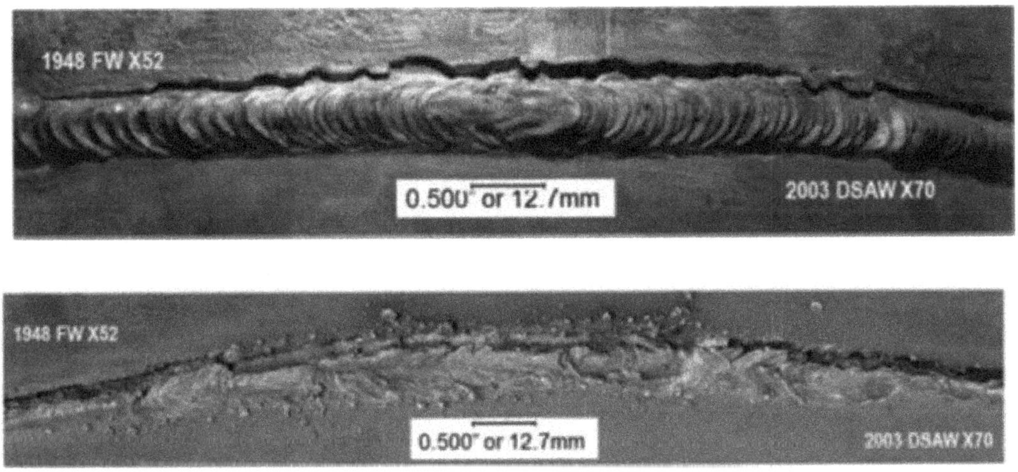

Figure 11-30. Overviews in the vicinity of the GW failure in Figure 11-29:

(a - top) OD view showing the bottom button centered in the image;

(b - bottom) ID view (not centered about the bottom button as above).

Settlement or support-related issues. Settlement and/or support-related issues can develop during construction and lead to failure that could be evident prior to or during the pre-service hydrotest or a benchmarking ILI, or could develop through in-service degradation. Such failures can be due to M&C issues, but also can develop if construction surveys and related planning underestimate or poorly quantify the actual loadings encountered along the RoW.

Figure 11-28b illustrated one typical differential settlement scenario that could lead to secondary quasi-static loads, with similar situations plausible at cased crossings, or circumstances where settlement or soil movement leads to outside forces acting on the pipeline. Such scenarios lead to external bending and tension whose effects localize at changes in cross-section that tend to localize the bending, or at stress raisers like weld toes. Figure 11-31 is shown in complement to Figure 11-28b to illustrate issues where inadequate support in design goes unnoticed during construction, leading in this case to dynamic excitation in service that initiated and grew a fatigue crack at a GW, which leaked as the cracking grew through-wall. Figure 11-32 presents the details of this failure.

Figure 11-31a shows a post-repair overview of this failure, wherein a heavy valve body sits atop an extruded tee that exits into a smaller diameter offtake, all of which is part of compressor station piping. High flow rates through the valve excited the valve and piping, creating the loading that initiated cracking along the GW at the exit from the tee, as shown and highlighted in the Figure 11-31b.

The inset to Figure 11-32 indicates that this cracking developed along the edge of an LoF defect that formed at the interface of the cap bead and the outlet of the tee, which initiated cracking over an arc of about 2.5 inches (64 mm). The scanning electron microscopy image in this figure captures this cracking, which advanced by fatigue crack propagation (FCP), doing so at a rather slow rate based on sequential bands of striations and blunting due to corrosion. Evidence of corrosion debris is apparent in this figure near the origin as the patchy features on the crack plane.

It follows from Figure 11-32 that this failure was due to workmanship related to intermittent LoF between the cap pass and the base metal. This LoF created a sharp notch that initiated a fatigue crack driven by periodic dynamic cyclic loading and continued via FCP, eventually growing through the wall causing a detectable leak. Absent the effects of this LoF defect or another of comparable significance, it is unlikely that this defect would have initiated.

Crossings, directional drills, and horizontal bores. Few pipelines can be built today without encountering geographic barriers such as rivers and lakes, and/or man-made barriers such as roads and railroads. Absent concern for interfering with the use of a waterway, rivers and lakes often could be crossed using traditional dam and cut methods. While effective, such methods are open to environmental concerns that today usually can be managed by using a directional drill, or by building various forms of support structures, like suspension bridges. While crossing many waterways does not involve concern for interfering with their use, such is generally not the case for roads and railroads, which increasingly are being dealt with by directional drills and horizontal bores. Shorter straight-line bores are usual in crossing roads and railroads, whereas directional drills are better suited to moving below obstructions, or in cases where high-value assets like optical cables lie in or near the path of a straight-line bore. Figure 11-33 is useful in illustrating the usual issues encountered where drills and bores emerge to meet open-trench construction.

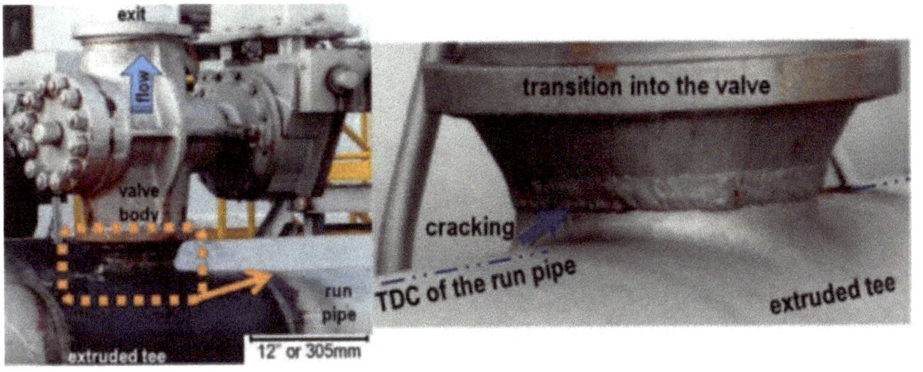

Figure 11-31. Illustrating inadequate support leading to GW cracking:

(a - left) overview of the tee and valve;

(b - right) detail local to the cracked GW at the exit of the tee.

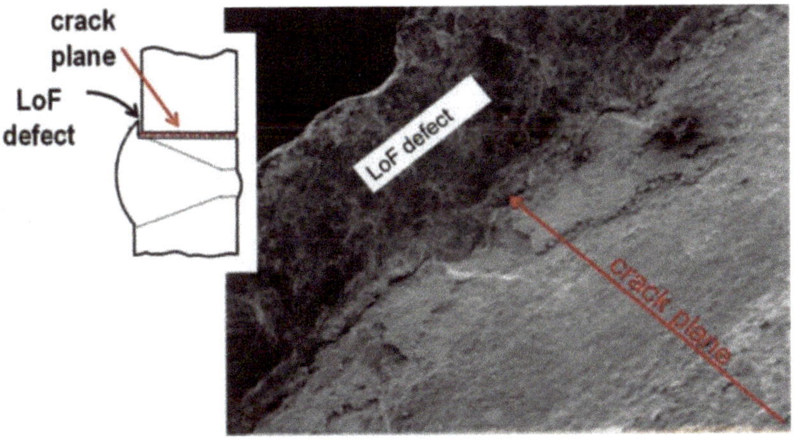

Figure 11-32. Illustrating the consequences of inadequate support and a LoF defect.

Figure 11-33. Horizontal road-bore and the open trench tie-in weld.

Inspection of Figure 11-33 shows the path of the subject pipeline through the clear-cut path evident on the horizon. A major highway can be seen as a barrier to this construction, which lies just above the dashed box that has been inset into this image. The expanded view of the area contained within that inset box lies in the lower right corner of this figure, wherein a second road also can be seen. As both roads had to remain in-service and without interference to traffic, given the terrain and other circumstances this barrier was traversed by a short horizontal bore, with the pipe being jacked from the far side of the road. Once the exposed ends were refaced and other checks completed the jacked pipe was tied into the pipe segment visible in the image. Thereafter, the same work was completed in the bore pit on the far side of the road. QR Code 11-11, Managing Geographic Barriers, contains images and a link to a video outlining aspects of crossings and drills that can be encountered during the construction sequence of some cross-country transmission pipelines, and provides video animation of a directional drill.

Aside from the issues unique to this horizontal bore, this phase of pipeline construction is largely the same as that elsewhere along the spread where tie-in welds are made. As the prior discussion makes clear, where quality practices are used by skilled welders, secondary loads and stresses are avoided or managed, the steels to be welded are understood, and the appropriate process and consumables are used, GWs do not pose a unique integrity threat. But, it must also be said, where any of this list is poorly dealt with, their integrity can be in jeopardy.

QR Code 11-11. Scan here for a Primer on Managing Geographic Barriers.

Commentary and summary. Review of the above discussion indicates that failures did not occur where the pipe body and the welds were sound and as specified, and the loadings were consistent with the design assumptions. It is therefore critical that the design process fully addresses the circumstances along the RoW and that acceptance of materials and the weld seams be in accordance with established specifications. Welds that are over-matched and transitions that are designed to fail in pipe rather than in the GW are equally important, as are materials and welds produced in accordance with proven manufacturing and welding procedure specifications. Regarding the failures discussed in long seams and the girth welds, workmanship issues and loadings other than pressure that lead to axial tension or local bending were recurrent themes as causes for the failures considered.

Survival of a pre-service high-pressure hydrotest can build confidence in system integrity, as can long-term service without evidence of issues. But care must be taken in this context, as random threats like mechanical damage can lead to delayed failures, whereas changing circumstances along a RoW can activate apparently stable / benign features. High-profile incidents like that at Ghislenghien (Belgium) or San Bruno (USA) drive these points home. Both failures involved key circumstances unknown to the operator. The first of these involved modern construction and relatively heavy-wall pipe – but unknown to the operator the line suffered mechanical damage while operating at reduced pressure associated with construction in the vicinity of the pipeline. While stable at the reduced pressure, as the pressure was being increased following the construction the damage grew stably through wall – and continued to grow after developing a small leak until it ruptured and caught fire. The second of these involved mid-1950s X42 construction operating at relatively low pressure the maximum for which was set in connection with the housing development in the vicinity of the pipeline. Its construction era was well prior to the advent of ILI – which made it difficult to pass such tools through the line. Unknown to the operator, this pipeline included what can be considered rogue pipe segments that were made with a partial-penetration long seam weld. Those rogue segments operated in the vicinity of work involving other services, which had the potential to activate near-critical otherwise latent features – such as the partial-penetration seam welds. Over time the partially completed weld failed, leading to a rupture and fire. The message in both cases is clear – regardless of their origin, features that are or appear to be stable can be activated. Care must be taken to fully understand the circumstances that define the integrity of the pipeline. The upside since the advent of ILI is that pipelines have been designed to facilitate the use of ILI tools.

11.6 Failure at defects by plastic collapse and fracture

The prior sections of this chapter have illustrated M&C defects and discussed briefly, at a high level, some pipeline loadings that can lead to failure at such defects. This section provides insight into the response of defects to pressure, and briefly considers ductile failure by net-section (plastic) collapse as well as by ductile and brittle modes of cracking. QR Code 11-12 provides some background concerning ductile and brittle modes of cracking, and the role of the mechanical properties in regard to collapse-controlled failure. This QR Code also considers the usual pipeline fracture properties in regard to fracture-controlled failure. Taken together these properties quantify the resistance of the line pipe to collapse and fracture at defects. While other approaches exist to quantify the fracture initiation and propagation resistance of line-pipe steels[e.g., 52], these are not considered herein.

11.6.1 Failure mechanisms and defect response to increasing pressure

Inspection of applicable codes and various regulatory requirements indicates that a significant margin of safety has by design been built into pipelines. Yet pipelines occasionally fail. Such failures are occurring in spite of efforts made to improve pipe steel resistance to failure and to develop inspection / quality management technologies to avoid, detect, and remove potentially injurious features, like M&C defects. While failure can be defined for pipeline systems in general terms that include metrics like efficiency or structural stability, failure for present purposes will be simply quantified relative to plastic collapse and fracture. The microscopic aspects of these mechanisms are considered next, prior to discussing the response of a defect to increasing pressure.

Plastic-collapse-controlled failure. Plastic collapse – which when it occurs through a reduced cross-section is termed net-section collapse (NSC) – is ductile, and as such shows evidence of plastic deformation in proportion to the material's uniform strain capacity. The underlying micro-mechanism involves void nucleation, growth, and coalescence. Collapse controlled failures occur transgranularly (TG); that is, they cut across the grains, which in turn can be thought of as the microstructural building blocks that grow integral to one another. In the early steels these building blocks typically range in size from about 100 microns (0.004 in), whereas for the more recent steels this size decreases to about 1/20 of that or ~5 microns (0.0002"). Finer grained steels are stronger and generally tougher than their coarser-grained counterparts – all else being equal. The voids nucleate at 'particles' within the grains, which can involve either desirable or deleterious constituents, depending on the circumstances. Given the size of these particles, this mechanism is often referred to as micro-void coalescence (MVC).

Figure 11-34a shows a magnified image illustrating fine equiaxed dimples within MVC. QR Code 11-12 illustrates the development of this failure process wherein it is apparent that the high stresses at the tip of discontinuity like cracks first lead to blunting of the discontinuity, the extent to which depends on the toughness. The voids nucleate subsurface to this blunted feature due to the hydrostatic stress field that develops there. As the local net-section decreases, more voids nucleate and eventually interact and coalesce locally and then across the net-section, leading to failure. Such coalescence typically leads to planar cracking unless other factors like geometry cause directional changes.

Fracture-controlled failure. Fracture-controlled failures can occur either in a ductile mode or in a brittle mode, depending on whether the service temperature is above or below the ductile-to-brittle transition temperature (DBTT). Ductile failures tend to occur along a shear plane such that shear-area (SA) serves as a measure of ductility. One-hundred percent SA (%SA) reflects fully ductile response whereas zero % SA reflects fully brittle response. Predominantly ductile response is evident above 85%

SA such that service at temperatures above the 85% SA DBTT the mode of cracking is predominantly ductile. QR Code 11-12 outlines testing methods that are used to quantify the DBTT, and to quantify the corresponding toughness. The underlying micro-mechanism of ductile fracture is as it was for collapse-controlled failure – that is MVC. And as for collapse controlled failures, blunting first occurs the extent to which depends on the toughness, where after TG cracking ensues – leading to generally planar cracking unless the geometry and/or other factors affect the cracking directions.

Fracture also can occur in a fully brittle mode, or it can occur in a mixed mode that couples the brittle and ductile modes. Brittle fracture at a micro-scale looks much different than MVC, as is evident in Figure 11-34b that illustrates quasi-cleavage which follows a TG path. Brittle fracture also occurs intergranularly (IG) – that is between the grains along the grain boundaries. Distinguishing between cleavage and quasi-cleavage both of which occur by TG cracking is by the presence of localized ductile response within the otherwise cleavage-dominated fracture surface. For Figure 11-34b this MVC is evident in the upper right corner of this image. Secondary cracking out of plane to the quasi-cleavage is also evident. Whether TG or IG such brittle cracking involves little to no evidence of plastic flow, is highly reflective if untarnished, and tends to be planar. But as QR Code 11-12 illustrates, such cracking can be fragmented or stepped.

This QR Code provides further discussion along with images of IG cracking – in terms of fracture surfaces and cross-sections through such cracking.

QR Code 11-12. Scan here for more on failure mechanisms and pipeline properties.

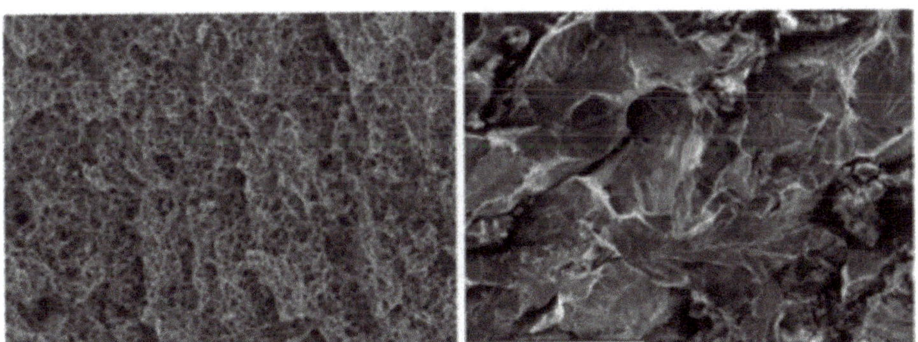

Figure 11-34. Illustrating the micro-mechanisms of typical pipeline failures:

(a - left) fine equiaxed dimples within MVC (plastic collapse and ductile fracture);

(b - right) quasi cleavage (brittle fracture).

Macroscopic images of ductile and brittle failure. The significant difference in the micro-appearance of the modes of failure is also evident at the macro-scale. This is evident in Figure 11-35 that contrasts views of typical axial ductile failure due to pressure loading with the sinusoidal propagation cracking paths that are typical of brittle axial failures due to pressure loading.

It is apparent from Figure35a that ductile axial failures present two characteristic 'looks' – both of which involve an axial component of cracking. The upper image in Figure 11-35a shows a split that in this case arrested without 'ring-off', wherein the crack path turns and begins to run with a circumferential component. Such splits are typical of long-seam failures, but this cracking also occurs in the pipe body well remote to the long seam. The lower image shows a ductile axial failure that has developed as a flap. The abutting edges along the path of ductile propagation lie at 45° in thinner-walled pipes (i.e., higher diameter to thickness ratio), with the failure occurring by void nucleation, growth, and coalescence. The origin for these ductile failures typically lies within the bulge formed as failure localizes due to the progression of through-wall collapse. Experience indicates that flap opening failure is more common for tougher steels. Both of these failures occurred in liquid (crude) pipelines such that the driving force to sustain axial cracking decayed almost immediately. Had these pipelines been transporting compressible or certain other fluids the circumstances associated with split would be such that it would have extended axially for at least several joints.

Figure 11-35b illustrates several traits typical of axial propagating brittle fractures, with other usual traits being less evident. This failure occurred in gas-transmission service at approximately 65% of SMYS being initiated through a corrosion defect whose length over which through- wall collapse occurred was in excess of the critical length for unstable axial propagation. It is apparent from this image that such brittle failures occur with limited energy dissipation, and track in-plane sinusoidal paths. The fact that the pipe in many places retains its curvature prior to failure indicates that brittle failures occur with limited energy dissipation remote to the fracture. It is also apparent that such brittle cracking involves multiple cracks that create 'shards' of pipe, with the shadows evident along the right margin of the image showing the characteristic sinusoidal cracking path. The stored energy and its release-rate is very high in such events – sufficiently so to eject the shards from the ditch and throw them well outside the pipeline's RoW. Where the propagation is brittle, such failures lie on a crack plane that is 90° to the pipe wall, with little associated through-wall thinning. Cracking occurs via cleavage in such a way to produce chevrons that point back to the origin, which when well-formed can be visible to the unaided eye. The inset to the bottom of this figure shows typical chevrons.

Figure 11-35. Macroscopic appearance of typical ductile and brittle failures in pipelines:

(a - left) ductile axial failures (both approx. 2006);

(b - right) brittle sinusoidal axial fracture (1972).

Defect response to increasing pressure. The prior sections illustrate the significant differences evident in the failure response of pipelines depending on the nature of the cracking. Not surprisingly these differences are clearly manifest in the pressure response of defects. This section briefly illustrates and discusses the differences first for ductile then for brittle initiation. QR Code 11-12 provides related background concerning ductile and brittle modes of cracking, and the role of the mechanical properties in regard to collapse-controlled failure. Further details can be found in Reference 53.

Figure 11-36 illustrates aspects of the response of a defect to increasing pressure under circumstances where the toughness of the steel supports fully ductile behavior. Figure 11-36a is a schematic of defect growth with increasing pressure, wherein the relationship between pressure and defect depth or crack-mouth-opening displacement (CMOD) is presented. The y-axis representing pressure is analogous to force imposed in the tension test, while CMOD on the x-axis is analogous to displacement in the tension test – with one key difference. Tension tests are typically done in displacement or strain control wherein the process remains stable beyond the UTS, whereas burst in a pipeline occurs under force control which tends to failure once the UTS is achieved. It is apparent from Figure 11-36a that the trend up through the ultimate capacity of the pipe is similar to that of a tensile test – the pressure rises, the pipe yields local to the defect, and plastic flow ensues – with some evidence of strain hardening which develops local to the defect.

Figure 11-36b presents the corresponding phenomenology. It is evident from this image that significant local thinning is possible, and that local crack-tip stretch and flank rotation occur with the extent for both dependent on current defect depth and the toughness of the steel. The zone labelled vee-notch and blunting reflects the blunting that occurs at the tip of this initially sharp vee-notch. Such blunting occurs prior to the onset of initiation of ductile tearing, as tough steels resist crack advance. Micro-voids nucleate and grow in the hydrostatic stress field that forms below the blunting tip of the defect. The failure at defects in tough steels is therefore controlled by plastic collapse through the remaining wall below the defect. Similar behavior occurs for initially blunt features like areal or pitting corrosion, but as there is no crack in such cases the x-axis CMOD must be replaced by a measure of thinning or local pipe expansion, such as bulge displacement or hoop expansion around the pipe and across the defect. NSC analysis is suitable to predict such response, as could be fully general nonlinear fracture mechanics (NLFM), although such would be much more complex.

As the pressure increases micro-voids continue to nucleate and grow, and coalescence occurs between them leading to stable tearing. As the net-section reduces, MVC continues at the peak pressure, being driven by stress-activated creep[e.g., 54, 55]. Such cracking can continue despite no increase in the pressure, since at the peak load this becomes an autocatalytic process. Unless the pressure is decreased, cracking continues toward the critical size for failure. Blunting thus is the first trend encountered as pressure increases in Figure 11-36a. It is followed by the onset of stable growth, then the onset of unstable growth, and finally by the trend corresponding to the failure pressure.

In contrast, for cases where the loading occurs below the DBTT brittle fracture begins to dominate the failure response. As the resistance to crack advance is low under such conditions, little to no blunting develops. Where brittle fracture controls, there is no blunting line in the context of Figure 11-36a, and under plane strain conditions initiation is immediately followed by propagation – such that all boundaries shown in Figure 11-36a become coincident with that denoted unstable propagation. Failure now is fracture-controlled and typically occurs at much lower pressures because the toughness is reduced.

Further discussion of fracture initiation and propagation can be found in Reference [53]. QR Code 11-13 expands the scope of this discussion whereas QR Code 11-12 provided background concerning ductile and brittle modes of cracking, and the role of the mechanical and fracture properties in regard to collapse-controlled and fracture-controlled failure.

QR Code 11-13. Scan here for a Primer on Defect Response to Increasing Pressure.

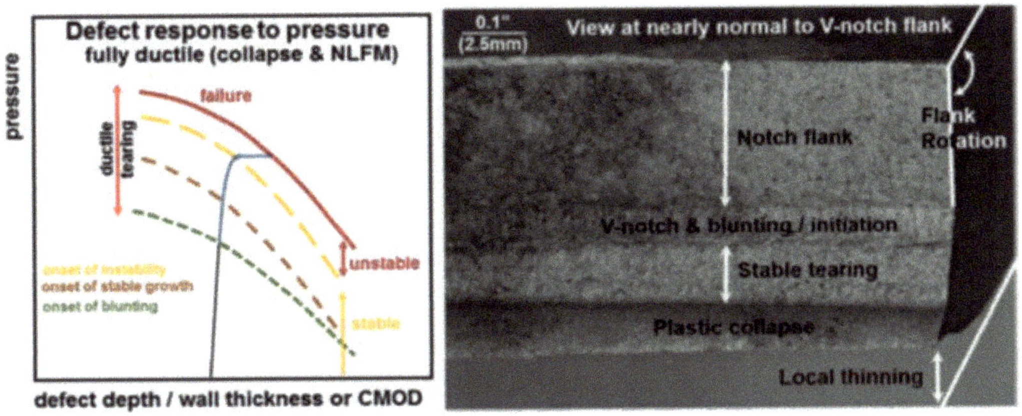

Figure 11-36. Defect response to pressure and its phenomenology:

(a - left) pressure vs growth response;

(b - right) fracture surface characteristics.

11.6.2 Commentary and summary

It is apparent from the above discussion that there are two extremes in the response of a defect to loading – fully ductile response – or fully brittle response. Such response is respectively evident at the micro-scale through the presence of MVC or cleavage. The influence of this micro-level response is clearly manifest at the macro-scale in terms of the phenomenology as well as the response of a defect to increasing pressure. Three regimes can be identified: I - linear-elastic opening of the crack, II - blunting and initiation at the crack tip, and III - growth – which is initially stable but becomes unstable as the cracking extends into the pipe wall leading to failure. Such response is typical of steels that are highly resistant to cracking, which is evident as high CVN energy, or high "toughness" – that as noted above supports failure under collapse-controlled conditions. Stable cracking over the course of Regimes II and III is often termed stable tearing, with the extent and nature of this response strongly dependent on toughness.

The effect of toughness is apparent in the extent of plastic deformation, and for pipelines is evident in the separation between the four bounding trends shown in Figure 11-36a. Tougher steels resist cracking and so sustain higher pressures. As toughness decreases the trend denoting the onset of stable tearing shifts down toward the trend denoting the onset of blunting (and initiation of cracking). Likewise, the trend denoting failure shifts down toward the trends denoting the onset of blunting (and initiation) and the onset of crack propagation. To this point this chapter has focused on the effects of pressure and failure in the context of defects of the effects of increasing pressure. QR Code 11-14 makes clear that by design pipelines are highly resistant to failure, as design provides for a significant MoS – with strain-hardening providing a hidden MoS. As has been illustrated, that MoS can be eroded by M&C defects. To this point the discussion of such failures has not considered the possibility of in-service degradation due to corrosion, environmentally assisted cracking (EAC), and other processes. QR Code 11-15 discusses defects that fail other than due to increasing pressure, whereas the QR Code 11-16 presents details of some common degradation and cracking processes and briefly considers their kinetics. These aspects also are considered elsewhere in this book.

QR Code 11-14. Scan here for Codes and Margins of Safety.

QR Code 11-15. Scan here for Illustrations of Some Pipeline Defects.

QR Code 11-16. Scan here for Degradation and Related Cracking.

11.7 Summary and conclusions

This chapter has considered material and construction defects, beginning with the background and related history. Sources and causes of such defects were considered with respect to the pipe body, as well as in or along the longitudinal and girth welds. Blisters, hard spots, laminations, laps, slugs, and plug scores were considered in the context of the pipe body. That discussion also considered breakdowns in QC/QA and issues with pipe acceptance testing.

The section on long-seam issues began with consideration of lap welded pipe, a seam type no longer in use for line pipe, and thereafter considered resistance welded pipe in regard to cold welds, paste welds, stitched welds, hook cracks, penetrators, pinholes, inclusions, and arc burns. Alignment issues also were considered in terms offset edges, excessive ID and OD trim, and inadequate trim. SAW pipe was considered in regard to single SAW, a seam type no longer in use for line pipe, and double SAW pipe, with a focus on toe cracking due to various issues.

Attention then turned to construction defects. This included extensive consideration of girth-welding issues, after which handling and shipping issues and damage during lowering-in were discussed in regard to bedding and/or padding. Typical girth-weld issues were discussed with a focus on tie-ins and transitions, and settlement / support-related issues. Finally, scenarios involving crossings, directional drills, and horizontal bores were considered.

The failure process at defects was considered next in terms of plastic collapse and fracture. The underlying failure mechanisms were introduced followed by consideration of the response of defects to increasing pressure. Aspects of plastic-collapse-controlled failure were then outlined, which was followed by consideration of fracture-controlled failure. This discussion first dealt with the microscopic aspects of these failure mechanisms, which was followed by parallel discussion in terms of the macroscopic response.

Important conclusions that can be drawn from the chapter include:

- Pipelines as designed include a significant MoS – with strain-hardening providing a latent MoS, particularly for the lower-strength Grades;
- The design MoS can be eroded due to defects introduced in the production of the steel and during completion of the long-seam weld (if used) and the girth welds used to join adjacent pipe joints;
- Defects take many forms that can generically be described as blunt or volumetric and sharp or planar;
- Failure localized at defects can occur either by plastic collapse or fracture, with the properties – particularly toughness – and the pipe geometry determining which of these controls the failure process;
- Tougher steels blunt initially sharp defects, ensuring collapse-controlled failure which occurs at a higher pressure as compared to fracture-controlled failure – all else being equal;
- Welds that are overmatched should ensure failure occurs in the pipe body, provided that they satisfy usual workmanship standards and other case-specific requirements;
- Tie-in welds tend to be hotspots along the pipeline, such that consideration should be given to better managing their placement and ensuring their quality;
- In-service degradation occurs by a wide range of mechanisms all of which act to further erode the design MoS – periodic condition assessment is key to managing this issue;
- Care must be taken to fully understand the circumstances that affect pipeline integrity, be they associated with the current physical condition of the line pipe, the systems put in place to protect it, and/or the RoW factors that can activate otherwise stable defects or lead to circumstances that further erode the design MoS.

11.8 References

[1] Duffy, A. R., McClure, G. M., Maxey, W. A. and Atterbury, T. J., "Study of the Feasibility of Basing Natural Gas Pipeline Operating Pressure on Hydrostatic Test Pressure," Catalogue No. L30050, American Gas Association, 1968.

[2] Anon., "In-line Inspection Systems Qualification," API Standard 1163, 2nd Edition, American Petroleum Institute, April 2013.

[3] Anon., "Specifications and Requirements for In-Line Inspection of Pipelines," Pipeline Operator's Forum, 2016.

[4] McClure, G. M., "The NG-18 Project: Historical Perspective," 4th A.G.A. Symposium on Line Pipe Research, Paper D, Dallas, TX, 18 November, 1969, A. G. A. (now PRCI) Cat. L30075.

[5] McClure, G. M., Eiber, R.J., Hahn, G.T., Boulger, F.W., Masubuchi, K., 1st A.G.A. Symposium on Line Pipe Research, Dallas, TX, March, 1961: A.G.A. (now PRCI) Cat. 40/PR (listed later as L00290), 2nd in Columbus OH in 1963, 3rd in Dallas TX in 1965, and thereafter in Houston TX in 1969 (4th), in 1974 (5th), in 1979 (6th), in 1986 (7th), in 1986 (7th), in 1993 (8th), and in 1996 (9th and last).

[6] Anon., History of the Engineering Research Station, 1964 -1993, Old Flames Website maintained by BG Technology, 12 May, 2009 (http://oldflames. org.uk/BGRS/)

[7] Anon., "Standard on Imperfection Terminology," API 5T1, R2010.

[8] Anon., "Specification for Line Pipe," API 5L, Edition and Date as applicable.

[9] Leis, B.N., "Managing an Aging Pipeline Infrastructure," Chapter 44 in Oil and Gas Pipelines: Integrity and Safety Handbook, R. W. Revie, Editor, Wiley, 2014, pp 609-633.

[10] Anon., "Potential Low and Variable Yield and Tensile Strength and Chemical Composition Properties in High Strength Line Pipe," US PHMSA Advisory Bulletin ABD-09-01, May 21, 2009: see also Anon., "Use of Substandard Steel by the U.S. Pipeline Industry 2007 to 2009," Plains Justice, 28 June, 2010.

[11] Leis, B.N. and Clark, E. B., "Implications and Trends in SMYS Built into Pipelines," 8th International Pipeline Conference CD, ASME, Calgary, 2010, IPC2010-31523.

[12] Leis, B. N., Clark, E. B., and Flamberg, S. A., "Fitness-for-Purpose of the 42-inch Diameter Segments in Aspects of Boardwalk's System," to Boardwalk Pipeline Partners, LP, October 2009: see also Leis, B. N., Kurth, R. E., Zhu, X-K., and Clark, E. B., "Fitness-for-Purpose of the 36-inch Diameter Segments of the Boardwalk System," to Boardwalk Pipeline Partners, LP, June, 2010. (available via FOIA request by Plains Justice, 02 September 2009)

[13] Kiefner, J. F. and Clark, E. B., History of Line Pipe Manufacturing in North America, ASME CRTD-Vol 43, 1996.

[14] Phillis, W. A., "The Manufacture of Modern Wrought Pipe," The Rose Technic., Vol XXV, No 6, March 1916, pp. 138-141

[15] Anon., "Hammer Welded Steel Pipe," Forging and Heat Treating, Vol 7, No 12, December 1921, pp. 601-603 (abstracted from Bulletin No. 13., "Hammer-Welded Steel Pipe," published and copyrighted by the National Tube Company, Pittsburgh, 1921).

[16] Anon, "Spellerizing," The National Engineer (Advertising Section), December 1914, p 127: see also Anon, "Spellerizing as Applied to Pipe - Details of the Process, Its Significance and Utility," Railway and Locomotive Engineering, September, 1917, pp 280-281.

[17] Clark, E. B., Leis, B. N., and Eiber, R. J., "Integrity Characteristics of Vintage Pipelines," Final Report to the INGAA Foundation, Battelle, October 2004. https://primis.phmsa.dot.gov/gasimp/docs/integritycharacteristicsofvintagepipelineslbcover.pdf

[18] Anon., Handbook of Electric-Weld Tubing, Steel and Tubes Inc., 1934

[19] Leis, B. N., "Time-Trending and Like-Similar Analysis for ERW-Seam Failures," Task 4.2 Contract DTPH56-11-T-000003, Pipeline and Hazardous Materials Safety Administration, July 2013.

[20] Anon., "Electric Welding Aids Pipe Industry," Pittsburg Press, 23 February 1930, p. 19.

[21] Neale, J. E., Youngstown Sheet and Tube, "Improvements In or Relating to Welding Methods and Apparatus," United Kingdom Patent GB369,431, 24 March, 1932.

[22] Mirfield, G. E., Lewis, R. E, Firth, C., and Neale, J. E., "Electrical Welding Apparatus," US Patent 2,003,118 (assigned to Youngstown Sheet and Tube), 28 May 1935.

[23] Scheil, M. A., Fratcher, C. E., Henry, S. L., Uecker, E. H., "Manufacture and Metallurgy of Flash-Welded Line Pipe," The Engineering Journal, February 1957, pp. 60-71.

[24] Bennett, E., (assigned to Wisconsin Alumni Research Foundation), "Method of Heating Electrically Conducting Bodies for Industrial Operations," US Patent 2,066,668, 5 January 1937.

[25] Heald, S. T., "Radio-Frequency Resistance Welding of Carbon Steel Pipe and Tubing," in Mechanical Working of Steel, AIME, 1963, pp 23-38.

[26] Body, A. C., "High Frequency Induction-Heating Apparatus," US Patent 2,654,019 (assigned to The Ohio Crankshaft Company), 29 September 1953.

[27] Rudd, W. C. and Stanton, R. J., "Induction Welding," US Patent 2,763,756 (assigned to The Magnetic Heating Corporation), 18 September 1956.

[28] Rudd, W. C. and Stanton, R. J., "Seam Welding by High Frequency Resistance Heating," US Patent 2,818,488 (assigned to The Magnetic Heating Corporation), 31 December 1957.

[29] Rudd, W. C. and Stanton, R. J., "Tube Welding," US Patent 2,857,503 (assigned to The Magnetic Heating Corporation), 21 October 1958 and "Welding by High Frequency Resistance Heating," US Patent 2,876,323 (assigned to The Magnetic Heating Corporation), 3 March 1959.

[30] Messier, R. W., Jr., Principles of Welding, Processes, Physics, Chemistry, and Metallurgy, Wiley-VCH Verlag, 1999.

[31] Leis, B. N. and Nestleroth, J. B., "Battelle's Experience with ERW and Flash Weld Seam Failures: Causes and Implications," Battelle Interim Report, Subtask 1.4, US DoT Contract No. DTPH56-11-T-000003, September 2013.

[32] Kiefner, J. F. and Kolovitch, K. M., "ERW and Flash Weld Seam Failures," KAI Interim Report on Task 1.4, , US DoT Contract No. DTPH56- 11-T-000003, September, 2012

[33] Nichols, R. K., "Common HF Welding Defects," Thermatool Corp., (www.thermatool.com/.../common-hf-welding-defects.pdf), undated

[34] Simion, P., Dia, V., Istrate, B., and Munteanu, C., "Controlling and Monitoring of Welding Parameters for Micro-alloyed Steel Pipes Produced by High Frequency Electric Welding," Advanced Materials Research, Vol. 1036, 2014, pp 464-469.

[35] Haga, H., Akoi, K., and Sato, T., "The Mechanisms of Formation of Weld Defects in High-Frequency Electric Resistance Welding," Welding Journal, Welding Research Supplement, June 1981, 1104s-1109s.

[36] Choi, J.-H., Chang, Y. S., Kim, C.-M., Oh, J.-S., and Kim, Y.-S., "Penetrator Formation Mechanisms during High-Frequency Electric Resistance Welding," Welding Journal, Research Section, January 2004, pp 27-S to 31-S.

[37] Rainwater, S., “Submerged Arc Welding: Then and Now - It’s not your father’s sub-arc process,” Practical Welding Today, The Fabricator.com, July/ August, 29 July 2008.

[38] Anderson, G. C., “Design and Operation of Electric Weld Pipe Mill at National Tube’s National Works,” Iron and Steel Engineer, May 1952, pp 96-101.

[39] Kaplan, D. and Murry, G., “Thermal, Metallurgical, and Mechanical Phenomena in the Heat Affected Zone” in Metallurgy and Mechanics of Welding (Ed. Régis Blondeau), ISTE Ltd / Wiley & Sons, 2008, pp. 89-132.

[40] Ashby, M. F., and Easterling, K., “A First Report on Diagrams of Grain Growth in Welds,” Acta Met., Vol. 30, 1982, pp. 1969-1978.

[41] Williams, H. E., Ottsen, H., Lawrence, F.V., Munse, W. H., “The Effects of Weld Geometry on the Fatigue Behavior of Welded Connections,” Civil Engineering Studies Structural Research Series No. 366, University of Illinois, 1970.

[42] Anon., “Recommended Practice for Railroad Transportation of Line Pipe,” API RP 5L1, 7th Edition, 2015: see also “Recommended Practice for Truck Transportation of Line Pipe,” API RP 5LT, 1st Edition, 2012.

[43] McClure, G. M., “Field-Failure Investigations,” 3rd A.G.A. Symposium on Line Pipe Research, Dallas, TX, November, 1965, A. G. A. (now PRCI) Cat. L30000, pp 127-138.

[44] Anon., Code for Fusion Welding and Gas Cutting in Building Construction, 1st Edition, American Welding Society, 1928.

[45] Anon., Welding of Pipelines and Related Facilities,” API 1104, Edition 1, 1953, or Edition and Date as applicable.

[46] Henderson, W. M., “X-Ray Examination of Pipe Joint Welds,” Proceedings, Pacific Coast Gas Association, Vol. 31, 1940, pp. 134-136.

[47] Wang, W., “The Great Minds of Carbon Equivalent: Part III: The Evolution of Carbon Equivalent Equations,” Published and Copyright, Edison Welding Institute, 2016.

[48] Anon., “Boiler and Pressure Vessel Code,” ASME, Edition and Date as applicable.

[49] Anon., “Gas Transmission and Distribution Piping Systems,” B31.8 Code, ASME, Edition and Date as applicable.

[50] Zhu, X.K. and Leis, B.N., “Plastic Collapse Assessment of Unequal Wall Joints in Pipeline Transitions,” Storage Tank Integrity and Materials Evaluation, PVP-Volume 490, ASME, 2004, pp. 253-260. (This work was available in detail circa 2002 in the form of a PRCI Report.)

[51] Merchant, S. Y., “An Overview on Effect of Preheating on Cold Cracking of Low Alloy steel and Stainless Steel Weld Joint,” International Journal of Application or Innovation in Engineering & Management, Vol 4, No 4, 2015, pp. 70-77.

[52] Zhu, X.K. and Leis, B.N., “Fracture Resistance Curve Testing of X80 Pipeline Steel using SENB Specimen and a Normalization Method,” Journal of Pipeline Engineering, Vol. 7, 2008, pp. 126-136.

[53] Leis, B. N. and Eiber, R. J., “Fracture Control Technology for Transmission Pipelines”, PR-003-084506, December 2014: currently PRCI Member Access only

[54] Leis, B. N., Brust, F. W., and Scott, P. M., “Development and Validation of a Ductile Flaw Growth Analysis for Gas Transmission Line Pipe”, (NG-18 Report No. 193), American Gas Association, A.G.A. Cat. No. L51543, June 1991.

[55] Brust, F. W., and Leis, B. N., "A New Model for Predicting Primary Creep Damage in Axially-Cracked Cylinders: Part I - Theory," Engineering Fracture Mechanics, Vol. 43, 1992, pp. 615-627: see also Brust, F. W., and Leis, B. N., "A New Model for Predicting Primary Creep Damage in Axially-Cracked Cylinders: Part II - Application," Engineering Fracture Mechanics, Vol. 43, 1992, p. 629-639.

11.9 Sources and acknowledgments to supplemental videos in Chapter 11

QR 11-1. Steps involved in pipe making

- **Introduction to pipe making**
  https://youtu.be/8vIoxzYnYXw
  HardHat Engineer – https://hardhatengineer.com/

- **Seamless pipe manufacturing**
  https://youtu.be/ztcEyel47Kg
  Tenaris

- **Hot-rolling**
  https://youtu.be/4LBkLCCjpM0
  Amazing Technology

- **Electric resistance welding manufacturing process**
  https://youtu.be/0x1uRR9Jb34
  Tenaris

- **Resistance welded pipe production**
  https://youtu.be/dOLPKSI3FaI

- **Mechanical cold expansion**
  https://youtu.be/2nfw8bPC_iM
  China Smart CNC

QR 11-2. Pipeline construction primer

- **Typical stages of pipeline construction**
  https://youtu.be/nC9c3D3VknM
  ProVideoDuluthMN

  https://youtu.be/_gW6EU0g6ys
  Spectra Energy (Enbrdge)

QR 11-10. DSAW Pipe Production

- **Introduction to pipe making**
  https://youtu.be/8vIoxzYnYXw
  HardHat Engineer – https://hardhatengineer.com/

- **Production of DSAW pipe**
  https://youtu.be/HmQYEJ52kbU
  Hebei Allland steel pipe manufacturing Co

- **DSAW process as applied to girth / tie-in welds**
  https://youtu.be/H6QGLGJ-BOE
  https://youtu.be/TC-JU-GAjQo

QR 11-11. Geographic barriers

- **Horizontal Directional Drilling Installation Animation**
  https://youtu.be/mdLCD6t6C-w
  Directional Technologies

QR 11-12. Failure mechanisms

- **Failure Analysis – Strength and Fracture Properties**
  https://vimeo.com/351295203/3f2a3e7daf
  ROSEN

QR 11-13. Defect Response to Pressure

- **Failure Analysis – Defect Response to Pressure**
  https://vimeo.com/351295221/4e23bf2fd2
  ROSEN

QR 11-15. Illustrations of some pipeline defects

- **Failure Analysis – Defects**
  https://vimeo.com/351295231/3c24a958bc
  ROSEN

QR 11-16. Degradation and related cracking

- **Failure Analysis – Degradation and Cracking**
  https://vimeo.com/351297850/254e1eb8d2
  ROSEN

CHAPTER 12

Pipeline protection

Marcelino Guedes Gomes

12.1 Introduction

In the 1980s, pipeline companies in different parts of the world became aware of theft ('illegal taps') of product from their facilities. These 'taps' (valves) are easy and quick to install.

Do you believe that this is a new problem in the world? It is not. Product thefts are becoming increasingly common, with very rapid growth since 2000. This illegal tapping is no longer a problem only affecting developing economies: it has been occurring in Europe, alongside more widely reported incidents in Asia, Latin America and Africa.

Pipeline thefts are relatively rare in the United States although there are reports that North American Pipeline operators are concerned about the rising number of tapping incidents. This is therefore a real-world problem. The most important pipeline activity is in USA, Russia and Canada, where the longest pipelines are located. However, we have no information from these countries. Zero incidents in those countries? Probably not. It is a global problem that has been costing billions of dollars per year and poses a high risk to safety, the environment and company image. Besides that, illegal tapping can damage the pipeline, causing pipeline shutdown and a production stoppage, or operating pressure reduction.

There are three causes of malicious damage to a pipeline:

- Vandalism;
- Sabotage; and,
- Theft.

Vandalism and sabotage will not be presented in this chapter, even though some countries have been facing these difficulties. On the other hand, theft, which was probably carried out by amateurs in the beginning, is now certainly controlled by organized crime.

Pipeline protection is absolutely vital to the security of the world´s energy supply. Disruption of energy supplies can have disastrous effects on the lives of people, as well as on businesses and governments. Pipeline protection cannot be guaranteed, but can be managed through the combination of smart initiatives and resources.

Before starting to read this chapter, it must be understood that there is a dichotomy. It is not possible to teach pipeline professionals the solution to the illegal tapping problem without the risk of notifying the gangs involved. It was decided to take the risk.

12.2 Motivation

There are various factors that cause increases in the criminal activities of pipeline product pilferage. The principal motivation for a clandestine diversion is the quick financial gain, with low risk.

There is minimal investment to make an installation on a pipeline: a simple connection of valves and hoses makes the connection viable. We know that these days the clandestine diversion systems are becoming more sophisticated, installed using professionals. They are therefore more expensive, but even so continue being low cost when compared to the potential gains (see Figure 12-1).

Figure 12-1. Illegal tapping.

The market demand for the stolen product also provides a motivation to increase thefts:

- Gasoline, naphtha or refined oil products, are in demand in the automobile market, principally in the Americas, Africa and Asia;
- Diesel is in demand for European cars and transport and agricultural vehicles everywhere; and,
- Crude oil has differing uses in different countries, including exportation to the black market or inefficient usage for local consumption.

Illegal tapping is easy, with installation taking just thirty minutes, and a speedy distribution of the high value product to its consumers. Gangs have moved from drug trafficking, smuggling, hijacking or other crimes, to the theft of oil products, because there is low investment, low cost and low risk.

12.3 Who is tapping?

When we try to identify where the illegal tappers originate from, we find a variety of places. Since the 2000s, the economic crises throughout the world have given rise to a shortage of employment available. Some ordinary professionals, who lose their jobs, are attracted by an easy way of earning money, where there are no honest job opportunities.

Considering the level and type of information in the hands of pipeline professional people, whether direct employees or outsourced employees (pipeline walkers, maintenance and right of way (ROW) crews, etc.), the gangs have contacted them to share the knowledge. In some countries, immigration is considered to be a contributory factor.

In certain countries, some policemen or authorities have also been approached to warn or notify the gangs of official movements, and to work for their protection, maybe by minimizing or hiding evidence, etc.

To collaborate with the gangs is relatively simple and probably lucrative; however, once somebody chooses this path, there is no return, and there will never be an opportunity or chance to go back. The gangs could threaten his/her family and friends, thereby destroying the person's character, credibility, relationships, dreams, and life.

QR Code 12-1. Scan here to watch a video: Mexican Oil and Drug Cartels: Cocaine & Crude.

12.4 How to tap

First identify the area, the position on the pipeline, dig and then install the illegal taps.

Pipeline companies throughout the world identify their pipelines, showing their position, size, depth of cover, and product carried. While this information is essential in the prevention of accidental damage caused by third-party digging, it is clearly useful to criminals in identifying and locating product lines. Nowadays, there is a worldwide discussion about if we have to keep this procedure to identify pipelines along the ROWs with so much detailed information.

Having identified the pipeline to be tapped, the exact point of access is selected before digging can begin. Ideally, the entry point will be secluded or hidden among trees, bushes or undergrowth in the countryside, or a house or other building in a built-up area.

'Hot tapping' is the technique used to install a valve to a pipeline in operation, by creating an opening in the pipeline by spiking, drilling, or cutting into the steel. Figure 12-2 illustrates a typical hot tap installation.

Remember, that a hot tap is a very dangerous operation, which should only be carried out by professionals. Each hot tap should be properly designed; the hot tap location must be inspected, and the installation procedures approved. Hot taps are successfully completed every day all over the world by pipeline companies, which use highly trained professionals with appropriate specialized equipment, considering the high risk inherent in the operation.

QR Code 12-2. Scan here to view a video about manual hand tap machine hot tapping.

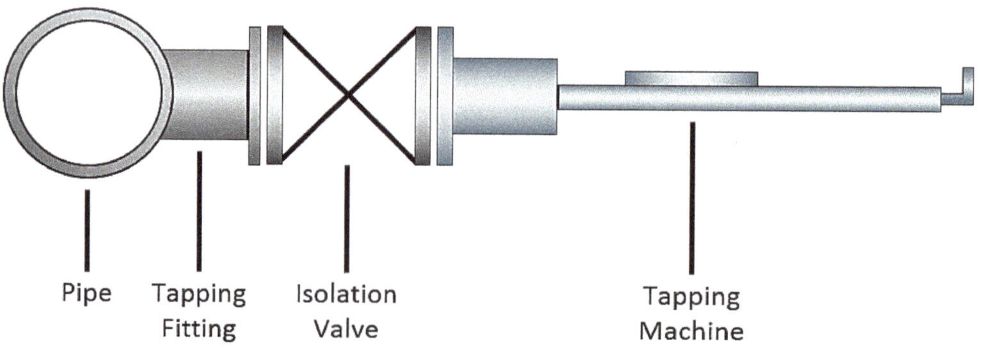

Figure 12-2. Tapping system.

12.5 Types of illegal tapping and connections

In this section, different types of illegal tapping will be shown.

12.5.1 Position

(a) Above ground

In certain countries, illegal tapping started in unprotected valve stations in the countryside. The process of tapping is simple, but it is easily identified by the line walkers. It can be disconnected quickly and only serves for short-term operations. Illegal taps later became installed on the actual pipelines running above the ground, for example, when a pipeline crosses a stream or a small river (see Figure 12-3).

(b) Underground

Pipelines are usually laid underground. The thieves prefer to steal from underground pipelines because the taps are hidden. Once the taps are installed, the soil is replaced covering the taps, making it more difficult for the line walker to identify. After the installation of a tap, the site may need to be camouflaged (see Figure 12-4).

12.5.2 Location

There are two principal classifications for the location of illegal tapping: rural; and urban / industrial. Each of these classifications has its own methods of installation and identification, but they are completely different:

Rural. In the countryside, the criminals will have much more time to install the taps and more natural camouflage to hide the installation. In these areas, there are fewer people to pay attention to the installation and notify the authorities or pipeline company (see Figure 12-5).

Urban/Industrial. On urban sites, there is no natural camouflage and there are always plenty of neighbours looking on. It is often necessary to employ wells and tunnels to access the pipeline in this situation. The local community will easily notice a sudden increase in traffic, especially oil tankers and trucks (see Figures 12-6, 12-7).

Figure 12-3. Pipeline crossing stream with illegal taps.

Figure 12-4. Buried pipeline with illegal tap.

Figure 12-5. Hidden tap installation in a rural area.

Figure 12-6. Different connections in warehouse.

Figure 12-7. Tunnel in urban site.

12.5.3 Types of tap

Metal spike. A spike into the pipeline is the most unsafe methodology used to tap pipelines, by criminals who are not concerned with the quality of this installation. There are two types of metal spike, with valve or without valve.

- With valve: This is simple. After digging to uncover the pipeline, the pipeline thickness is reduced by manual grinder, and the thinned wall is perforated by penetrating a spike into the pipeline. There is a high risk of leaks because of the low level of sealing between the spike and the pipe wall; therefore, this illegal tapping is used on few occasions.

- Without valve: After digging, the wall pipeline thickness is reduced by grinding. A pit and/or ditch is dug, just to keep the product contained to be collected immediately or later. People can steal the product in buckets, big cans, or in some cases with trucks. Their behaviour is like looters. There are some cases of huge accidents in the world, mainly in Africa, where hundreds of people have died. This method can only be used once.

QR Code 12-3. Scan here to view a video.

Spike – Nipple. It is not common to find this kind of illegal tapping, but it is safer than the previous methodology. In this process, it is not necessary to use a hot tapping machine. After grinding to reduce the wall thickness, a simple nipple is welded on to the pipeline surface. A spike (steel bleed nipple screw) with a valve is screwed into this nipple and it perforates the pipe wall allowing the product to flow. A hose is attached to this spike-screw so that the product may be transported for storage (see Figure 12-8)

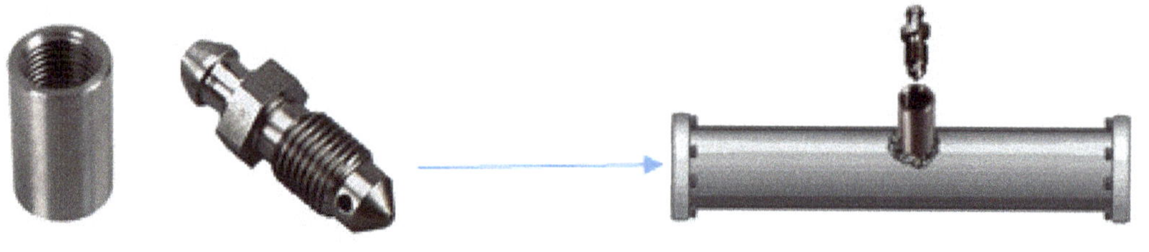

Figure 12-8. Spike – nipple (left); Nipple in place (right).

Clamp. This is the usual method of installing illicit valves. There are different types of clamp, and they have been found tapping several active pipelines belonging to companies all over the world.

Generally, they are applied to pipelines of up to 20 inches diameter (due to the difficulty of installation in locations having difficult access), and there being no necessity for electrical equipment, welding or lifting equipment. The type most commonly found is a clamp made from steel plates with a valve pre-welded in one of the parts. They may have one or two parts, but there are different installation techniques:

- The two parts are bolted together using O-rings to seal, Figure 12-9;
- There is one part which is secured to the pipe using bars or ties, Figure 12-10; and,
- There is only one part secured to the pipeline using chains. Figure 12-11.

A high risk of a leak is caused by the possibility of low-quality sealing between the interference clamp and the pipe-wall, and the material and type of sealing joint utilized.

Welded. After the clamp, welded taps are the most common type of illegal tapping found. The additional difficulty is the necessity for welding equipment to make the connection between the illegal devices and the pipeline. Generally, direct welding is utilized for theft from pipelines of over 20 inches diameter, because any clamp would be too heavy to be practical. Most illegal welded taps have been installed at the 12 o'clock position on the pipe (see Figure 12-12). The welded taps demand great care in installation on pipelines in operation (hot taps). There are several cases of pipeline perforation followed by explosion, fire and similar consequences.

12.5.4 Instrumentation and connections

In some illegal taps, we have found instrumentation systems to assist with the installation of the connection, and to supply the parameters for the operation of product withdrawal (see Figure 12-13).

When we have found instrumentation, all had pressure meters, with some systems being a little more sophisticated and having flowmeters. They were generally non-intrusive, ultrasonic flowmeters to measure the flow rate of the product removed. In one of the most sophisticated set-ups, a supervisory system was found, that controlled the product withdrawal, and compensated with the injection of water.

Another important consideration is the generation of hydrocarbon vapor, and in some installations, there are pieces of equipment and devices to minimize the generation of vapor. Pipeline operators distribute guidelines to local residents on how to live near pipelines safely, and one point raised is that if hydrocarbon vapor is seen, it is dangerous and should be reported to the pipeline operator.

Figure 12-9. Common clamp.

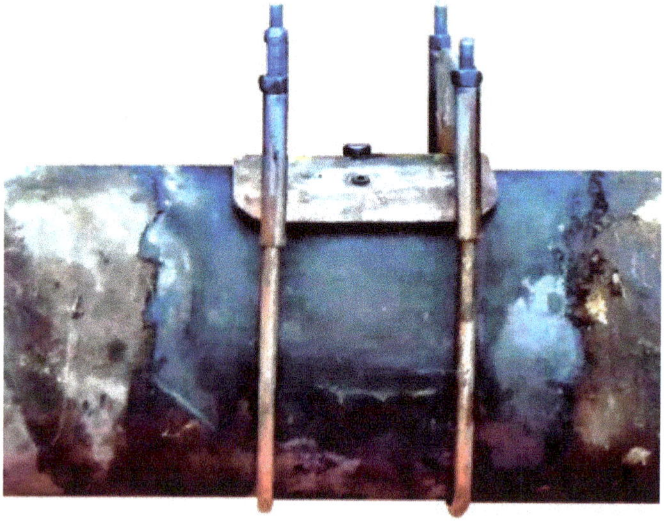

Figure 12-10. Illegal tap with ties.

Figure 12-11. Illegal tap with chain.

Figure 12-12. Illegal tap welded.

Figure 12-13. Illegal tap with pressure gauge.

It is not common to find the dissipation of the hydrocarbon vapours in clandestine storage installations. Only a few gangs are aware of this problem and they do not want to attract attention to their activities. These vapours often bother the neighbours of the installations, who, therefore, generally communicate the fact to the police authorities or pipeline operators.

There are two types of connections between the illegal tap (hole) and the withdrawal point (valve) of the product:

- High-pressure hose with crimped ends; and,
- Galvanized tubing.

In some cases, low-pressure garden hoses with simple hose clamps have also been found.

12.5.5 Number of holes (see Figures 12-14, 12-15, 12-16)

It is easier to drill small holes due to the weight of the equipment necessary. So, three or four small holes close together can remove a volume similar to one large hole, but they are easier to open.

A simple hole can be of any size but the smaller the hole, the more difficult it is to detect by the pipeline control room. Criminals often begin with one small hole and, if not detected by the pipeline operators, they can open the second hole/valve near the first. It is common to find holes from 3/8 in (10 mm) to larger than 1 in (>25 mm). There is illegal tapping with multiple holes from 2 up to 10 holes.

12.5.6 Withdrawal point

The withdrawal point is where the product leaves the pipeline through a storage system. This may occur within the right of way where the withdrawal point is directly on the pipeline, or where there is a short hose from the pipeline to the withdrawal point. In either case, there is a risk that the connection between the withdrawal point and the storage system may not be perfectly secure. This would allow telltale drips to become visible on the ground to both the line walker and sniffer dogs. The tap would be more difficult to identify using this method if the hose were underground, and at least fifty meters from the pipeline outside the right of way. With a simple tap on the pipeline itself, it could be easily detected by the line walker or sniffer dogs.

Figure 12-14. Tap with one hole – one valve.

Figure 12-15. Tap with two holes.

Figure 12-16. Tap with three holes.

12.6 Types of storage

When speaking about "storage", we are talking about the types of recipient we find on the other side of the connection. Or, in other words, how the stolen product will be physically held or transported. In very rare cases, there is a consumer connected directly to the pipeline. In most cases, however, there is a tank or vehicle receiving the stolen product. In the case of tanks, small fixed metal tanks with volumes from 4 m^3 up to 50 m^3 have been found in residences or warehouses. Presently, the criminals are widely using small containers known as IBCs - intermediate bulk containers, which are reusable, multi-use industrial containers that have been engineered for the mass handling, transport and storage of liquids.

The vehicles used for transporting the stolen product are tanker-trucks and/or cars, trucks or vans camouflaged to facilitate passing police control points. There are all types of camouflage: water transport trucks that are carrying hydrocarbons, skips for stones or sand, trucks for the delivery of drinks or supermarket vehicles, etc. In the same way, there could be school transport, ambulances or private vehicles, etc. (see Figures 12-17, 12-18, 12-19, 12-20)

12.7 Detection and dealing with illegal tappers

In this section, we are presenting the initiatives to deal with and combat clandestine tapping activities. We can broadly divide the operations and applications into two groups:

- Remote installations, such as valve stations and pumping stations; and,
- ROWs.

12.7.1 Valve stations (see Figure 12-21)

There are various options:

- Metallic fences with barbed wire;
- Metal cage;
- Sensors for the opening and closing of doors and sensors for presence of people;
- Monitoring cameras, all of these systems connected to the Pipeline Protection Control Centre; and
- Welding of the line accessories and secondary connections.

Figure 12-17. A plastic water tank.

Figure 12-18. A set of IBCs.

Figure 12-19. Minivan with tank.

Figure 12-20. A truck with steel tank.

Figure 12-21. Valve station with metallic fence.

12.7.2 ROW solutions

ROW aerial monitoring. These are devices to detect the arrival and movement of interlopers in the ROWs before starting digging activities. We may have images from satellites, aerostats, helicopters, drones with rotors, and fixed wing drones.

Satellite images. Satellite images are tools that may greatly facilitate the work of monitoring ROWs. This solution of monitoring changes along and around the ROW is an interesting option to observe constructions and alterations in the vegetation, and even digging. It is not continuous, because satellites are orbital and accordingly we receive discrete images (photographs). It continues to be a high cost solution and the interpretation systems for illegal tapping are still being perfected. The images may be acquired from several suppliers due to the different constellations of satellites, and have a resolution to identify any object of 1.5 meters to 0.25 meters.

Aerostats. Aerostats are different from airplanes and helicopters: these surveillance balloons do not need pilots on board, and in comparison to aerial photography drones, they offer much more versatility. Even the best drone for aerial photography cannot stay in the air for the long periods of time usually needed to capture the perfect shots.

Surveillance balloons can shoot photos and stream video of the ROW scene to meet specific missions, in order to detect intrusions, and they also provide night vision surveillance. These balloons can carry a payload of up to 18 kgs. (40 lbs), provide surveillance from an altitude of up to 300 meters (1000 ft) with a radius of 5 km (3.1 miles). There is a portable ground control station where the technician can analyse and process the transmitted surveillance photos, videos and data. This tool is recommended for surveillance on ROWs located in flat areas (see Figure 12-22).

Pipeline patrol helicopters/airplanes. Pipeline Patrol is the most useful inspection tool in ROWs to verify leaks, encroachments, or other abnormal conditions, such as illegal tapping. It is commonly performed by using helicopters or small airplanes. They normally focus on the observation of changes that take place along the ROW. The main aim in these inspection services is to discover damage or criminal activities to access the pipeline, and so it is possible to mitigate damages and losses. It is however very expensive in comparison with other techniques.

Drones. Drones are being used by pipeline companies, more and more to carry out ROW inspection. Drones can help in detecting leaks, illegal tapping, vegetation changes, land surface feedbacks, and other potential problems associated with a pipeline. Commercial drones provide additional flexibility in that they can be scheduled to cover long distances systematically. Drones can be used to carry out thermal inspection of pipelines to detect liquid leaks caused by illegal tapping, which can be captured by thermal imaging due to temperature differences between the fluid and the surrounding soil. They can also detect the presence of interlopers.

Drones can dramatically reduce the high costs, safety risks, and time involved with conventional surveillance methods. As with any technology of this type, there are rules and regulations in each country that control commercial and industrial use of drones. All operations near the ground need to be conducted carefully. Policies such as those related to intrinsic safety have long been in place to protect people and facilities.

The main limitation of small drones is in their flying time. You can expect between 10 and 30 minutes in the air. Despite the extraordinary benefits and numerous applications of domestic drones, it is not recommended to hire domestic tools. It is also not recommended to fly drones in rain or even in drizzle or strong wind. There are different types of drones and this fantastic application is just beginning (see Figure 12-23)

Intrusion detection. Detection and tracking of interlopers is a major problem in this application. When intrusion occurs in large terrains, such as ROWs, it is difficult to detect.

Fibre optics. This equipment has been used for more than 10 years to monitor the pipeline ROWs. The basic idea is that it uses fiber optics as a sensor to collect signals produced by soil vibration near pipelines, and then focuses on intelligent processing and smart recognition of soil vibration signals. The model is calibrated for each sector of pipeline. Experiment results show that this system can fully satisfy the real time requirement. Furthermore, the level of training for the operators of this system is an important factor in minimizing the false alarm rate. After six months of calibration, the accuracy level in determining a real threat will be higher than 98%. The cost of installing a fiber optics cable to be used only for monitoring the presence of outsiders can be prohibitive; however, when the cable is also used for other communications this solution becomes viable (see Figure 12-24).

Acoustic devices. The Intrusion Detection System draws upon more than thirty years' experience in the field of acoustic event detection and location. The methodology operates on the basis that sound waves propagate at velocities that are characteristic of the surrounding soil. The difference in arrival times of an acoustic signal at a set of fixed detectors, geophones, whose position is known, defines the location of the source of disturbance.

Seismic sensors are installed at intervals along the ROW of the area under surveillance. There is no limit to the length of the perimeter; it can range from tens to tens of thousands of meters. The sensors are buried at a depth of 0.25 – 1.0 m to pick up ground vibrations, and then the acquired data is processed by a computer to obtain the location of the ROW invaders (see Figure 12-25).

Figure 12-22. Balloon – Aerostat: for surveillance.

Figure 12-23. Drone operating over a ROW through a swamp.

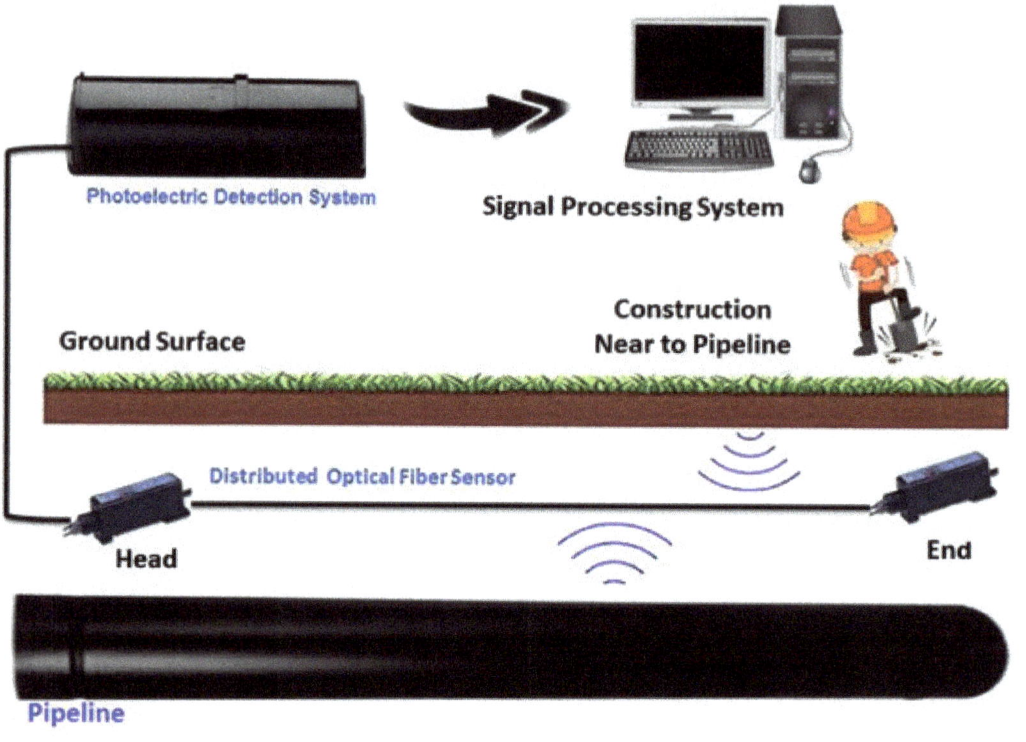

Figure 12-24. Schematic diagram of fibre sensors.

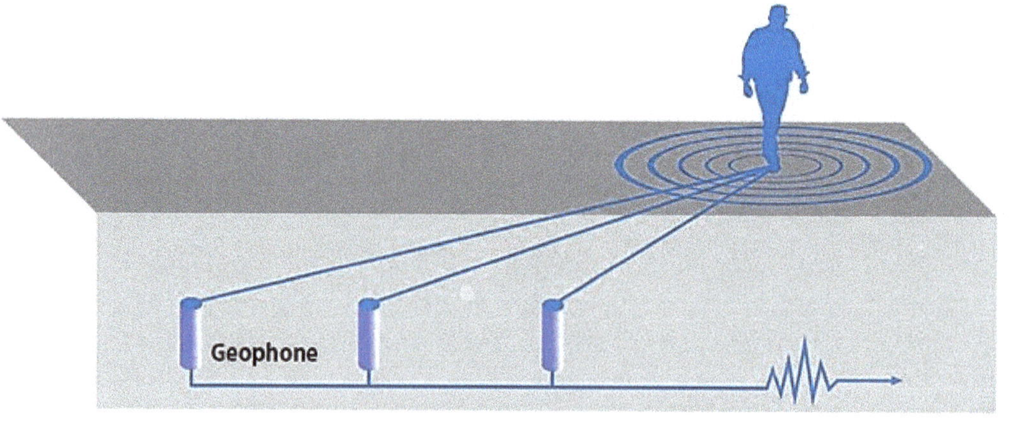

Figure 12-25. Schematic diagram – acoustic sensors.

12.7.3 Tap detection and location

It is very important to differentiate between detection and location. Existing taps can be detected by measurement of product loss, and drops in pressure; however, to pinpoint the exact place of theft is more difficult.

Pipeline Control Rooms: These are facilities from where pipeline operations are monitored using Supervisory Control and Data Acquisition - SCADA systems. Trained operators monitor flow, pressure and many other parameters specific to the operations of a pipeline, 24/7, 365 days a year. Sometimes it is not easy to identify the differences between leaks and illegal operations. The operators need to be highly trained.

There are two methodologies applied to detect illegal taps.

1. Based on software (computational algorithms, which can analyse data of product transported with a view to identifying theft).
   a. Mass balance. The volume of liquid entering a system is compared constantly with the volume leaving the system. Any differences arising could be either a leak or a theft. Temperature and pressure are used to calibrate the model.
   b. Alteration in pressure. Using a simulator, the drop in pressure is compared with the expected result. If a significant difference is found this could indicate a possible theft, or leak.

2. Based on hardware (sensors throughout the pipeline which can detect the theft).
   a. Negative pressure wave – NPW. Pressure sensors are installed at both ends of the pipeline section to be measured. A negative pressure wave on the monitor will demonstrate a loss of liquid and an indication of the exact point of loss. The opening and closing of an illegal tap will provoke small variations in the pressure line at each end. The timing difference between the effects at each end will allow an estimation of the location of the loss. The synchronization between the end sensors is critical.
   b. Acoustic methodology – hydrophones or acoustic devices are installed at both ends of the pipeline section to be measured. They can measure the vibration caused in the pipe wall and the fluid due to the opening or closing operation of an illegal tap. The timing difference between the effects at each end will enable an estimation of the location of the escape.

In-line inspection ('pigging'). Due to the increasing number of illegal taps, several pipeline service companies have been developing simplified MFL (magnetic flux leakage) inline inspection tools to inspect pipelines with high frequency. These MFL 'pigs' are tools applied worldwide and have increasingly lower costs. A pig for finding clandestine operations has to be inexpensive to allow many runs. The use of these smart pigs is a more efficient technique in finding the exact locations of leaks or taps. One important recommendation is to maintain secrecy in these operations. The deadline for receiving the pig reports should be a maximum of three days. This will allow the intelligence groups to work and act before possible changes in the configuration of illegal taps along the pipeline.

Ground-penetrating radar – GPR. This is a geophysical method that uses radar pulses to image the subsurface. This methodology uses electromagnetic waves (UHF/VHF frequencies) of the radio spectrum and detects the reflected signals from subsurface structures. GPR can be used to detect tunnels, pipes, cables, metals, rocks, water and different structures underground. Without GPR, it is difficult to detect the tunnels, which have been used increasingly by the gangs. So, whereas a line walker cannot identify tunnels, the GPR can.

Sniffer dogs – K9s. An unusual smell, such as hydrocarbon odor will sometimes accompany pipeline leaks or an illegal tapping. Sniffer dogs are one of the most efficient methods to pinpoint illegal taps along a pipeline and inside the ROW. These dogs can do tap detection surveys on in-service pipelines, and they can detect all kinds of events with hydrocarbon smell at ppb (parts per billion) and can outperform modern technology available that only detects at ppm (parts per million). This technique is more recommended for locating illegal taps within the ROW. It is often applied to locate the tap accurately when the general area is known. This method is highly efficient in rural areas but less so in urban areas. One team comprises two dogs, a line walker, one vehicle and driver. Each dog can work an average of 3.5 to 4 km per task, so one team can inspect up to 8 km per day. It is necessary to keep the dogs focused and dig manually when a possible tap is found. The safety of the team is paramount. (see Figure 12-26).

12.7.4 Pipeline protection control room – PPCR

When a pipeline operator decides to structure a dedicated group to combat criminal activities, one of the most important initiatives is to have a control room. The control room receives all data and images collected and distributes to interested parties. From here, it is possible to accompany all initiatives being undertaken by the company at any point in time. For example, the line walkers, inspection cars, images from CCTV, drones, satellites, etc., pig reports and denunciation and social media reports.

The principal role of this centre is to integrate the information received and activate any remedial action necessary to prevent theft in real time.

12.7.5 Investigation-intelligence

Just as there are specialists in pipeline integrity, smart pigs, pipeline operations, and hydraulic analyses, there are also experts in the intelligence area. To combat illegal activities in the pipeline industry, it is fundamental to create a secret group to survey the data from the field, information coming from neighbouring communities, dedicated investigations and other sources to prepare reports to be sent directly to the Police Department for appropriate legal action. It is recommended to establish a formal agreement with the Police Department to facilitate this liaison. We cannot forget that in most cases of illegal acts, organized crime is involved, and therefore efficient intelligence is mandatory.

Figure 12-26. Sniffer dog near pipeline marker.

12.7.6 Denouncement channel - public awareness

One important strategy to engage the public, who may wish to denounce something they have seen or heard anonymously, is to establish a direct and easy link with them. To file a report, they would only need to call the denouncement channel. A three-figure telephone number is recommended because 0800 numbers are too long to memorize. The chosen number must be divulged strongly along the ROW. Regular campaigns help to reinforce this number. Baseball caps, refrigerator magnets, pens, and small gifts have all been used (see Figure 12-27).

Denouncements could include any activities related to fuel theft, illegal fuel storage and the sale of stolen fuel. This channel is in addition to the social media channels of WhatsApp, Facebook, etc. It is easy to photograph or film any illegal event and forward it to the pipeline company or the authorities.

The relationship with communities is very important because their eyes are the eyes of the pipeline companies. In general, pipeline companies have been investing in technology and other expensive methods, but to engage the local people is more efficient. To establish social programs, such as the kitchen garden program, helps the community to identify more closely with the pipeline company.

In addition to liaising with the Police Department, it is important to liaise with community leaders. These leaders could suggest initiatives and social programs to improve safety and relationships along the ROW. The benefit is increased safety for both the community and the pipeline.

Figure 12-27. Oil pipeline sign with 168 free call.

12.8 Found an illegal tap? What next?

Once an illegal tap has been found, it is necessary to make a series of decisions, and their sequence is critical (see Figure 12-28).

1. Is the tap or connection leaking? If yes, repair immediately. If no, go ahead.
2. Is there any risk to life or health? If yes, repair it immediately. If no, go ahead.
3. Is there any risk to the environment? If yes, repair quickly. If no, go ahead.
4. Is the illegal tap active? If no, repair it. If yes, go ahead.
5. Call the intelligence group. This group will be responsible for establishing a contact with police and security groups. They will determine the methodology applied in the tap.

These decisions, to go ahead to next questions, are not easy because the location cannot be altered as this would alert the criminals that the tap has been found. They would easily notice if the pipeline company decided to dig around the pipeline. It is very important to clarify that the final target of the pipeline companies is to eradicate the organized group involved in illegal tapping, and not just to arrest the truck drivers or ordinary people involved.

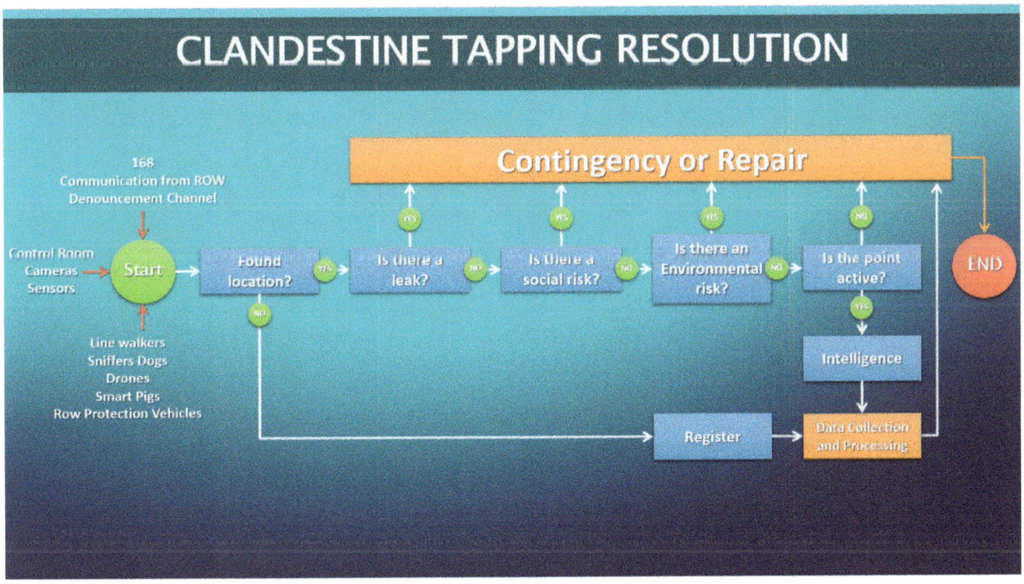

Figure 12-28. Diagram for resolving illegal tapping.

12.9 Law, legislation and local regulations

The Federal Legislative Powers of some countries have proposed specific laws to prevent the illegal tapping in pipelines from being so profitable. Specific laws are necessary because the old laws did not envisage this type of crime. Despite the difficulties in passing a new law, this initiative would help judges determine the appropriate penalties for this crime. The guilty parties in this crime should include all those who participate. The gangs involved are extensive from technicians, drivers, property owners and those who finance the activities.

12.10 How to manage: daisy methodology and dashboard methodology

Illegal tapping is a complex issue and the methodology for combating it will necessarily differ in different countries due to the legal system, people and culture. Accordingly, there is no one procedure to be followed universally, which will resolve everything. Further, the methodology to be followed will vary in degree according to the country situation and boundary conditions.

We have been using a combination of Daisy Methodology and Dashboard Methodology.

Daisy methodology (see Figure 12-29). We have noticed that in several countries this methodology has been put into use to some extent. Each petal represents a different approach, which taken together help to resolve the problem. The petals may be classified into three separate groups:

- Security, intelligence and official/external relationships;
- Engineering-technology, operational procedures and HSE matters; and,
- Legal, social and communication actions.

Dashboard methodology (see Figure 12-30). This is the weighting given to each petal in the daisy. Some of these are applicable in certain situations but not others. So, we must measure our efforts for each parameter, as we would use a control panel. What is good for France or the UK may not work in Spain, Brazil or Mexico. For example, a change of law may be necessary in one country but not necessarily in all. Social initiatives will work in some countries but are totally inapplicable in others.

Figure 12-29. Daisy methodology considerations.

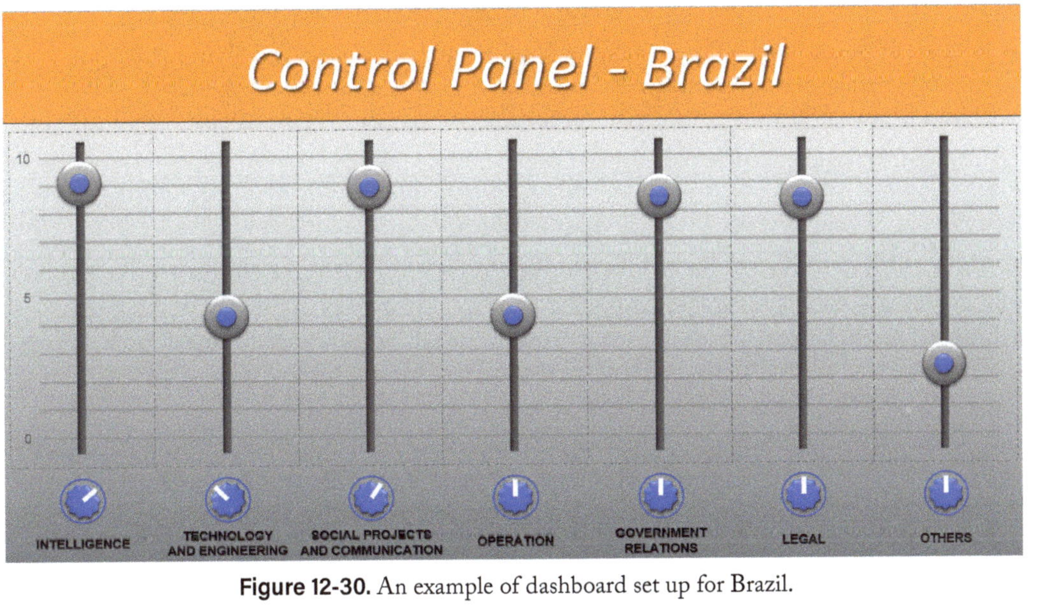

Figure 12-30. An example of dashboard set up for Brazil.

12.11 Final considerations - the role of the pipeline community

Although pipeline operators have an Illegal Tapping Management System, and some prevention technologies implemented, these are events where they cannot control either the timing or location. Occurrences reported in Mexico, Colombia, Nigeria, Spain, Turkey, France, UK, Italy, Brazil, China, India and Russia have a higher incidence than others.

The last 20 years have brought an increase in illegal pipeline tapping. This has fundamentally changed the role of the pipeline industry and will continue into the future. It is a new discipline for us. It is a challenge that causes us to use new skill sets involving several aspects. It is not only a technological challenge, because it involves the law, social programs and interaction with all types of people from ROW communities to criminals. These problems have not been faced before. They are not comparable with technical challenges because it is new ground.

We must now work together to solve a problem which is common to all of us. It will not be easy because the subject is wide ranging, and it is time to share our knowledge and experiences.

12.12 References

[1] I M Ralby, 'Downstream Oil Theft – Global Modalities, Trends and Remedies', Atlantic Council – Global Energy Center, January 2017.

[2] B Corlett,'Using the latest ILI Technologies to identify pipeline pilferage', Technology for Future and Ageing Pipeline Conference, Gent, Belgium, 11-12 April, 2018.

[3] A. Ambituuni, P. Hopkins, J. M. Amezaga, D. Werner, J. M Wood, 'Risk assessment of petroleum product pipeline in Nigeria: the realities of managing problems of theft/sabotage', 6th International Conference on Safety and Security Engineering. Opatija, Croatia. Wessex Institute. 6-8 May 2015

[4] Anon., 'Report 6/18 - CONCAWE 2018, Report Performance of European cross-country oil pipelines Statistical summary of reported spillages in 2016 and since 1971', CONCAWE, March 2018.

[5] R Landstorfer, 'Hot tapping and product theft on pipelines: A way to detect and locate these spots during normal operation', 11th Pipeline Technology Conference 2016, May 2016.

CHAPTER 13

Pipeline defect assessment basics[1]

Dr Phil Hopkins

13.1 Introduction

Structures are made from materials, and materials fail. A failure means a structure cannot carry out its function or cannot meet its specification or expectations. These structures can fail for a variety of reasons ranging from being overloaded beyond their design limits, to developing a defect (such as a crack) during service.

Defects in structures can decrease their ability to function, and can lower their strength; consequently, we need to assess the effect of any defect in a structure (see Figure 13-1).

This Chapter covers the effect of defects in structures, and methods for assessing these defects, but first we need to understand why structures fail, and what we mean by a defect.

13.1.1 Why do structures fail?

Structures have to withstand forces; for example, tall buildings have to withstand forces created by the wind, and road surfaces have to withstand the forces from passing traffic. These forces can lead to failures.

1 This Chapter is based on the e-learns available at the Competence Club (https://competence.rosen-group.com/) and is courtesy of Rosen Group.

What is stress in a structure? Why are structures sensitive to these forces? A force is a push or pull on an object. The force causes the object to stretch or compress. This stretching or compression creates a strain in the object. We can see this strain as deformation of the object. Stress is created in the object by this straining: it is the resistance to the straining. Stress is always within the object and is caused by the deformation.

These stresses can be so high, that they fail a structure. Obviously, structures cannot withstand higher and higher stresses: they are limited by the strength of the materials the structure is made from. These materials have a limiting strength. This limiting strength is a material property called the yield strength or tensile strength, and when this strength is exceeded, the material (and hence the structure) fails.

Cause of failures. We have all seen failures, ranging from broken hinges on the doors in our house, to cracks in our highways. Structures fail due to a combination of:

- the forces (or loads) on the structure (the forces on a structure cause a stress in the structure);
- any defect in the structure (a material or design defect will weaken the structure, or elevate stresses, and accelerate failure);
- the properties of the materials that make up the structure (strength, shape, etc.);
- time the force operates (the longer a structure operates, the more likely it is to fail (for example, due to repeated loads called fatigue));
- the environment the structure operates in (e.g., temperature, corrosion, or irradiation, can, over time, weaken the material).

This Chapter is concerned with defects, such as dents and cracks, failing structures.

13.1.2 What is a defect?

We have many definitions, or alternative words, for a defect in a structure (see Table 13-1). Most describe a defect as something that should not be there or exceeds a limit that has been specified or agreed for that structure.

It is often easy to see defects in structures. Most of the homes we live in will have cracks in the walls and if we own a bicycle, we will often experience a flat tire caused by a hole in its inner tube (see Figure 13-2). These defects are not going to cause an immediate disaster, but we do need to know if they pose a longer-term problem, and if we need to replace or repair the defective component.

Table 13-1. Definitions of defect, etc. [1]

Term	Definition
defect	"imperfection of a type or magnitude exceeding acceptable criteria".
imperfection	"flaw or other discontinuity noted during inspection that may be subject to acceptance criteria during an engineering and inspection analysis".
flaw	"imperfection that is smaller than the maximum allowable size".
discontinuity	"interruption of the typical structure of a material, such as a lack of homogeneity in its mechanical, metallurgical, or physical characteristics".
anomaly	"a possible deviation from sound... material or weld".

QR Code 13-1. Scan here for two case studies of pipeline failures caused by defects.

Figure 13-1. Defects in structures can lower their strength.

Figure 13-2. Defects, and their effects, are a common sight.

13.1.3 Defects in pipelines

Pipelines can contain defects such as corrosion and gouges (see Figure 13-3). Defects can be caused during the manufacture of the line pipe (our pipeline is constructed using tubular steel sections called line pipe), fabrication of the pipeline on site, or operation of the pipeline.

Most of these defects are introduced during the operation of our pipeline by threats to the pipeline; for example, buried pipelines can be damaged by earth moving equipment that is not aware of the presence of the pipeline. The damage caused by this external interference can puncture, gouge, or dent the pipeline. The threats to a pipeline are summarised in various publications (for example [2]). Table 13-2 gives a simple summary.

These threats create differing types of defects in the pipelines; for example, corrosion pits, dents and gouges in the pipe, and cracks in welds in the pipe. The threats can also fail the pipeline; for example, Figure 13-4 shows the main causes of failures in European natural gas pipelines (from 2007 to 2016).

It should be noted that some threats are on the increase: theft of product from pipelines (by illegally tapping (drilling) into the pipe) is on the increase in many countries; for example, Figure 13-5.

Figure 13-4 and Table 13-2 show that pipelines may contain defects such as corrosion and cracks; therefore, we must ensure we detect these defects before they can fail the pipeline, and then assess their significance when we detect them.

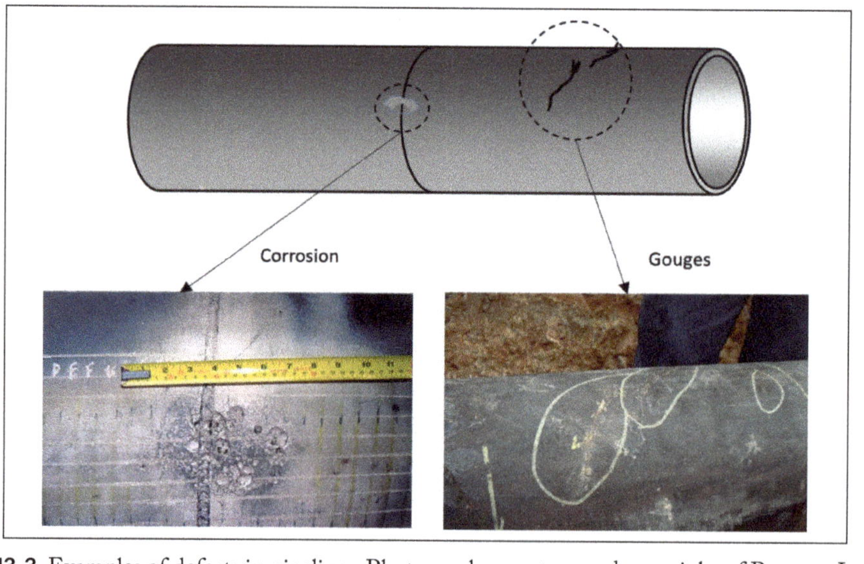

Figure 13-3. Examples of defects in pipelines. Photographs courtesy and copyright of Penspen Ltd, UK.

Table 13-2. Threats to a pipeline.

Threats
Corrosion (on the internal or external surfaces of the pipe).
External interference (impacts on the pipeline (e.g., from earth-moving equipment (onshore pipelines) or anchors (subsea pipelines) causing defects such as dents, gouges, or punctures).
Material defects (in the pipe and its welds).
Cracks in the pipeline (e.g., caused by fatigue).
Natural environmental force/hazard (e.g., cold weather or floods damaging the pipeline).
Man-made environmental force/hazard (e.g., mining subsidence).
Design errors.
Equipment error (e.g., problems with valves, flanges).
Operation/operator (human) error.
Vandalism/theft/sabotage/terrorism.

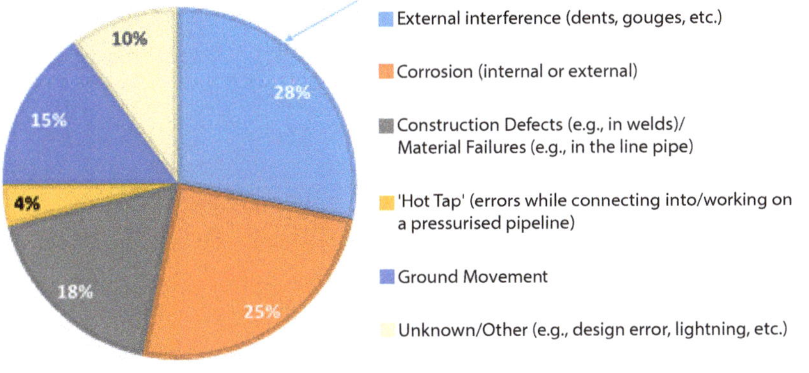

Figure 13-4. Causes of failures in onshore European gas pipelines [3].

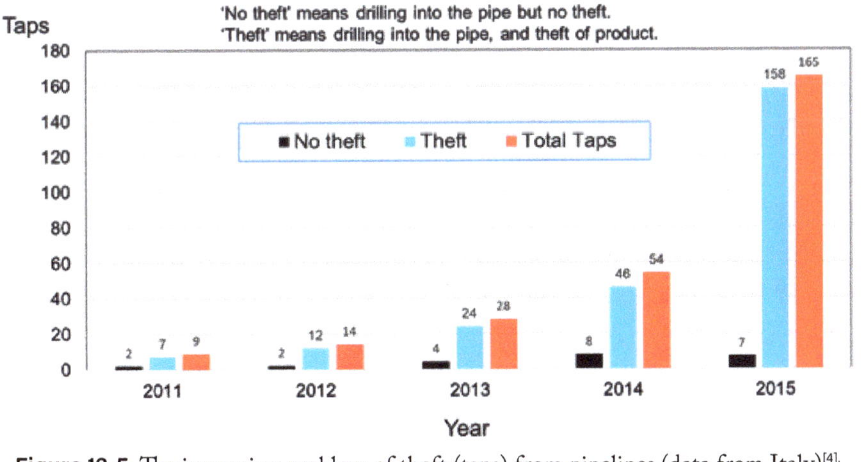

Figure 13-5. The increasing problem of theft (taps) from pipelines (data from Italy)[4].

13.1.4 Assessment of defects: fitness for service

A defect normally means something not within a requirement in a standard (see Table 13-1) so you might ask 'Why assess a defect? All defects should be repaired?' This is a reasonable question, but an unreasonable solution, as all structures contain defects. Nothing is perfect; therefore, defects in new structures are necessarily limited by quality control levels. These quality levels are usually based on past experience of building the structure and are the limits of good workmanship. When you buy a new house, you are expecting good workmanship, but you would not expect perfection.

As a structure ages we find defects in excess of these original quality control levels (e.g., Figure 13-2). A house is a good example of aging defects, as you cannot compare a very old house with a new house. Old houses have many defects which would not be accepted in a new house, but you would not repair every defect, or replace the old house. You would evaluate the defects and determine if they are a threat to the safety of the house.

We need to determine if these larger, or more numerous defects, need to be repaired or not. We do this by assessing the defects using engineering calculations. A British Standard[5] gives the reasoning behind these assessments:

> 'In circumstances where it is necessary to examine critically the integrity of new or existing constructions… it is also necessary to establish acceptance levels for the flaws revealed. The derivation of acceptance levels for flaws is based on the concept of fitness for purpose. By this principle, a particular fabrication is considered to be adequate for its purpose, provided the conditions to cause failure are not reached. A distinction has to be made between acceptance based on quality control and acceptance based on fitness for purpose.'

> 'Quality control levels are usually both arbitrary and conservative but are of considerable value in the monitoring and maintenance of quality during production. Flaws that are less severe than such quality control levels as given, for example, in current application standards, are acceptable without further consideration. If flaws more severe than the quality control levels are revealed, rejection is not necessarily automatic. Decisions on whether rejection and/or repairs are justified may be based on fitness for purpose, either in the light of previously documented experience with similar material, stress and environmental combinations or on the basis of an "engineering critical assessment" (ECA).'

In summary:

- fitness for service, or fitness for purpose assessments are quantitative engineering evaluations that are performed to demonstrate the structural integrity of a component containing a defect[6];
- these evaluations are often called an engineering critical assessments (ECAs), and are detailed in international standards (e.g., [5, 6]); and,
- these evaluations use theories and models based on fracture mechanics.

The evaluation will tell us how much the strength of a structure is reduced by the presence of a defect. We will cover fracture mechanics later in this Chapter.

13.1.5 Perspective on defect assessment

Defect assessment methods provide a quantitative measure of the acceptability of a defect in a structure – they will calculate the stress the structure will fail at, due to the defect, and we can then decide if this stress is too low. The methods are both useful and powerful, but they must not to be used in isolation: they are part of an overall process for the management of defects, and the wider integrity management plan for the structure or system[5]. This process needs to address many factors, including[5]:

- the cause of the defect, and remedial action to prevent further occurrences or growth;
- whether a previous inspection failed to detect this defect, and if the defect has not been detected, what are the reasons;
- the previous history of the structure and whether it can explain the occurrence of the defect (its nature, location and size);
- whether an inspection indicates that the defect has grown, and the observed growth is consistent with assumptions about loading and defect growth rates, after allowing for uncertainty in the inspection results;
- the implications for other structures of the same or similar design and whether modifications to the structure, or a change in the service conditions, might be required; and,
- whether there is a pattern of the defect being detected in other structures of the same design.

Clearly, there is a need to conduct a full engineering evaluation of the cause and effect of any defect detected in a structure and manage the whole process. The next section gives an example of what this management of defects entails.

13.1.6 Example of defect management

Cracks can be created during the manufacture of materials and products, and the fabrication of structures. This Chapter is concerned with assessing the significance of defects, such as cracks, but the assessment is just one part of the management of a defect. This section briefly explains how pipeline operators manage pipelines containing cracks.

Cracks can be present in pipelines. They could be present in the line pipe used to construct the pipeline, or introduced during the construction of the pipeline, or during operation.

Pipelines operate under a safety management system (for example [7]). These systems identify and manage risk and improve safety. A key element of a safety management system is the integrity management program (for example [8, 9]), where the policies, processes, and procedures to prevent failures are documented.

These systems give an overall process to manage a wide range of defects in a pipeline, but it is better to have a special process for some defects that are both common and dangerous. This is the case for cracks in pipelines: a specific management process is available[1].

Crack management involves:

- identifying the effect of the cracking on pipeline integrity;
- conducting a risk assessment on the pipeline to determine the segments of the pipeline at most risk; and,
- establishing reassessment (by inspecting the pipeline or testing the pipeline) frequencies.

This management will require[1]:

- determining the necessary pressure reductions needed in a cracked pipeline;
- presenting suitable methods for assessing the condition of the cracked pipeline;
- establishing criteria for inspection and testing, to detect the cracks and prove integrity;
- calculating the remaining lives of cracks remaining in the pipeline; and,
- developing preventive and mitigative measures.

This process shows that defect assessment is just one part of defect management.

We will now start to explain how we assess defects, but first we need to explain the basis of defect assessment methods. This basis is made up of:

- structural mechanics (the effect of loads and deformations on materials); and,
- fracture mechanics (the effect of a defect such as a crack on materials).

13.2 Structural mechanics

All structures are subjected to loads and deformation. Structural mechanics is the study of stresses (loads) and strains (deformation) in engineering structures.

13.2.1 Early days and the strength of structures

We have many magnificent, beautiful, ancient engineering structures (e.g., Egypt's pyramids, or the Roman viaducts) that have survived for thousands of years. We know these structures must have been designed, but we do not know how. We suspect it was mainly by trial and error, or design and failure... but the results are impressive. Remember, these early pioneers drew with quills and ink, worked under candlelight, and had a life expectancy of 30 years....

Our ancestors understood the importance of the strength of a structure: the famous Roman architect Marcus Vitruvius (70 BC - 15 BC) wrote that an architect should focus on three themes when designing a building:

- firmitas (strength);
- utilitas (functionality); and,
- venustas (beauty).

His ideas of beauty were based on symmetry and proportions.

The modern development of structural mechanics began with an increasing demand for the structural analysis of domes in medieval cathedrals. The best examples are the dome of Santa Maria del Fiore in Florence (1415 -1436), and the dome of St Peter's Basilica in Rome (planned in 1506, but completed in 1590). The only available design method at that time was to build a small-scale model of a dome; however, this method was not very reliable, because of the scale effects.

What are scale effects? We can easily build small models of structures. It is easy to build a small boat to model a large boat. It is easy to model a large building with a small-scale model, but these models will not use the same materials, the same thicknesses, the same joining processes, etc., and can only present a model that will not represent the full-scale structure. A good example was designing obelisks: small obelisks were simple to build, difficult to fail, and lasted many years. When an engineer wanted to create a much taller, heavier obelisk, it was very difficult to build, fragile, and would fail after a few years. The scale effect was evident....

The main motivation in the early development of structural mechanics was to understand the strength of materials, and therefore control the fracture of materials used in new and larger buildings. There was an increasing demand for bigger buildings (castles, cathedrals, bridges, etc.), and these bigger buildings (higher, wider, deeper, and heavier) created huge loads on the supporting structures.

Many great engineers and scientists were attracted to designing and understanding structures; for example, Michelangelo (1475- 1564) was involved in the design of the Basilica dome in Rome. Galileo Galilei (1564-1642) had an interest in many things, including the strength of materials (see Figure 13-6). He investigated the breaking loads of rods under tension and correctly concluded that the load was:

- proportional to the cross-section area of the rods; but,
- independent of the length of the rods.

This does not seem a major finding today, but hundreds of years ago engineers were relying on intuition, experience, models, and trial and error to build bigger structures. They did not have structural theories, or detailed data on the strength of materials. Their work was primarily heuristic.

The world had to wait for Leonhard Euler's theory of column buckling (1757), Thomas Young's work on the strength of materials (1807), Louis Navier's theory of beams (1826), and Otto Mohr's work on calculating forces on a structure (1874). These works allowed engineers to design and build bigger and better structures.

13.2.2 The effect of defects

Before Galileo, Leonardo da Vinci (1452-1519) had sketched in his notebooks a test of the tensile strength of a wire and investigated metal failure (see Figure 13-6). He observed a fact: the strength of a material was dependent on the presence of defects.

Leonardo da Vinci was the first engineer to document an investigation into the fracture of materials. He was assessing the strength of materials, using long iron wires, and (correctly) assumed that the strength of a wire was constant (for a specific length of wire tested). He tested iron wires of various lengths, loaded by baskets of sand, and was surprised to observe that the strength of the wires was inversely proportional to the length of wire: short wires failed at higher loads than long wires. Da Vinci correctly concluded that the long wires had more chances of containing a defect, than the shorter wires. The presence of these defects lowered the wire's failure load: long wires failed at low loads, short wires failed at high loads.

Defects in a structure greatly decrease strength, so, for a structure to be strong – and withstand high loads – it must also be defect-free. Unfortunately, nothing is perfect, and all structures/bodies contain defects: the secret of a strong structure/body is to keep defects to a minimum.

13.2.3 Can giants exist?

Galileo is thought to have merged:

- his own thoughts of cross-sectional area being related to strength; and,
- Leonardo da Vinci's observation of small defects weakening a structure,

to note that giants (should they exist…), would be very fragile creatures. Why? Well, we can expect a giant's strength to be greater than ours, since the cross-sectional areas of their skeleton and muscles would be larger, but their bones would be long and wide, giving a greater volume (their bone volume would be huge compared to ours).

Also, their weight (and the loads the bones must support), would also increase; therefore, the chances of defects in their bones would greatly increase, as would the loads on their bones. This means a simple fall may break their giant bones. Conclusion? Do not be afraid if you are confronted with an aggressive giant. Just push him/her over….

QR Code 13-2. Scan here for more understanding of structural mechanics.

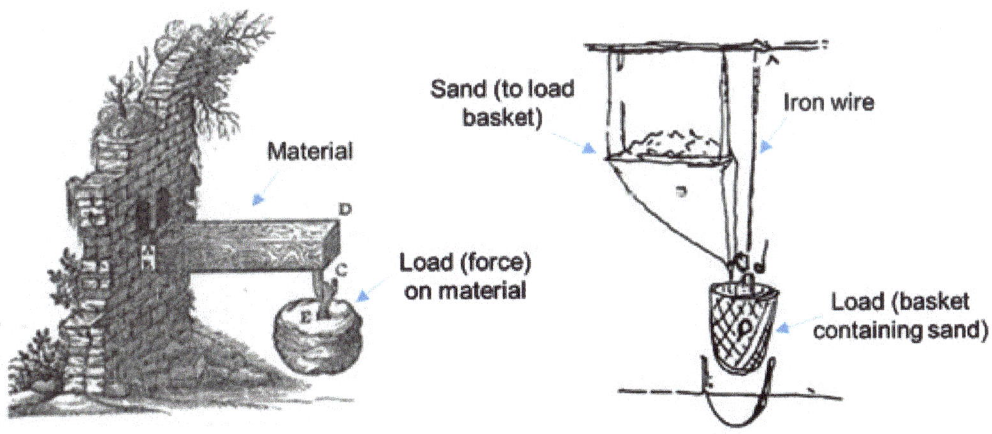

Galileo's tests on strength of materials

da Vinci's tests on the fracture of materials

Figure 13-6. Galileo's and da Vinci's tests.

13.3 Fracture mechanics

All structures will contain defects; for example, Figure 13-7. Fracture mechanics is the study of the effect of defects on structures subjected to forces and deformations. Our early scientists and engineers such as Galileo and da Vinci were interested in how defects affected a structure's performance. Early studies into fracture mechanics were limited, as there was no great demand for it – if a structure cracked, or failed, you simply replaced it, or ensured the next design of the structure was thicker and stronger. Also, very small defects such as cracks could not be seen or detected, so little could be done to control them.

The 20th century witnessed bigger, longer, heavy, and faster structures, such as ships, being built. Additionally, new structures, such as jet aircraft, were put into service. These increases in size, weight, and speed put greater forces and deformations on the structures, leading to failures, and a need to understand why these structures failed, particularly from cracking.

Cracks were originally thought to be small and insignificant, and not able to be a threat to these new, large, elegant machines. This was not the case. Ships and aircraft experienced failures due to cracks. Two famous examples were:

- The Liberty ships (see Figure 13-8) were ships built in the USA, during the Second World War. They were used to transport goods across the Atlantic to Europe and bring the region liberty. These ships experienced serious fractures: cracks ran through the ship's hull, causing dramatic failures. The cracks in the ship's hull propagated rapidly and were not arrested by the hull steel, as the steel had a low toughness (see later).
- The world's first passenger jet aircraft was de Havilland's Comet, Figure 13-8. It first flew in 1949.
    - ◊ Passenger aircraft in the 1940s were often uncomfortable converted slow wartime bombers, and all were noisy (because of piston engines and propellers), with unpressurised cabins (meaning no air to breathe at high altitudes, so they had to fly low where there is plenty of air), and this low-flying meant they had to fly through, rather than above, the weather with all its discomforts and reduced speeds. The Comet was high flying (12 km, avoiding bumpy and noisy weather), with a pressurised cabin, comfortable, with many windows, quiet (jet engines), and it was fast (500 mph (807 km/h)).
    - ◊ Three fatal crashes of Comet aircraft in 1953-54 were caused by cracks growing from corners of the square fuselage windows. These square corners served as stress risers, accelerating crack formation, and the cracks subsequently grew in a fuselage stressed by the necessary pressurisation

during high altitude flight. This pressurisation of the fuselage is repeated many times for passenger comfort: it is pressurised at the start of each flight and depressurised when the engines shut down. This repeated increase and decrease of pressure causes fatigue. Fatigue is caused by repeated stresses on a structure and assists cracks to grow in size.

Defects such as cracks create very high stresses in a structure and can cause a material to fail below its yield strength. Clearly, the high stresses created at a crack tip are a problem.

Figure 13-7. Crack in a pipeline.

Figure 13-8. Examples of cracks causing failures (left, a Liberty ship, right, the Comet aircraft).

13.3.1 Applying fracture mechanics

In structural mechanics, we compare a remotely applied stress on a defect-free structure, to the yield strength of the material of the structure. If applied stress>yield strength, I assume my structure is overloaded, and liable to fail (see Figure 13-9). In effect, we are comparing load (applied stress) with material resistance (yield strength).

It is similar in fracture mechanics: we need a measure of the local stresses at the crack tip caused by the remotely applied stress. We need a measure of the material's resistance to these local stresses (see Figure 13-9). We will discover that the resistance to these stresses is provided by the material's toughness: the higher the material's toughness, the more resistance to the presence of a crack. And finally, we need an equation that links this load and resistance, to give us a failure condition (see Figure 13-9). These equations are published in standards (for example [5, 6]), and they allow us to assess defects in any structure.

Therefore, to be able to use fracture mechanics to predict the behaviour of the defect, we will need information about (see Figure 13-9):

- the defect (e.g., its size);
- material properties (e.g., toughness): and,
- the applied stresses on the structure.

We input this information into an equation that predicts if the defect will fail, or not (see Figure 13-9).

Organisations, such as British Standards Institute (BSI) and American Petroleum Institute (API) have produced documents (BS 7910[5], and API 579[6]) that give a full fracture mechanics methodology. They give the necessary equations, procedures, etc., and help with the necessary assumptions and safety factors. API 579 considers the following systematic approach to defect assessment:

- defect and damage mechanism identification (cause?);
- applicability and limitations of assessment procedures (can I assess it?);
- data requirements (have I the data?);
- assessment techniques and acceptance criteria (what method will I use?);
- remaining life evaluation (how long can the defect survive?);
- remediation;
- in-service monitoring; and,
- documentation.

QR Code 13-3. Scan here for more understanding of fracture mechanics.

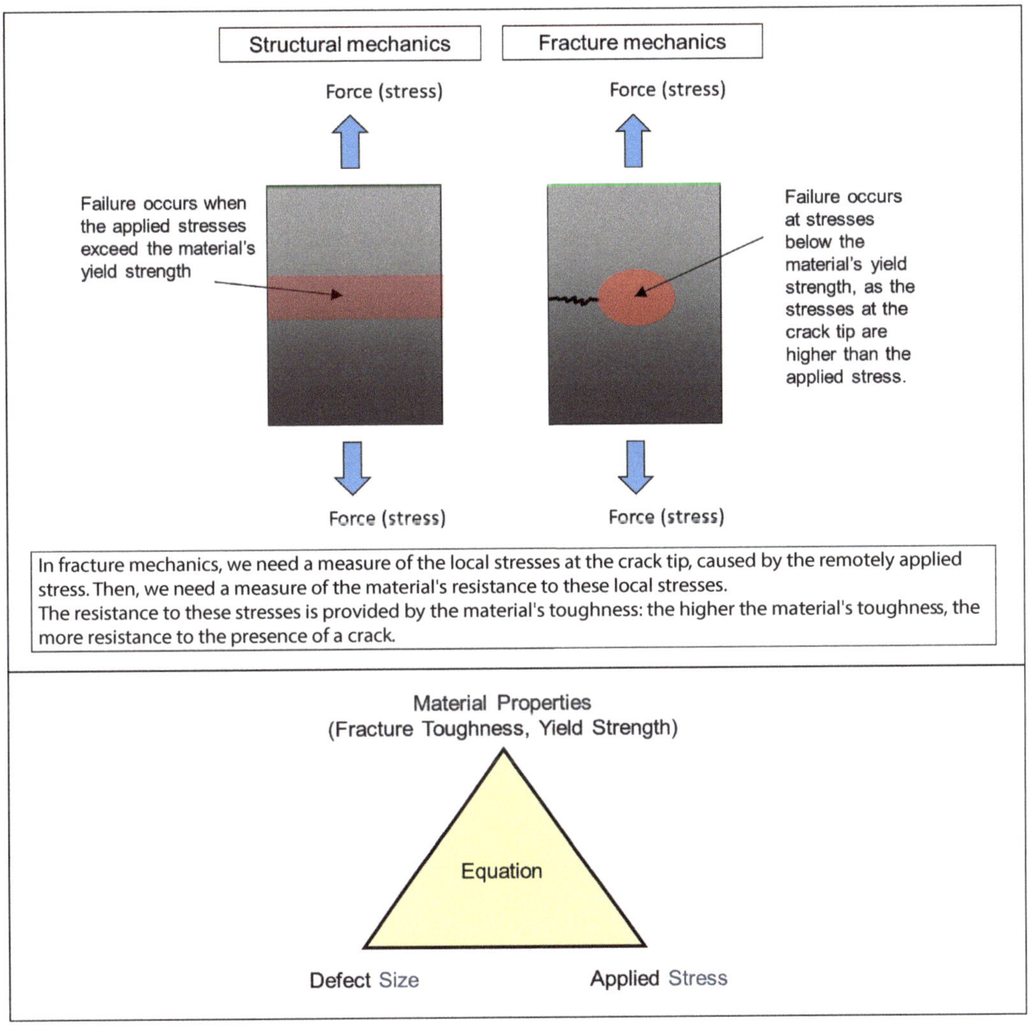

Figure 13-9. Applying fracture mechanics.

13.4 Defect assessment in pipelines

13.4.1 The early days

Most oil and gas pipelines were built between 1950 and 1980, and this was a period when fracture mechanics was still developing, and defect assessment was relatively new. Some pipelines were showing signs of age in the 1960s (e.g., corrosion), particularly in the USA.

13.4.2 The pioneers

The research organisation Battelle in Columbus, Ohio, USA did the first major fracture mechanics work on pipelines. Fracture mechanics methods had been available since the 1950s, but in the 1960s they were still new and not well-developed. Methods did exist in the 1950s for brittle materials, and thick-walled structures (such as nuclear pressure vessels), but these methods were not directly applicable to thin-walled structures such as pipelines, which were made from relatively tough and ductile materials. Additionally, the pipeline industry did not measure the fracture toughness (in terms of quantitiative measures such as stress intensity factor (K), J-integral (J), or crack tip opening displacement (CTOD)) of its pipelines, to allow it to use the various fracture mechanics methods. The only toughness measure the pipeline industry had was the Charpy toughness, which is a qualitative measure of toughness, and not easily used in fracture mechanics.

Battelle and other workers conducted full scale tests on defective pipeline sections to overcome limitations in fracture mechanics for pipelines, and the absence of quantitative measures of toughness. They developed semi-empirical (a combination of theory and experiments) methods to predict the failure pressure of defects such as corrosion and cracks in pipelines. Battelle produced fracture mechanics equations to predict how defects would behave, but they had to use the Charpy toughness as their toughness.

13.4.3 The problem with Charpy

The pipeline industry still uses Charpy toughness and does not have fracture toughness data in terms of a quantitative measure of toughness (i.e., K, J or CTOD). Charpy is not a quantitative fracture toughness; therefore, it has to be correlated with quantitative fracture toughness measures (e.g., the stress intensity factor, K).

This Charpy-K correlation is a particular problem when we assess cracks in pipelines. API 579 and BSI 7910 use recognised correlations, but these correlations introduce inaccuracies. The toughnesses used in these standards may be only one third of the actual toughness. This leads to low predictions of failure stress for cracks, and the problem of multiple repairs.

These correlations are not a problem when we assess corrosion. Corrosion is a blunt defect and is not sensitive to toughness, and its failure is mainly governed by the strength of the pipe. This is not the case for a crack – the failure of a crack in pipelines can be governed by the toughness. This means that corrosion assessment is not a problem: the fracture mechanics models we use are reliable.

13.4.4 Assessment of corrosion, gouges, dents. and cracks in pipelines

Corrosion. Pipelines inevitable corrode as they age. The Battelle workers proposed an equation to predict the failure pressure of corrosion in pipelines (see Figure 13-10). The equation only needs inputs of the corrosion length and depth, pipeline radius and wall thickness, and the strength of the line pipe steel.

These types of equations have been adopted in standards; for example, ASME B31G[10]. Figure 13-10 is an acceptance curve obtained from ASME B31G, for a specific pipeline geometry and strength. The equations given in the standard can be used to construct these acceptance curves, which allow a simple and quick method for determining acceptance.

Note that acceptance curves are not failure curves:

- If the defect is below the acceptance curve, it is acceptable and does not need repairing. This acceptable defect will not fail.
- If the defect is above the acceptance curve, it is not acceptable and needs repairing. This unacceptable defect will not necessarily fail the structure, as the acceptance curves contain safety factors.

Failure curves do not contain safety factors – if the defect is above the failure line, it will fail the structure.

There are many methods for assessing corrosion in pipelines, but the most popular methods are detailed in ASME B31G[10]. ASME B31G is the benchmark standard for assessing corrosion but is limited to internal pressure loadings only. It cannot be applied to pipelines subjected to external loads; for example, from ground movement.

Gouges. Pipelines are usually either buried below ground or are subsea. They can be damaged by earth excavations, anchors, etc. This can lead to damage such as gouges or dents. Therefore, a gouge is surface damage to a pipeline, usually caused by foreign objects removing part of the pipe wall.

Gouges can be assessed using similar equations to the ones used with corrosion (see Figure 13-10). Failure curves for gouges can be developed using these equations (see Figure 13-11).

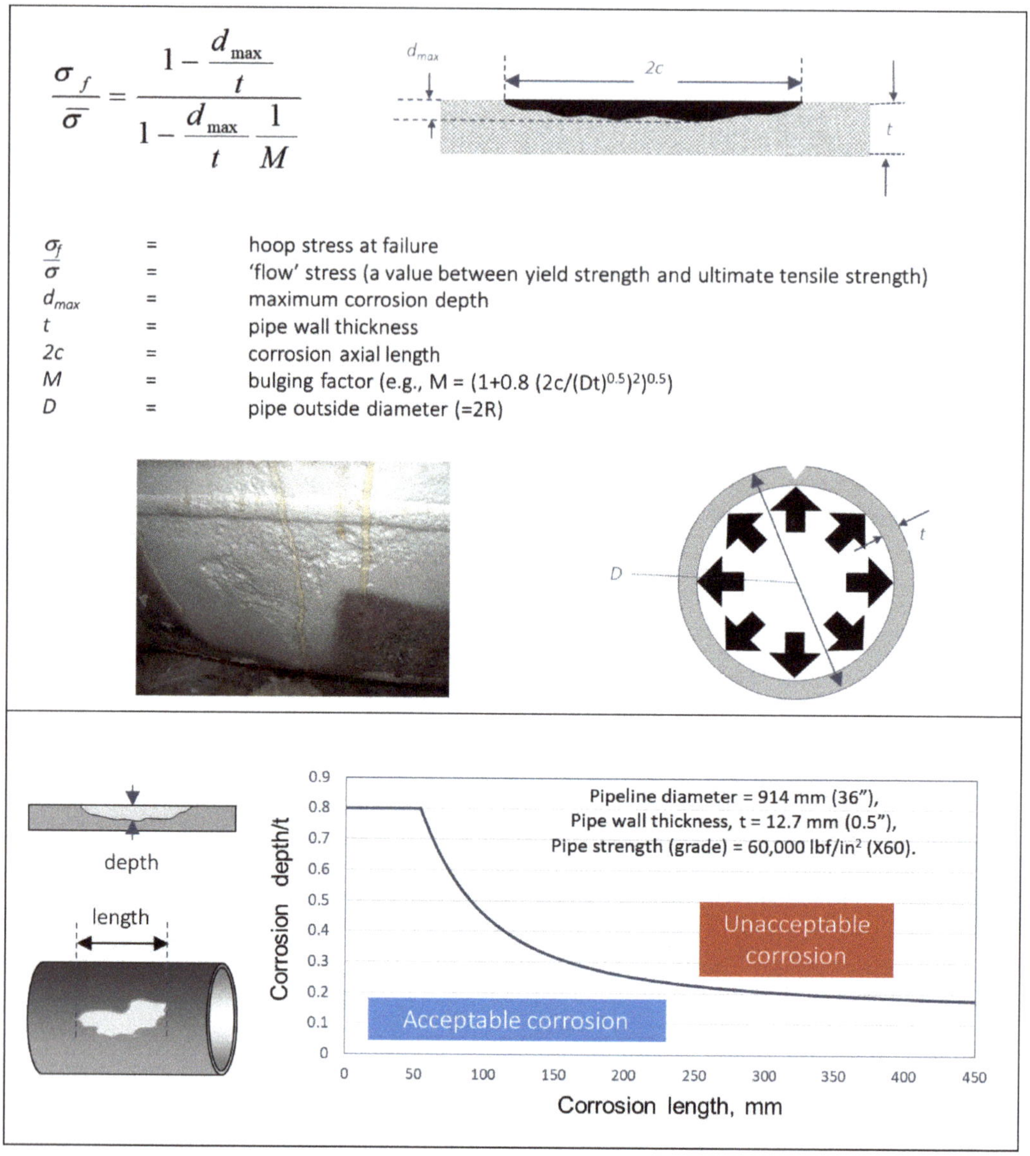

Figure 13-10. Example of a corrosion assessment equation, and an acceptance curve (Photograph courtesy and copyright of Penspen Ltd).

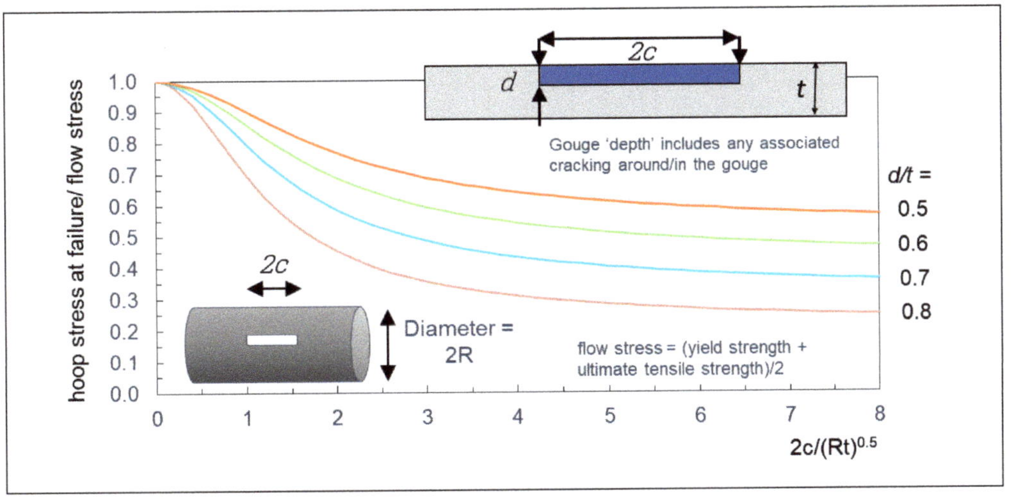

Figure 13-11. Example of failure curve for gouges in pipelines under internal pressure.

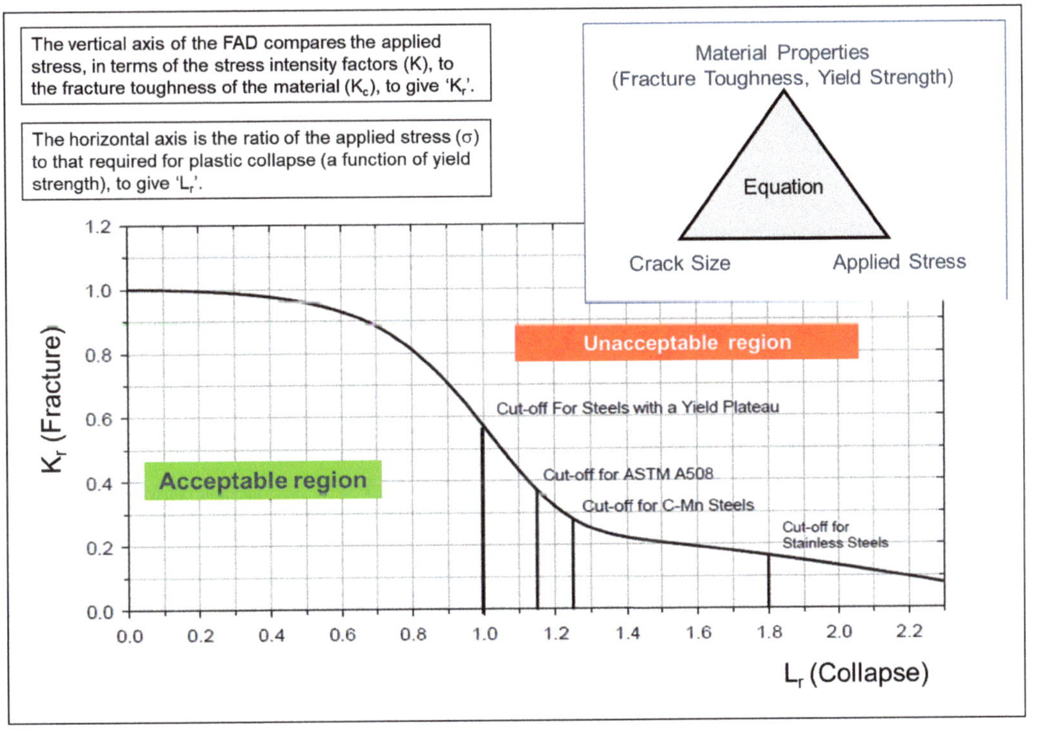

Figure 13-12. The failure assessment diagram (FAD) in API 579 and BS 7910, used for the assessment of cracks.

Dents. Pipelines can be dented during construction and/or operation. These dents can be introduced by accidental impact from earth moving equipment when the pipe is buried, anchor damage when the pipe is subsea, etc.

The pipeline industry has always adopted a pragmatic view on dents in pipelines and accepts that dents will be present in operational pipelines. Pipeline standards detail acceptance levels for dents; for example, Table 13-3 shows the size/type of dents requiring removal or repair or an engineering evaluation.

Cracks. Cracks can be created during the manufacture of our line pipe (for example, by poor steel making processes), or its construction (for example, by poor welding). These cracks are inadvertently 'built into' a pipeline. Quality control, with its inspection and testing, can eliminate most of the manufacturing and construction cracks: they need to be detected and repaired, as cracks can cause pipeline failures.

Cracks can also be created during pipeline operation. Corrosion, and fatigue can create and grow cracks in a pipeline; therefore, we can sometimes find cracks in pipelines, and they will require an assessment.

Cracks are usually assessed using published procedures. The two most popular procedures - API 579 and BS 7910 - use fracture mechanics methods. The standards use a failure assessment diagram (FAD) to calculate the failure stress of defects in a structure (see Figure 13-12). The FAD was originally developed by the nuclear industry in the 1970s, but it now used in many other industries.

Table 13-3. Dent acceptance levels for operational pipelines[11].

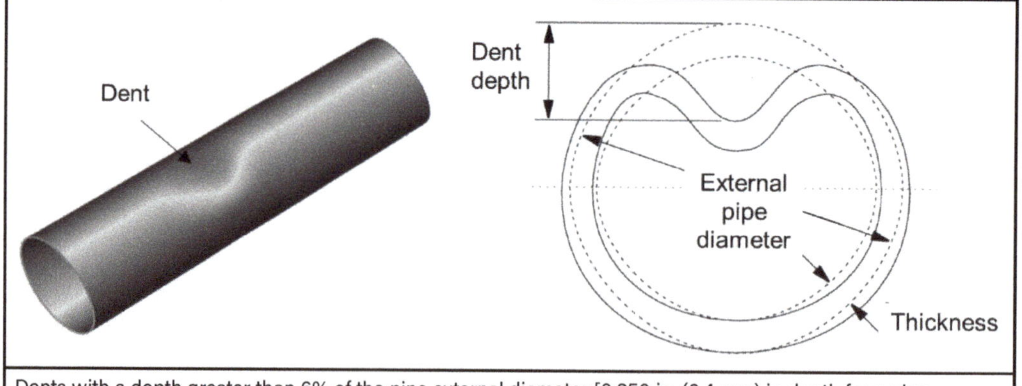

Dents requiring removal or repair or an engineering evaluation
Dents with a depth greater than 6% of the pipe external diameter [0.250 in. (6.4 mm) in depth for a pipe diameter 4.5 in. (114.3 mm) and smaller].
Dents containing gouges, grooving, scratches, cracking or other stress riser.
Dents containing metal loss resulting from corrosion or grinding where less than the minimum wall thickness permitted by the component specification remains.
Dents that affect pipe curvature at a girth weld or a longitudinal seam weld.

The FAD needs three main inputs (data) (see Figure 13-12):

- crack size (length, depth, shape);
- applied stresses on the cracked structure;
- material properties (fracture toughness, and yield/ultimate tensile strength).

The standards detail all the necessary equations to determine if the crack is acceptable. These calculations are in two parts:

- one calculation (focussed on fracture - K_r) requiring the fracture toughness of the cracked material; and,
- the other calculation (focussed on overload (collapse) of the material around the crack - L_r), requiring the strength of the material around the crack.

The results of these two calculations are then plotted on the FAD, to determine if the crack is acceptable or unacceptable (see Figure 13-12).

QR Code 13-4. Scan here for more understanding of pipeline defect assessment.

13.5 Summary

This Chapter has covered how we assess defects, such as corrosion, gouges, dents, and cracks in a structure.

We have been introduced to fracture mechanics, and discovered it is the application of structural mechanics to the problem of predicting the behaviour of a crack in a structure. Fracture mechanics defines the local conditions under which a defect such as a crack will extend, by using more global parameters such as load and geometry. We have discussed these parameters that model the conditions at the crack tip (K, J, and CTOD), and have discovered that when these values of K, J, or CTOD reach a critical value (the material's fracture toughness, K_c, J_c or CTOD), the crack fails the structure. The analogy is the stress in a defect-free structure reaching the yield strength of the structure's material, and failure ensuing.

It is important to note that defect assessment is just one part of the management of defects in a pipeline. This management includes inspection criteria, repair methods, etc. We must never treat defect assessment in isolation – it is part of our safety management system.

Finally, we will end with some advice. We can assess many defects in pipelines, and there are many documents/software to help us in these assessments, but whatever defect we assess, we need to remember… the 5 Rules of Defect Assessment:

- always understand the cause of any defect you are assessing;
- understand and use the best assessment practices;
- use all relevant data (e.g., inspection data, operations records, maps, etc.);
- check calculations, inputs, outputs, and assumptions; and,
- always appreciate the consequences of any failure of the defect.

And… we have one final piece of advice. You will be dealing with pipelines containing defects. You may have to assess a pipeline containing many thousands of defects. But remember… sometimes the pipeline contains so many defects… its condition is so poor… it is beyond defect assessment. Advice? Never assess a mess….

13.6 References

[1] Anon., 'Recommended Practice for Assessment and Management of Cracking in pipelines. American Petroleum Institute. API 1176. July, 2016.

[2] Anon., 'Pipeline Risk Modelling Overview of Methods and Tools for Improved Implementation', Pipeline and Hazardous Materials Safety Administration, USA. Draft 1. May 9, 2018.

[3] Anon., 'Gas Pipeline Incidents: 10th Report of the European Gas Pipeline Incident Data Group (period 1970 – 2016)', EGIG. Doc. number VA 17.R.0395. March 2018.

[4] http://www.ukopa.co.uk/wp-content/uploads/2016/06/June-2016-Newsletter-Final-Version.pdf

[5] Anon., 'Guide to methods for assessing the acceptability of flaws in metallic structures'. BS 7910:2013. British Standards Institution. UK. 2013.

[6] Anon., 'Fitness For Service', API 579-1/ASME FFS-1. American Petroleum Institute. June, 2016.

[7] Anon., 'Pipeline Safety Management Systems', ANSI/API Recommended Practice API 1173. July, 2015.

[8] Anon., 'Managing System Integrity of Gas Pipelines', American Society of Mechanical Engineers. New York, USA. ASME B31.8S-2018. Supplement S of ASME B31.8. 2018.

[9] Anon., 'Managing System Integrity for Hazardous Liquid Pipelines', American Petroleum Institute. API Recommended Practice 1160. Third Edition. 2019.

[10] Anon., 'Manual for Determining the Remaining Strength of Corroded Pipelines', American Society of Mechanical Engineers. New York, USA. ASME B31.G-2012. 2012.

[11] Anon., 'Pipeline Transportation Systems for Liquids and Slurries', American Society of Mechanical Engineers. New York, USA. ASME B31.4-2016.

CHAPTER 14

Pipeline integrity management

Dr Phil Hopkins

14.1 Introduction

Engineering plant is expensive, complicated, and needs constant attention. Asset management[1] is the management of physical assets such as engineering plant (see Figure 14-1), including their selection, maintenance, inspection and renewal.

Physical assets are one of the five broad categories of assets that have to be managed in order to achieve an organizational strategic plan[1]. The other categories are:

- human assets;
- information assets;
- intangible assets (reputation, etc.); and,
- financial assets.

This Chapter is limited to the management of physical infrastructure assets.

14.1.1 Process safety management

The oil and gas industry has been using asset management principles for many years, with a focus on process safety management[2]: this is the safety aspect of asset management, aimed at preventing major incidents. But what is process safety?

Safety is managed by identifying and eliminating threats (hazards). Safety in organisations is often divided into two categories:

- Occupational safety (safety, health, and welfare of people at work, focused on the person): these are high frequency, low consequence events affecting workers.

- Process safety (the engineering and management associated with preventing major accidents and near misses in hazardous industries such as chemicals and petroleum products): these are low frequency, high consequence events in facilities affecting workers and potentially members of the public.

The main concerns in process safety are major accidents (e.g., an explosion): there is less interest in occupational hazards (such as slips or falls)[6]. This is important: many companies record occupational hazards and incidents, and might claim a good safety record, but in the oil and gas industry there are much bigger concerns, as the consequences of a major hazard could be an explosion killing many. A good example of a correct focus on safety is the airline industry: they are very concerned about an aircraft crashing. The aircraft industry would never use worker safety (trips and slips) as a guide to the safety of their aircraft.

Hazard, threat, risk [3, 4, 5]

A hazard is anything that may cause harm, such as chemicals, electricity, working from ladders, an open drawer, etc. A hazard may be defined as a source of potential harm.

Factors that can trigger hazards are often termed 'threats', but 'threats' and hazards' are often taken as having the same meaning.

'Risk' is the chance, high or low, that somebody could be harmed by these and other hazards, together with an indication of how serious the harm could be.

Risk = probability x consequence.

'Probability' is the chance an event (the hazard) will occur (for a pipeline it could be corrosion). It has a value between 0 and 1.

'Consequence' is the effect of the event (the hazard) occurring (e.g., a pipeline fire).

The major objective of process safety management of hazardous plants is to prevent unwanted releases of hazardous chemicals, especially into locations which could expose employees and others to serious hazard[7]. Process safety management is aimed mainly at major hazard organisations, but the models it uses for establishing a performance measurement system applies to other organisations requiring a high level of assurance that systems and procedures continue to operate as intended[8].

Figure 14-1. Managing the physical assets of a pipeline system (photographs courtesy and copyright of Penspen Ltd).

QR Code 14-1. Scan here for more on 'probability'.

14.1.2 Integrity management

Integrity management is part of process safety management. In the pipeline industry, it includes the many and varied activities pipeline operators must undertake to ensure that releases of products from their pipelines do not occur.

BSI PD 8010-4[9] defines integrity management as a: 'set of processes that ensures incident-free transportation of fluids through a pipeline system.' The management of reliability and maintainability in operations is closely related to integrity management but remember, that integrity management is not solely concerned with operations: BSI PD 8010-4 states: 'Integrity management is a complete process for a pipeline system, covering its life from design to abandonment'.

14.1.3 Integrity management systems

The day-to-day safety of a pipeline system is managed through a pipeline integrity management system. But... what is a... pipeline integrity management system?

- Integrity means a pipeline system maintains its structural integrity and does not leak or rupture (i.e., prevent a release of product)[10].
- Integrity management is the activities pipeline operators must undertake to ensure that releases do not occur[10]. It is a program of integrity work, within a written plan, within a system.
- A system is a group of interacting, interrelated, or interdependent elements (components) that carries out a specific activity, performs a duty, or solves a problem.

A management system helps us manage a particular aspect of our business; e.g., plant safety or pipeline integrity, and is often required by regulators of pipelines [e.g., 9, 11, 12, 13]. It will need (see Figure 14-2):

- a clear policy;
- staff organisation;
- a plan, setting targets and objectives, and identifying hazards;

- assessment of risks, and establishing standards against which you can measure performance (and show continuous improvement); and,
- periodic audit and review.

An integrity management system will need:

- leadership and management commitment;
- stakeholder engagement;
- a clear policy;
- staff organisation;
- risk management;
- operational controls;
- safety assurance;
- incident investigations, lessons learnt; and,
- emergency preparedness and response.

Most management systems are based on the continual improvement cycle of 'Plan, Do, Check, Act' (Figure 14-2):

- Plan: plan the work to be done;
- Do: complete the work;
- Check: evaluate and monitor the work; and,
- Act: improve and integrate lessons learned.

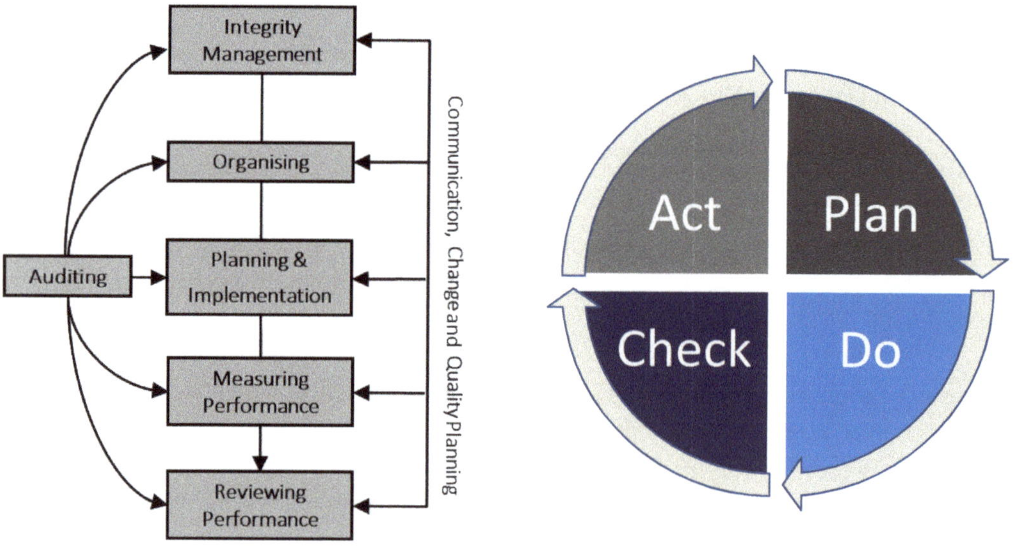

Figure 14-2. Pipeline safety is controlled by a pipeline integrity management system.

The system will be a comprehensive and integrated management program, providing safety improvements to systems by allowing an operator to effectively allocate resources to prevent/detect/mitigate threats to their pipeline.

It should be noted that managing safety is not just about having a safety management system[14]. The success of whatever process or system is in place depends on the attitudes and behaviours of people in the organisation. This is safety culture.

'Accident' versus 'Incident'

There are many different definitions of 'accident' and 'incidents'.

An 'accident' is usually considered an unplanned event that causes death, injury, or property damage.

An 'incident' has many meanings in the literature, ranging from an unplanned event causing death, injury, or property damage, to an event not causing harm or damage, but with the potential to cause harm.

QR Code 14-2. Scan here to understand why 'incidents' occur.

QR Code 14-3. Scan here for more on 'safety management systems'.

QR Code 14-4. Scan here for more about 'safety culture'.

14.1.4 Risk management

We manage the safety of engineering assets, such as pipelines, by managing the risks associated with operating the structure[1, 15]. Risk management is the process of selecting appropriate risk reduction measures and implementing them in the on-going management of the activity.

API 1160[11] considers risk management: 'An overall program consisting of: identifying potential threats to an area or equipment; assessing the risk associated with those threats in terms of incident likelihood and consequences; mitigating risk by reducing the likelihood, the consequences, or both; and measuring the risk reduction results achieved'.

Examples of threats to pipelines, are:

- activities (e.g., earth moving) around a pipeline;
- adverse weather conditions;
- sabotage, theft, vandalism;
- poor coatings on the pipeline leading to corrosion;
- poor quality materials or construction;
- operating faults;
- equipment faults;
- internal corrosion caused by a corrosive product.

Examples of consequences of pipeline failures are:

- casualties;
- environmental damage;
- contamination;
- collateral damage;
- loss of revenue;
- repair costs;
- loss of public image;
- loss of shareholder confidence;
- litigation;
- lower staff morale;
- etc.

Risk management is an important element of asset management.

QR Code 14-5. Scan here for a comprehensive list of threats to pipelines.

14.2 Pipeline integrity management in practice

Pipeline regulators and regulations expect and require pipeline operators to manage the integrity of their pipelines. Operators have historically relied on pipeline maintenance and inspection to ensure the pipeline was safe[16]. This approach is focused on the pipeline itself, and the supporting systems, and is administered through an operator's inspection program. The bias is ensuring a pipeline does not fail (i.e., it is concentrated on probability of failure), and pays little attention to any consequences of failure.

Integrity management takes a broader view, as it encompasses the environment around the pipeline, as well as the pipeline. This means that operators are required to know more about:

- the areas their pipeline traverses;
- the nature of the population in the area; and,
- the existence of environmentally sensitive areas near the pipeline.

Integrity management seeks to understand and control both the probability of failure, and the potential consequences of failure of a pipeline in a particular area along its route[16]. It sets priorities for inspection and operations and maintenance based on whether people, property, or the environment might be at risk should a pipeline failure occur, rather than simply following an agreed inspection and maintenance plan.

14.2.1 High consequence areas

The sections of the pipelines that must receive the most scrutiny and attention are the sections where a failure would have the greatest impact. These areas are called high consequence areas (HCAs): operators are required to devote additional focus, efforts, and analysis in HCAs to ensure the integrity of the pipeline in that area.

The consequences of natural gas and hazardous liquid pipeline releases differ; therefore, the criteria for their HCAs also differ[17]:

- HCAs for natural gas transmission pipelines focus on populated areas (e.g., ASME B31.8's Class Location 3 or 4[18]; or, Class Location 1 or 2 with a

potential impact radius > 660 feet (200 m) containing ≥20 occupied buildings, or sites such as a school or camping ground).
- HCAs for hazardous liquid pipelines focus on populated areas (e.g., 10,000 people/sq mile), drinking water sources, busy commercial navigable waterways, and unusually sensitive ecological/environmental areas (e.g., where critically imperilled species can be found, or where migratory water birds concentrate).

When the HCAs have been identified, a risk assessment is conducted to identify all threats and mitigate all high risks. Additionally, the operator is required to conduct an integrity validation (through inspection, testing, and analysis) for all their HCAs segments. The risk assessment, identification of HCAs, integrity validations, etc., are all included in a documented integrity management program (see Figure 14-3)[12].

We will now cover these integrity management programs, threat assessments, risk analyses, and integrity validations.

14.2.2 Integrity management program

An integrity management program[11]:

- identifies and analyses all events that could lead to failure;
- examines likelihood and consequences of potential pipeline incidents;
- examines and compares all risks;
- provides a framework to select and implement risk mitigation measures;
- establishes and tracks performance, with the goal of improvement.

The program must be comprehensive, systematic, and integrated, to provide the means to improve the safety of pipeline systems. It must provide information for an operator to effectively allocate resources for appropriate prevention, detection, and mitigation activities that will result in improved safety and a reduction in the number of incidents[12].

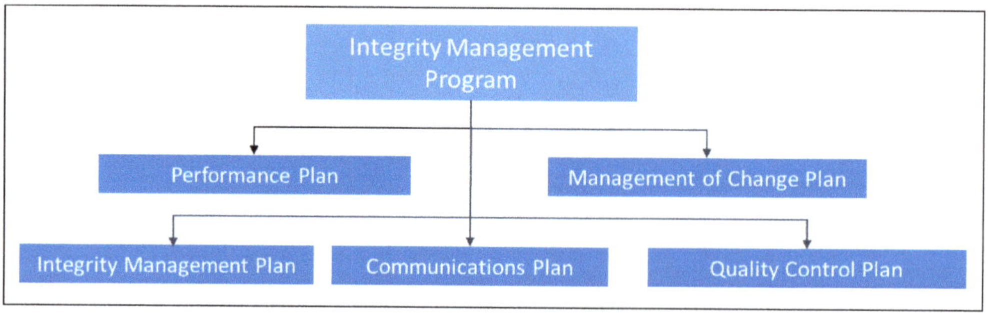

Figure 14-3. An integrity management program with its plans.

The integrity management program can be either performance-based or prescriptive[12]:

- a performance-based integrity management program utilizes risk management principles and risk assessments to determine prevention, detection, and mitigation actions and their timing;
- a prescriptive integrity management program follows pre-set conditions that result in fixed inspection and mitigation activities and timelines.

The prescriptive process will provide all the inspection, prevention, detection, and mitigation activities necessary to produce a satisfactory integrity management program, but a performance-based integrity management program uses more data and more extensive risk analyses, which enables the operator to achieve a greater degree of flexibility in order to meet or exceed the requirements of standards[12].

Risk: how often... not when

Remember... risk analysis will tell us **how often** something will happen, but it will not tell us **when** it will happen....

QR Code 14-6. Scan here for more on prescriptive versus performance-based programs.

14.2.3 Evaluating threats in HCAs

The integrity of a pipeline is first managed by identifying and mitigating threats in HCAs. A review[19] of pipeline threats grouped them into 13 categories (see Table 14-1).

The threat, or hazard, has a potential consequence. The risk associated with this threat is high, if the threat has a high probability of occurring, and the consequences of it occurring are high. An integrity management system needs to assess all these threats and consequences, and mitigate any associated risks if the risks are high.

These threat assessments and risk assessments are included in the integrity management plan in the integrity management program (see Figure 14-3). A schematic of an integrity management plan is given in Figure 14-4. The risk assessments use the data from the integrity management plan.

Table 14-1. Threats to a pipeline.

External Corrosion.
Internal Corrosion.
Environmental Cracking (including SCC).
Structural/Material Degradation (non-steel pipe).
Manufacturing-related Defects (includes defective pipe and seam acted on by fatigue or other failure mechanisms).
Instruction-, Installation-, or Fabrication-related Defects (includes defective girth weld, fabrication weld, wrinkle bend or buckle, stripped threads, broken pipe, etc.).
Equipment Failure (includes failure of control/relief equipment, pump, compressor, seal/pump packing failure, etc.).
Excavation Damage (includes damage by operator, contractor, or third-party).
Other Accidental Outside Force Damage (includes causes such as vehicles, other fire or explosion, electric arcing).
Intentional Damage / Vandalism / Sabotage.
Incorrect / Improper Operation (includes human errors).
Geohazards / Weather / Natural Force Damage.
Other / Uncategorized / Emerging Threat.

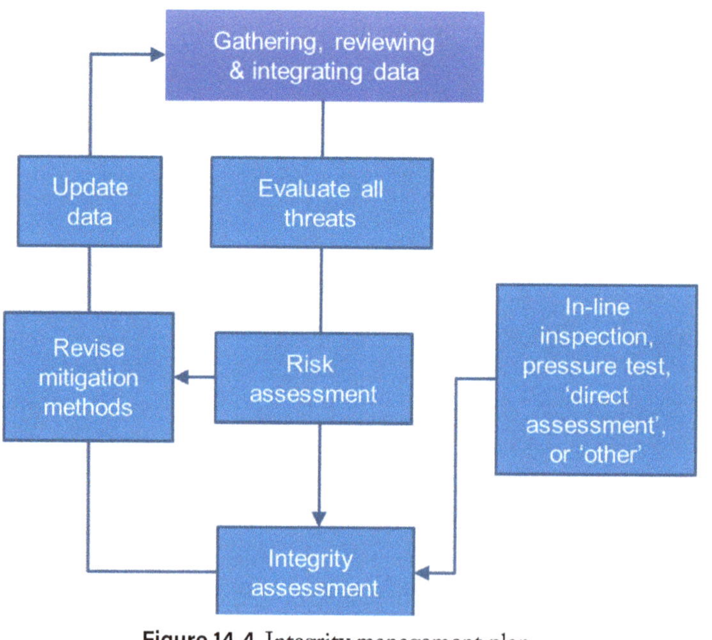

Figure 14-4. Integrity management plan.

14.2.4 Evaluating risk in HCAs: risk analysis

Threats must each be subjected to a risk analysis (see Table 14-2), using a risk model, where the scale of the threat is quantified. The quantified risk must be compared (assessed (see Table 14-2, Figure 14-5)) to an acceptable or tolerable level of risk:

- if the risk is not acceptable, the risk must be mitigated;
- if the risk is acceptable, the risk can be accepted and monitored.

Risk (R) is a combination of the probability (P) of an event occurring and its consequences (C), and is expressed mathematically as R=P x C. A technical description of risk is: risk is the likelihood of an event occurring in conjunction with the potential consequence of such an event[3]. Safety is linked to risk: safety is the absence of risk, or the reciprocal of risk.

A risk analysis is the starting point for a risk assessment, and can either be expressed qualitatively (e.g., there is a high chance of rain tomorrow) or quantitatively (e.g., there is a 70% chance of rain tomorrow); consequently, and in simple terms, two types of risk analysis can be performed[21, 22]:

- qualitative: a relative risk, or a ranking number, is calculated, which allows for a comparison with another risk (e.g., risk is high for a specific pipeline, compared to others); or,
- quantitative: a risk based on numerical estimates of probability and consequence is calculated (e.g., 1 fatality every 1000 years from a failure of the pipeline).

This simplification can be expanded into four types of risk model[19] (see Table 14-3).

Risk models differ in their approach, inputs, outputs, and algorithms, but the overriding principle in employing any type of risk model/assessment is that it supports risk management decisions to reduce risks and satisfies regulatory requirements[19]. The models to be used in future should be more quantitiative[19]: '... the continuing occurrence of significant pipeline incidents points to a continuing need for operators to upgrade [to risk assessment processes using quantitative or probabilistic risk models] their tools for risk assessment and risk management.'

Table 14-2. Explanations of risk terms.

Term	Meaning [20]	Action
Risk analysis	The estimation of risk from an activity using a risk model.	Identify threats and consequences. Perform risk calculations.
Risk assessment	A review of the acceptability of risk, based on comparison with risk standards or criteria, and then trial of various risk reduction measures.	Compare with acceptable risk levels.
Risk management	The process of selecting appropriate risk reduction measures, and implementing them in the on-going management of the activity.	Perform cost benefits. Decisions, actions, mitigation

Table 14-3. Types of risk model.

Model	Description	Output	Algorithms
Qualitative	Expressed in descriptive terms (e.g., high, medium, low), not quantitatively or numerically.	Qualitative, with no units.	Matrix mapping Inputs to Outputs
Relative assessment/ index	Scoring rules, or algorithms, defining a risk index (e.g., 5 or 10 on a scale of 1 to 20), based on input information. The scoring rules do not strictly adhere to the laws of probability but capture subject matter expert judgement of the relative importance of various risk factors.	Quantitative, but no units.	Risk Index Scoring
Quantitative	Expressed in terms of a numerical quantity or involving the numerical measurement of quantity or amount.	Quantitative (e.g., probability, frequency, expected loss) with units.	Quantitative Model
Probabilistic	A type of quantitative model adhering to laws of probability, with inputs that are quantities or probability distributions, and with outputs that can be expressed as probability distributions.	Probability distributions with units.	Quantitative Model

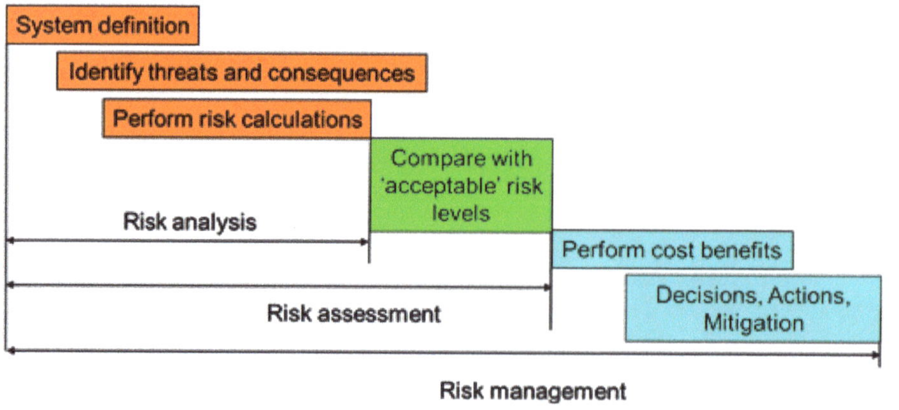

Figure 14-5. Risk analysis, assessment, and management[15, 20].

Qualitative analysis. A qualitative analysis can be based on a decision made by one or more subject matter experts considering the likelihood of threats to an asset and ranking those likelihoods. An extension to this process is to include potential consequences, and evaluate the resulting risk using a simple matrix with alphanumerical scales to represent the two risk components (see Figure 14-6). This type of matrix is often called a 'Boston Square': it was originally developed by Boston Consulting Group (BCG) in the 1970s to help analyse business units within companies. Again, the risk result is a scale value giving a relative value. These relative risk rankings can be compared with other risk rankings calculated using the same matrix. Qualitative risk analyses do not calculate an absolute risk (e.g., 1 fatality every 1000 years), or an acceptable risk.

Results evaluated using simple matrix, ranking or index scoring methods have proven effective as a screening tool to identify potentially high-risk areas/scenarios that may require a more detailed quantitative risk assessment, or intervention. Qualitative risk analyses provide a systematic approach that is easy to understand and apply.

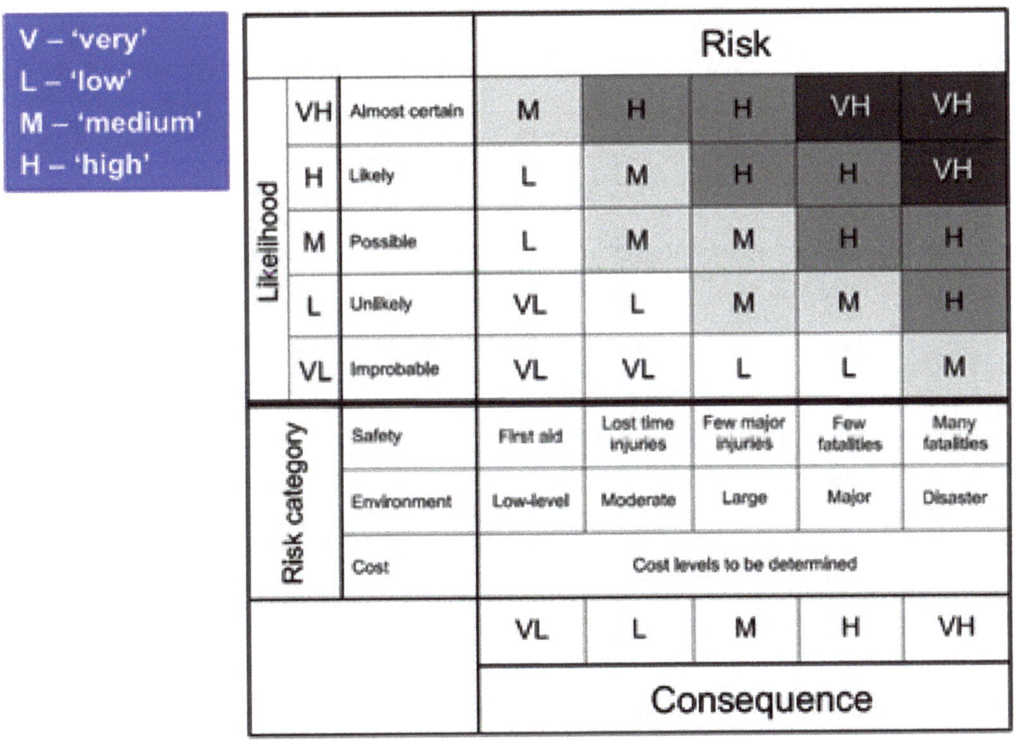

Figure 14-6. A risk matrix [9].

Semi-quantitative analysis. Risk analyses are rarely purely qualitative or quantitative. The term semi-quantitative is often used in risk analyses, and these semi-quantitative analyses are viewed as smarter models[23]. Semi-quantitative means that some real (not relative) numbers are being used in the analyses; for example, quantifying the area around a nuclear power station that would be affected by a radiation leak. Probabilities and consequences should be defined in quantitative terms where possible[9].

Many types of risk can be assessed using this type of matrix, for example:

- safety;
- environmental;
- financial;
- reputational.

Semi-quantitative risk analysis is popular in the pipeline industry (e.g., [11, 12]), and risk matrices are used in pipeline standards (e.g., [9, 24, 25]).

Quantitative analysis. Qualitative risk analyses are popular in the pipeline industry [26]. These analyses are straightforward to use, but only provide a risk ranking/priority, which may be difficult to justify, and cannot be used to optimise risk reduction measures. The alternative to qualitative risk analysis is quantitative risk analysis: examples of quantitative risk methodologies are detailed in References 27 and 28.

Quantitative risk analysis involves quantifying all aspects of the risk assessment; therefore, it will involve detailed calculations, and often the type of risk methodology used will depend on the regulatory regime applicable to the industry/asset.

A quantitative risk analysis presents actual risk profiles around an asset which can then be compared with acceptable levels. Quantitative analysis requires defined input parameters, relevant data and an agreed methodology. The risk values can be compared to risk acceptance criteria and used to assess risk reduction measures. The detailed nature of many of these calculations, and particularly the sparse nature of some of the input data, can lead to the potential for technical disagreements and inconsistencies in the quantitative risk analysis results, which in turn can adversely affect the decision-making process.

Another problem we have with the quantitative label is that it can imply an accuracy that does not exist[29] due to, for example, the availability and quality of input data not matching the complexity of the quantitative model. This is important as many quantitative risk analyses are investigating rare events: this means data for the risk variables are likely to be sparse and limited, and there may be some variables for which there will be no previous history.

It must be acknowledged that input data issues will lead to uncertainty in the results but identifying data gaps is good and allows important data to be generated. Risk analyses can be conducted using limited data: as Charles Babbage (1792-1871) said, 'Errors using inadequate data are much less than those using no data at all.'

The results obtained from a quantitative analysis are not definitive: they are useful guides to policy making, but their limitations must be made clear to policy makers [30]. Accordingly, any type of risk analysis should be used in conjunction with knowledgeable, experienced personnel (subject matter experts and people familiar with the facilities) that regularly review the data input, assumptions, and results of the risk assessments[12].

Finally, experts like numbers, and like the probabilities and risks generated by a quantitative risk analysis, but the general public do not like numbers that tell them they are at risk of injury or death. Figures such as a 6 in 100,000 chance per annum of being a casualty from a pipeline failure may neither be understood, nor accepted. This means that the output of a quantitative analysis may, in itself, create a problem, regardless of how low the risk number. This is why qualitative risk analysis is useful, and still has a role, even when quantitative risk analysis is used.

QR Code 14-7. Scan here for examples of qualitative and quantitative risk analysis.

QR Code 14-8. Scan here to read about the importance of data management.

14.2.5 Evaluating acceptable risk: risk assessment

Risk analysis generates a risk value by identifying and evaluating all potential credible hazards likely to impact an asset and the consequences of any adverse effect caused by those hazards. A risk assessment compares this risk value with a target or acceptable risk, to aid decision making[31].

Health and safety regulations do not expect all risk to be eliminated, but industry is required to protect people as far as is reasonably practicable. Risk assessment is simply a careful examination of what could cause harm to people, so that we can assess whether we have taken enough precautions or should do more to prevent harm. The goal of risk assessment is to support decision making on how to respond to real and perceived risks.

There are many reasons for conducting a risk assessment, including:

- ensuring compliance with legislation and standards;
- mitigating the potential impact of an accident (cost, casualties, public relations, etc.);
- minimising insurance costs; and,
- improving the effectiveness of an integrity or safety management program for an asset.

A simple risk assessment identifies hazards and consequences and calculates a risk level, then compares this calculated value with an acceptable risk level, Table 14-2. This means the risk is analyzed then assessed (Figure 14-5).

Defining 'acceptable' risk. 'Acceptable' is a complex word... acceptable to whom, and at what cost? Many questions surround the word acceptable when used in risk assessment, in particular:

- If we accept a risk, what do we gain, other than continued exposure to a risk?
- Who pays for any risk reduction and how much are we prepared to pay?

For example, if I calculate that the risk of death from driving an automobile on lit highways is reduced by 10% compared to unlit roads, would you pay for the lighting? The cost of installing and powering the lighting could be very high. How do I determine if the 10% reduction in fatalities is acceptable, and how do I determine if the cost is reasonable, and... who pays?

'Acceptable' is also very subjective:

- Is a risk of death/year from a specific hazard of 1 in 100 acceptable? Obviously not.
- How about... 1 in 1,000,000? Perhaps... but not if I am the '1'.

How much would you be willing to pay to reduce the risk from 1 in 100 to 1 in 1,000,000? $100? $1,000,000? And... again... who would pay?

Finally, 'acceptable' has to be defined in the context of the type of risk we are considering; individual or societal:

- Acceptable individual risk is a risk level at which the general public will accept the probability of an individual fatality.
- In contrast, societal risk is generally expressed in terms of the probability of multiple fatalities (e.g., an aircraft crash) and although it is possible to

determine an acceptable societal risk level, it is a more complicated process than that for an acceptable individual risk.

The remainder of this Chapter relates to individual risk.

'Acceptable' risk at a 'reasonable' cost? We need an acceptable risk number with a reasonable cost. Government regulators or company management decide what is a broadly acceptable risk to the general public or a workforce. A broadly acceptable risk should be arrived at by consensus among all stakeholders, including the general public, but it is impossible to convince every person that a particular risk is acceptable[20, 32].

We all have a different view of risks, based on how they relate to us and things we value. Our perception is dictated by: how well the process (giving rise to the risk) is understood; how well we can control the risk; and, whether we volunteer to take the risk (e.g., we volunteer to smoke, but we do not volunteer to be harmed by a nuclear power plant incident)[33]. This means that a risk calculated by an expert may not be viewed in the same way by the general public whose perception is complex and also affected by cultural, ethical, and social aspects that are not easily incorporated into a risk analysis[33].

A public view. The general public's concerns can be summarised as[33, 34]:

- Individual concerns: how individuals see the risk from a particular hazard affecting them and things they value personally. Individuals may be prepared to engage voluntarily in activities that often involve high risks (e.g., sky diving), but usually are far less tolerant of risks imposed on them, and over which they have little control (e.g., flying).
- Societal concerns: risks from hazards which impact on society with multiple fatalities. These incidents often generate a lot of media attention and give rise to often unthinking responses from politicians and the public at large, such as incidents involving nuclear power plants or railway travel. Society has a greater aversion to a single accident killing 10 people than to 10 separate accidents killing one person each[35].

The general public, the media, and politicians may well ask for zero risk; therefore, the desired acceptable risk level becomes zero.... But zero risk is not achievable - even at huge costs. We must therefore seek a consensus to establish acceptable levels of risk.

Inevitably when an incident does occur, particularly if it results in many fatalities or significant infrastructure damage (raising societal concerns), there will be people and organisations (e.g., business leaders, politicians, the media), who will claim (with hindsight) that it is an unacceptable risk. These stances are understandable, but in reality, all risk reductions have a price, and a pragmatic approach needs to be adopted by all parties[20].

Proposing an 'acceptable' and 'unacceptable' risk. It will not be possible to propose a level of risk that is acceptable to everyone, (other than zero), but there are various approaches we can take in reaching a definition of acceptable; for example:

- state an arbitrarily defined probability (e.g., 1 in a 1,000,000 chance of death per annum);
- quote an existing risk already tolerated by society (e.g., the risk/annum of being killed by lightning in the USA is 1 in 11,000,000[1]);
- use values quoted in standards or regulations;
- experts propose a value;
- the general public gives a value;
- politicians give a value;
- determine a monetary value such that the cost of reducing the risk does not significantly exceed the costs saved.

Can we determine an acceptable risk that can satisfy most of the people? One way is to look at the risks that we are exposed to in our working lives. We know that all workers are exposed to risk to a greater or lesser degree, and the level of risk depends on the industry they work in (see Table 14-4)[33]. These workers' risk could be considered acceptable by the general public because the workers derive a benefit from being employed. The risks in Table 14-4 can be compared with an annual risk of death from cancer (1 in 397), annual risk of death in a road traffic accident (1 in 16,800), and the annual risk of death from being struck by lightning in the UK (1 in 18,700,000)[33].

The safety regulator in the UK, the HSE, has used the type of numbers in Table 14-4 to conclude that an individual risk of death of 1 in a million per annum for both workers and the general public corresponds to a very low level of risk[33]. They also note that each individual in the UK is subject to a background level of risk of approximately a 1 in 100 chance of dying each year, so a risk level of 1 in a million chance of a fatality is seen as a broadly acceptable guideline.

Using similar statistics from across industries, the HSE concluded that a risk level greater than 1 in 1,000 was unacceptable for employees. For members of the public exposed to risks from activities they have no control over, and gain no benefit from, the HSE defined a risk level of 1 in 10,000 or greater, as being unacceptable.

1 1 in 11,000,000 is a small probability, but if we apply this probability to the world's population (7.6 billion), we predict 7,000 deaths per year from lightning. The actual figure is much higher (24,000 deaths per year), as many people, in the Developing World, are in exposed areas more often than in the Developed World.

Table 14-4. Annual chance of death in various industries.

UK Industry	Annual risk	Annual risk per million people
All employees	1 in 125,000 (8 x 10^{-6})	8
Mining and Quarrying	1 in 9,200 (1 x 10^{-4})	109
Construction	1 in 17,000 (6 x 10^{-5})	59
Agriculture, forestry	1 in 17,200 (6 x 10^{-5})	58
Service industries	1 in 333,000 (3 x 10^{-4})	3

Proposing a 'tolerable' risk. The previous section noted that risk levels of 1 in 1,000,000 were broadly acceptable, and risk levels of 1 in 10,000 were unacceptable, but what about risk levels between these two values? Risk levels that lie in the region between unacceptable and broadly acceptable levels of risk are defined as tolerable risk levels[33].

A tolerable risk is not necessarily an acceptable risk. In the UK there is a legal requirement to demonstrate that a tolerable risk has been reduced to a level 'as low as reasonably practicable' – the 'ALARP' concept[33, 36].

Tolerable [33]

'... 'tolerable' does not mean 'acceptable'. It refers instead to a willingness by society as a whole to live with a risk so as to secure certain benefits and in the confidence that the risk is one that is worth taking and that it is being properly controlled.'

The ALARP process, represented schematically in Figure 14-7, involves:

- assessment of the risk to be reduced or eliminated;
- evaluation of the sacrifice (in money, time and trouble) involved in taking measures to reduce or eliminate that risk;
- financial comparison between this risk and the sacrifice[37].

The ALARP approach is one possible approach to determining an acceptable risk value. We need to compare the calculated risk with an acceptable risk level[31] and judge if the cost of controlling that risk is reasonable; i.e., there will be a balance between the desired risk reduction and available investment. In a normal cost-benefit analysis (CBA) the risk reduction measure would only be adopted if the costs (C) of implementing the reduction (e.g., cost of putting concrete slabs over a pipeline to protect it from excavating machinery that could damage it) did not exceed the benefits (B) of the measure –in this case a reduction in potential of damage incidents.

In the ALARP approach the mitigation measure is implemented unless the costs (C) are significantly greater than the benefit (B); i.e., implement unless C >> B[37, 40].

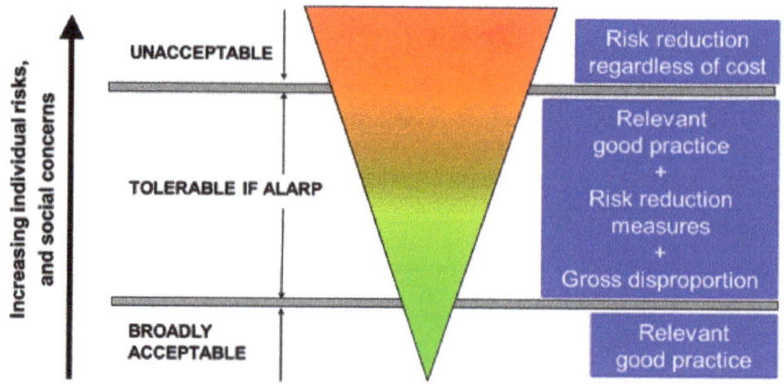

Figure 14-7. Levels of risk [33, 38, 39].

High uncertainty and the precautionary principle. Risk analysis results will contain uncertainties; e.g., uncertainties in the risk model, and the input data. Costs and benefits used in CBAs will also have uncertainties. Experts want certainty, and will look for it, wait for it, or work on developing it, but lack of certainty is not a reason for not taking preventive action.

When there is uncertainty, and in particular where the potential consequences are high, we can take a precautionary approach[41]: exercise caution in high uncertainty situations. The United Nations defines this principle as[42]: 'where there are threats of serious or irreversible environmental damage, lack of scientific certainty shall not be used as a reason for postponing cost-effective measures to prevent degradation'.

The precautionary principle[33] deals with situations of high uncertainty where the evaluation did not allow the risk to be determined within an acceptable degree of certainty. The principle is only applied in very special circumstances and is not applied to commonly recognised hazards.

QR Code 14-9. Scan here for more on 'cost of life'.

QR Code 14-10. Scan here for more on 'zero risk' and 'zero failures'.

QR Code 14-11. Scan here to learn more about uncertainty and the 'rare event'.

14.2.6 Risk management

Risk analysis and assessment is incorporated in an organisation's risk management process: risk management is the process of selecting appropriate risk reduction measures and implementing them in the on-going management of the activity, in a cost-effective manner (see Figure 14-5). The risk management process will conclude how a risk can be reduced, controlled or eliminated.

There are four ways to avoid/control risk:

- Risk Avoidance/Elimination: a considered decision to avoid completely a particular risk by discontinuing the operation producing the risk (e.g., locating a pipeline away from any human dwellings).
- Risk Retention/Acceptance: risk is retained in the organisation, with any consequent loss being financed by the company.
- Risk Transfer/Outsource: transfer the costs of certain potential losses from one party to another; for example, to an insurance company.
- Risk Reduction/Mitigation: risks are systematically reduced through control measures.

Pipeline risks need to be identified, quantified, and – where necessary – reduced to a reasonable level. The probability and consequences of a pipeline failure depend on many parameters (see Table 14-5). The control of the probability and consequences also depends on the same parameters.

Table 14-5. Examples of parameters affecting the probability and consequences of a pipeline failure.

Probability	Consequence
threats	explosion, ignition, and fires
fluid in pipeline	fluid in pipeline
pipeline type, design, and location	pipeline type, design, purpose, and location
environment	environment
time (of day, and duration)	time (of day, and duration)
company operating practices	dispersion, accumulation, or pool spread
age and construction quality	toxicity
maintenance and inspection practices	leak detection equipment
local social, economic and political environment	emergency response

14.2.7 Integrity assessments

We ensure these threats are managed and controlled by regularly testing, surveying, or inspecting our pipelines. These are called integrity assessments or integrity validations; for example (see Figure 14-8):

- We can test our pipelines by removing the product from the pipeline, and then filling it usually with water (we can use a gas at low pressures) (see Figure 14-9). We then pressurise this water (a pressure test or hydrotest) to a pressure well above the pipeline's operating pressure. If the pipeline survives this pressure test, we know it can then operate safely at the (lower) operating pressure of the pipeline.
- We can survey our pipelines for changes to its condition or surroundings. We can walk along the pipeline looking for changes in ground condition, and we can also do inspections on the pipeline to check the condition of the anti-corrosion coating. We can uncover the pipeline in areas where we have a concern with its condition, and directly inspect it: this is sometimes called 'direct assessment'.
- We can send robots (smart pigs) down the pipeline (see Figure 14-10). These robots flow with the product and can inspect the pipeline as they move through the pipeline (in-line inspection), detecting damage such as dents, or deterioration, such has corrosion.

Subsea pipelines, or water crossing, can be surveyed by either (see Figures 14-11, 14-12):

- divers;
- hydrographic surveys (these surveys can use acoustics, seismic, magnetic, and electrical surveys, on a variety of vessels);
- remotely operated vehicles (ROVs) (Figure 14-11); or,
- smart pigs.

QR Code 14-12. Scan here to learn more about inspection and surveillance of pipelines.

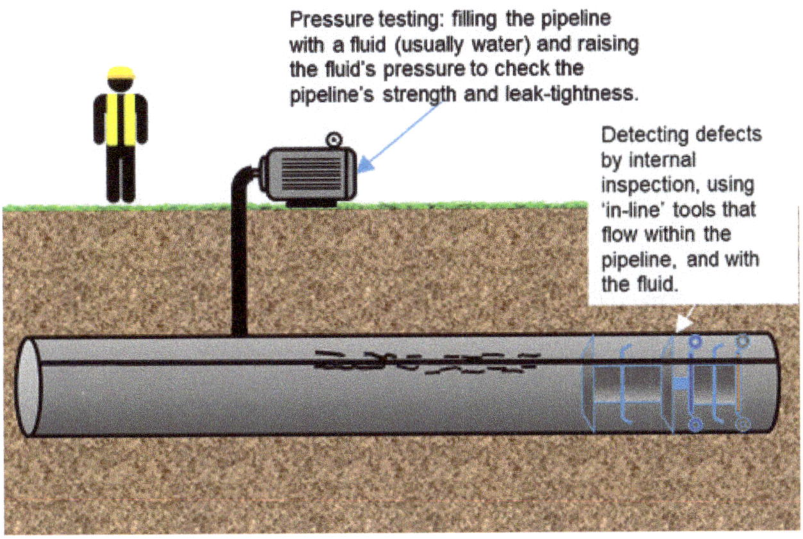

Figure 14-8. Pressure testing and in-line inspection.

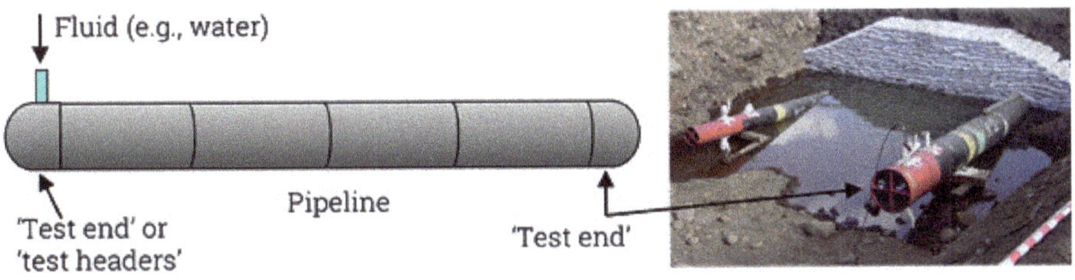

Figure 14-9. Pressure testing a pipeline.

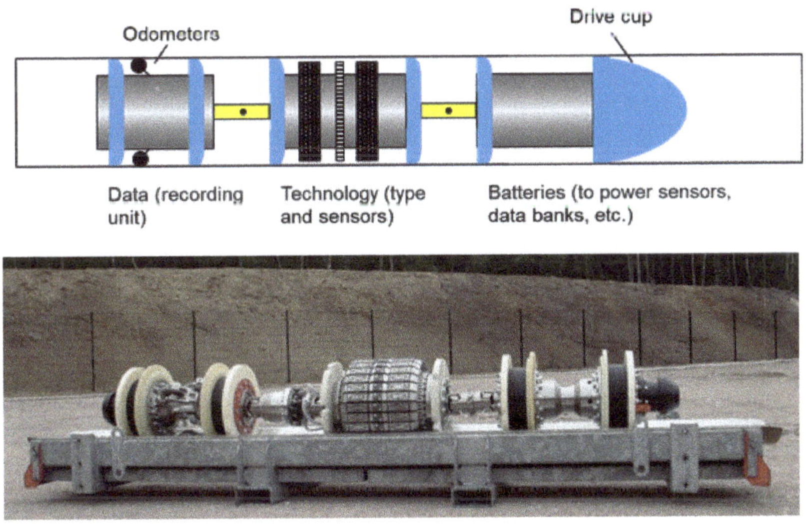

Figure 14-10. In-line inspection vehicle (smart pig). Photograph courtesy and copyright of Rosen Group.

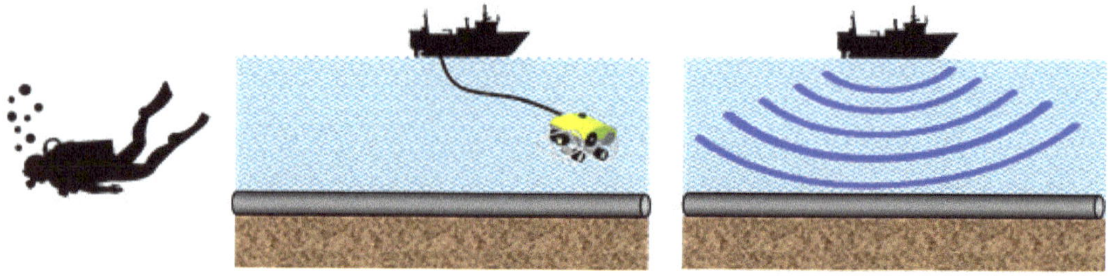

Figure 14-11. Diver, remotely operated vehicle, and hydrographic surveys.

Figure 14-12. Remotely operated vehicle for underwater inspection.

QR Code 14-13. Scan here to learn more about new threats to pipelines.

14.3 Summary

1. Asset management is the management of physical assets such as engineering plant, including their selection, maintenance, inspection and renewal.
2. Integrity management is part of asset management. It includes the many and varied activities pipeline operators must undertake to ensure that releases of products from their pipelines do not occur.
3. Standards bodies (such as in America (ASME), the UK (BSI), and Norway (DNV)) have all published standards to assist in the integrity management of pipelines.
4. Historically, pipeline operators relied on pipeline maintenance and inspection to ensure the pipeline was safe. Integrity management takes a broader view, as it encompasses the environment around the pipeline, as well as the pipeline. This means that operators are required to know more about: the areas their pipeline traverses; the nature of the population in the area; and, the existence of environmentally sensitive areas near the pipeline. This means that integrity management seeks to understand and control both the probability of failure, and the potential consequences of failure of a pipeline in a particular area along its route.
5. Integrity management starts with collecting pipeline data and determining which areas along the pipeline route would be most affected by a failure. Examples of these high consequence areas (HCAs) are where the pipeline passes through a highly populated area, or an environmentally sensitive region.
6. All the threats to a pipeline in these HCAs must be evaluated through a risk assessment of each threat, and high risks must be controlled and mitigated.
7. Regulations now require pipeline operators to conduct a pipeline integrity validation (through inspection, testing, and analysis) of pipelines that run through/near HCAs.
8. The threat and risk assessments and validations are all controlled using a pipeline integrity management system. These systems include integrity management programs which provide the means to improve the safety of pipeline systems and seeks to understand the potential consequences of a pipeline failure.

14.4 References

[1] Anon., 'Asset Management: Management Systems Requirements', ISO 55001:2014. International Organization for Standardization. Switzerland. 2014.

[2] Anon., 'Process Safety Management', U.S. Department of Labor Occupational Safety and Health Administration. OSHA 3132. 2000.

[3] http://www.hse.gov.uk/risk/controlling-risks.htm

[4] Anon., 'Integrity Management of Submarine Pipeline Systems'. Recommended Practice. DNV-RP-F116. Det Norske Veritas. Norway. October 2017.

[5] Anon., 'Efficient updating of risk assessments', Recommended Practice. DNVGL-RP-G107 Edition November 2016.

[6] A Hopkins, S Maslen, 'Risky Rewards', Ashgate Publishing Company. USA. 2015.

[7] A Herman, C Jeffress, ,'Process Safety Management', U.S. Department of Labor Occupational Safety and Health Administration. OSHA 3132. 2000. See also 'Process Safety Management of Highly Hazardous Chemicals', OSHA USA. Regulation 1910.119.

[8] Anon., 'Developing process safety indicators', HSG 254. UK's Health and Safety Executive. 2006. ISBN 978 0 7176 6180 0.

[9] Anon., 'Pipeline systems Part 4: Steel pipelines on land and subsea pipelines – Code of practice for integrity management', BSI PD 8010-4:2012. British Standards Institution. UK. 2012.

[10] http://primis.phmsa.dot.gov/comm/Inspection.htm

[11] Anon., 'Managing System Integrity for Hazardous Liquid Pipelines', American Petroleum Institute. API Recommend] ed Practice 1160. Second Edition. September, 2013.

[12] Anon., 'Managing System Integrity of Gas Pipelines', ASME B31.8-2018. American Society of Mechanical Engineers. New York, USA. 2018.

[13] Anon., 'Integrity Management of Submarine Pipeline Systems'. Recommended Practice. DNV-RP-F116. Det Norske Veritas. Norway. October 2017.

[14] http://www.hse.gov.uk/managing/health.htm

[15] Anon., 'Risk Assessment Techniques', Institution of Gas Engineers, IGE/SR/24. Communication 1655. 1999.

[16] https://www.phmsa.dot.gov/faqs/phmsa-and-pipelines-faqs

[17] https://primis.phmsa.dot.gov/comm/FactSheets/FSHCA.htm

[18] Anon., 'Gas Transmission and Distribution Piping Systems'. ASME B31.8-2018, American Society of Mechanical Engineers. New York USA. 2018.

[19] Anon., 'Pipeline Risk Modelling Overview of Methods and Tools for Improved Implementation', Pipeline and Hazardous Materials Safety Administration, USA. Draft 1. May 9, 2018.

[20] Anon., 'Marine risk assessment', UK Health and Safety Executive Offshore Technology Report 2001/063. HSE BOOKS. 2002. ISBN 0 7176 2231 2.

[21] M Nessim, 'Pipeline Risk Tutorial', International Pipeline Conference, ASME, Calgary Canada.

[22] www.hse.gov.uk/risk/faq/htm

[23] K Muhlbauer, 'Pipeline Risk Assessment. The Essential Elements', Pipeline and Gas Journal. May 2012.

[24] Anon., 'Pipelines - gas and liquid petroleum - design and construction', AS 2885.1-2018. Standards Australia, 2018.

[25] Anon., 'Oil and gas pipeline systems', CSA Z662-15, Canadian Standards Association. 2015.

[26] PHMSA Glossary. https://primis.phmsa.dot.gov/comm/glossary/index.htm ASTMInternational

[27] Anon., 'Pipeline systems Part 3: Steel pipelines on land. Guide to the application of pipeline risk assessment to proposed developments in the vicinity of major hazard pipelines containing flammables – Supplement to PD 8010-3:2009', BSI PD 8010-3: 2009. British Standards Institution. UK. 2009.

[28] Anon., 'Application of pipeline risk assessment to proposed developments in the vicinity of high pressure Natural Gas pipelines', Institution of Gas Engineers and Managers. IGEM/TD/2. Communication 1737. 2008.

[29] K Muhlbauer, 'Pipeline Risk Management Manual Ideas, Techniques and Resources', 3rd Edition. Elsevier.

[30] Anon., 'Societal Risk: Initial briefing to Societal Risk Technical Advisory Group', HSE Report RR703. 2009. http://www.hse.gov.uk/research/ rrpdf/rr703.pdf

[31] Anon., 'Five Steps to Risk Assessment'. UK Health and Safety Executive. Report INDG163, Revision 3. 2012. http://www.hse.gov.uk/pubns/ indg163.pdf

[32] N. Pidgeon et al., 'Risk Perception, in Risk Analysis: Perception and Management Report of a Royal Society Study Group, The Royal Society. London. pp.89-134. 1992.

[33] Anon., 'Reducing risks, Protecting people: HSE's decision-making process', HSE Books. R2P2, 2001.

[34] Anon., 'Pipeline systems – Part 1: Steel pipelines on land – Code of practice', British Standards Institution. Published Document. PD 8010-1:2015. 2015.

[35] Anon., 'Chapter 4: Government Policy in Practice'. Select Committee on Economic Affairs Fifth Report. UK Parliament. https://publications. parliament.uk/pa/ld200506/ldselect/ ldeconaf/183/18308.htm

[36] Edwards v. The National Coal Board (1949) 1 All ER 743.

[37] HSE principles for Cost Benefit Analysis (CBA) in support of ALARP decisions'. http://www.hse.gov.uk/risk/theory/alarpcba.htm

[38] http://www.hse.gov.uk/risk/theory/alarpglance.htm

[39] http://www.hse.gov.uk/risk/theory/alarp2.htm

[40] Anon., 'Internal guidance on cost benefit analysis (CBA) in support of safety-related investment decisions'. UK Office of Rail Regulation. February 2016. Crown copyright 2016. www.orr.gov.uk.

[41] http://www.hse.gov.uk/risk/theory/alarpcheck.htm

[42] Rio declaration on environment and development, made at UNCED 1992, ISBN 9 21 100509 4.

CHAPTER 15

Where next?

This Chapter comprises contributions invited from several industry leaders who were each asked to reflect on the subject 'Where next?'

15.1 Future technology

Hermann Rosen
Rosen Group, Stans, Switzerland

AS ADDRESSED IN THE FOREWORD to this book, pipelines are a key element in energy transportation as well as being the safest, most environmentally friendly way to transport large quantities of oil and gas. But they are also important in other industries, such as water and they will retain their important role into the future, for instance regarding the transportation of hydrogen. As this new book has been written by the authors especially with the young professionals in mind that have entered our industry, let me briefly re-emphasize a few points relating to pipelines.

Pipelines are highly valuable assets, in a strategic as well as an economic sense, that need to be protected. Their safe and economical operation must be ensured at all times. Technology, continuous enhancement and future developments and their application play a vital role in this context.

Managing pipeline threats, the theme of this new book, requires an understanding of what these threats are and assessing which effect they may have on the integrity and reliability of a given pipeline system. Important questions arising are how these threats can be identified and assessed, resulting in the need for accurate and robust information. Information of critical importance for decision making support.

Two major goals go hand in hand: to ensure the safety of a pipeline, protecting life and the environment on the one hand and process optimization of operating the pipeline, achieving best possible balance of efficiency and effectiveness of use, on the other. Achieving this, in short, requires knowing everything potentially affecting the status of the pipeline, for every location along its linear extent and at any time. A simple statement to make, but with considerable consequences when you start thinking about it – current technology, as sophisticated as it already is today, does not meet these requirements.

In my opinion we can only achieve this goal successfully if:

- We always deploy Latest Technologies: advanced sensors and data collection internally and externally with required resolutions, accuracies and threshold levels,
- We make advanced use of the data collected: turning data into useful information and visualizing it for easy use and decision-making support,
- We deliver Advanced Services: deploying technology and information gathering embedded in professional services for all types of pipelines including harsh and challenging environments.

Let me explain.

There is indeed already a wide range of "tools" available today to screen, monitor or inspect pipelines. Just consider all the internal and external inspection techniques, including in-line inspection, the various direct assessment methodologies or the information provided by SCADA systems or leak detection systems and the associated real time transient models used. But there still are gaps. Gaps in the information available, as well as gaps in the mind-set we use. The way forward is to be more comprehensive and more holistic.

Let us start with in-line inspection tools (ILI). These tools were originally developed to look for defects in the pipe wall. Different tools and non-destructive testing technologies were and are used to identify different defects such as dents, metal loss or cracks. Data collected provides information on the geometry, i.e. size and shape, of these features, which in turn is used for integrity assessment purposes. But defects can interact. Areas of corrosion can be treated as isolated from each other or interacting depending on their size and vicinity to each other. However, a specific defect category, such as a metal loss for instance can also interact with other defect categories, say cracks. But they can not only interact, there can also be a causality between them, i.e. the dent initiating a crack.

This has resulted in the need for ever higher resolution and measurement accuracies in tools as well as the use of multiple data sets, either from combined inspection runs or correlation of consecutive runs. The ILI industry is addressing these needs and will continue to develop even more refined tools.

But the potential threats not only relate to the steel wall of the pipe. The coating, applied for protection, or the cathodic protection system, installed to ideally prevent corrosion from taking place, can also deteriorate or malfunction. You need to know about it when it happens and for the entire line.

Defect assessment, however, does not only need input regarding the presence and geometry of defects, but also requires knowledge of the actual loading conditions of the pipeline as well as the true material properties. Loading, say the internal pressure or also external forces acting on the pipe, induces mechanical stresses in the pipe wall. Now, these stresses can be significantly influenced by macro or global forces, for instance due to pipe movement or by localized effects such as caused by stress concentrations. Technologies are being developed going beyond current capabilities of tools to measure stresses directly rather than deducing them from strains.

Material properties are sometimes not known, for instance if records are not complete or have been lost. This may result in the need to estimate values or introduce appropriate safety factors during an integrity assessment, often leading to very conservative results. Internal and external inspection technologies are already being introduced and will be further developed in the future to address material properties such as yield strength, ultimate tensile strength and especially fracture toughness.

These are just some examples of collecting more data about a pipeline or the pipe wall to be precise. All the chosen examples relate to data that can be collected internally or externally during the operation of the pipeline. But there is more information that can be used based on data collected during the steel and pipe manufacturing process, as well as pipeline construction.

Even more data is available through the SCADA system of a pipeline and by using data relating to the environment of the pipeline, for instance, optical cables to detect external mechanical interference with the pipe. But in future we need to use even more of the information available, such as weather-related data, or any kinds of geohazards, from earthquakes to flooding. All of that is what I mean with being more comprehensive.

This brings me to the next point. Data. It is not only a matter of collecting more and more data, but the challenging part is to make sense of that data and translate it into useful information. This relates not only to analyzing data collected but to correlating all that data from different sources to identify trends, recognize patterns.

A major part of future technologies in our industry will therefore be the use of big data analytics, artificial intelligence and machine learning to make the most use of the data available. The industry has already started this process, but the effort must and will intensify.

All this needs to be done for pipelines being inspected and also for pipelines that have already been inspected in the past. This will lead to better use of data, a better understanding of defect histories and mechanisms, and also faster availability of information.

We must use all available sources to collect data and then intelligently make use of it. This is what I mean by taking a more holistic approach.

Collecting data in order to derive useful information also requires that the data can be collected for all relevant locations along the length of a pipeline. Pipelines are linear in nature and often run through different terrain. In addition to new pipelines still to be built, most global pipelines are in operation and not all of them are ideally suited for inspection or monitoring equipment, including use of in-line inspection tools.

It is therefore often a matter of how to get the inspection device, the sensor that takes readings, to the location of interest. The actual propulsion of an inspection device is thus also of greatest importance, and future technologies will address this. Development of inspection devices incorporating their own drive all the way to fully autonomous devices for onshore and offshore use is under way, and the near future will see fascinating new tools being introduced.

But when we think about future technology and trends in our industry, we must also address delivery. Future technology and using it must also include the development and execution of advanced services. Achieving the above goals of knowing everything, everywhere and at any time is not a matter of developing and selling a technical component or product, but of embedding that technology within a professional service.

Such a service as a pre-requirement needs competent staff and optimized processes and procedures. This of course means that our industry must attract well qualified young professionals, train them, mentor and enable them to build relevant experience.

Let me sum up.

These are exciting times to be in our industry. In order to meet the future needs of our industry regarding the safe and efficient operation of pipelines and managing pipeline threats, the major areas of development have been introduced and can be categorized in the following way:

- Development and use of Latest Technologies: e.g. sensor technology, propulsion devices, data storage, energy supply.
- Advanced use of data collected: e.g. big data analytics, artificial intelligence, machine learning.
- Advanced Services: e.g. specialized processes and procedures to collect data, also in harsh and challenging environments.

My thanks to the editor, John Tiratsoo, for the opportunity to share these thoughts with you in this new book.

15.2 Pipeline integrity

Dick Williamson
TD Williamson, Tulsa, OK, USA

AS OPERATORS CONTINUE TO RISE to the challenge of managing our energy lifelines through ever-changing times, threats will evolve. Those of us who are passionate about pipelines will confront these with a duty of care that will call together additional expertise, wider experience and enabling technology.

There will be no reprieve from traditional, time-dependent threats such as internal and external corrosion, stable threats such as material and fabrication defects, or time-independent threats such as third-party damage, whether accidental or intentional (including pilfering). Some of these threats, including intentional third-party damage, may rise and incorporate new digital dimensions in the future. Other threats will become more prevalent by the introduction of unconventional product requirements, new pipeline materials and ever-more extreme environments.

Many of these factors have been traditionally considered as beyond the control of operators. This has led to many sleepless nights. Digital connectivity, new measuring devices, data analytics, more accurate inspection technologies and new materials will have a profound enabling impact, if correctly converged. Such convergence is essential for operators to be sufficiently confident in their data in order to make robust data-driven decisions. To borrow from Sir Isaac Newton, "if I have seen further, it is by standing on the shoulders of giants." Since previous sections of this book have covered traditional threats and the means of mitigating them, this section focuses on the future. Advances in cleaning methods, inspection technologies, corrosion monitoring systems and new sensors will continue to improve the operator's ability to detect and identify potentially injurious defects. These diverse technologies will benefit from data analytics and machine learning — enabling operators to correlate vast amounts of information gathered during cleaning events, in-line inspections, and external non-destructive examinations, as well as from corrosion monitoring systems and other data sources connected through the industrial Internet. Even today, operators are working to better understand integrity threats for each pipeline segment and to determine the combined impact threats have on the integrity of the segment within its operating environment. Whether the threat is a mill defect, mechanical damage, microbiologically induced corrosion or damaged coating, further development of data analytics will assist in providing a comprehensive view of these threats to help the operator best quantify risk, prioritize response and determine the most appropriate mitigation measures.

Monitoring systems permanently installed on or along entire pipelines will play an increasingly relevant role — with adoption rates increasing as the cost of deployment gets cheaper. These real-time systems will enable monitoring of environmental conditions, security and local erosion/corrosion. Additional soil condition monitoring, ground movement and other environmental data will provide key details in assessing the susceptibility of each pipeline segment to damage, and support the mapping of potential impacts on pipeline integrity. Again, correlation will be key, adding live data that comes from the system as a whole to the mix: Our future may see the concept of "self-thinking pipelines" that are able to flag potentially harmful conditions before they become a threat.

Inside of the pipeline, integrity threats associated with pipeline products will continue to evolve as pipeline systems are tasked with carrying a multitude of products. As we look ahead, such products will result largely from a diversification of energy sources — to include hydrogen, biogas and unconventional oil and gas production. The intermittent power-generation dilemma associated with renewable energy sources, namely wind power, has resulted in the development of a process by which hydrogen is produced and subsequently stored and transported in the gas grid to generate electricity at times when there is no wind. Similarly, in order to curtail methane emissions that occur during storage of organic waste as well as for the generation of power, there is growing interest in capturing and transporting biogas in pipeline systems. As unconventional oil and gas production becomes the norm, produced water, multiphase oil and gas flow and higher occurrences of sour gas production will create continued integrity challenges for operators. Add to this the need to transport and store ozone-damaging anthropogenic carbon dioxide — and it is easy to see there are challenging times ahead.

Outside the pipeline, threats associated with oil and gas production in extreme environments will pose new challenges. As onshore and shallow-water reserves become depleted, exploration and production companies will venture into arctic and extreme deep-water environments, subjecting pipelines to cold temperatures and extreme water depths to support such production efforts. Conversely, low-lying coastal areas will be impacted by rising sea levels and severe weather events. In these areas, land that is dry today may become temporarily or permanently submerged should sea levels continue to rise. Impact to pipelines in these areas must be evaluated for proper mitigation measures.

For integrity threats driven by either internal pipeline product or the external pipeline environment, common solutions that leverage new materials will emerge. For example, non-metallic and metal matrix composites provide corrosion resistance and flexibility, making them appealing candidates for extreme conditions. Furthermore, steel grades beyond X90, X100, and even X120 are becoming feasible with improvements in steel production techniques — providing sufficient fatigue performance and fracture toughness for extreme deep water applications. While both composite materials and high-strength steels provide solutions for a variety of pipeline product and operating environment challenges, new integrity threats associated with composite joints as well as field welding high-strength steels will emerge. Furthermore, new non-destructive examination technologies will be needed as traditional pipeline inspection techniques will not be suitable for long-runs composite pipelines.

News outlets have a seemingly constant stream of stories associated with acts of terror or malicious attacks. Pipelines continue to be the target of such attacks — and there are no signs of this threat subsiding. Furthermore, as pipelines become more connected to the industrial Internet, byproducts of this connectivity include additional potential entry points for those who wish to cause harm through digital means. Both physical threats and cyber threats to pipeline integrity from malicious third-parties will largely be mitigated through technology. Sensor technologies and algorithms will be utilized to mitigate threats from malicious third-parties. To further enhance the security of their assets, pipeline operators are exploring systems to help monitor the areas around their pipelines. Technologies such as fiber optic cables or seismic sensors buried in the pipeline right-of-way coupled with computer systems wielding trained algorithms are beginning to be utilized to monitor activity around the pipeline. This approach helps operators pinpoint the location and discriminate the nature of the threat such as footsteps in restricted areas (human or wildlife), vehicle movement or any other activities common along the right-of-way that may pose physical damage to the pipeline.

Cyber security threats are not restricted to the theft of information. Cyber threats that potentially jeopardize the integrity of the pipeline system will continue to evolve. As more online monitoring and remotely controlled systems are realized, digital entry points increase the potential for malicious attack as well the potential manipulation of hydrocarbon inventory to mask product theft. While algorithms trained in part through machine learning will be leveraged to detect physical and cyber-based threats, the human element in detecting and mitigating such threats will continue to be necessary. The women and men charged with protecting pipeline assets from these malicious third-party threats will retain a vital role in deciding and deploying appropriate responses to mitigate such threats, as well as in communicating to all stakeholders to ensure proper awareness and action throughout and beyond the pipeline system.

With respect to integrity threats in this forward-looking view, some high points have been captured. But this view should not be considered exhaustive. The next decade is expected to yield an abundance of pipeline integrity solutions leveraging data, sensor technologies and various tools deployable internally as well as external to the pipeline. As stewards of these energy lifelines that the world depends on, we have a duty of care to understand and share our concerns, pool our lessons learned and apply our expertise, experience and passion for pipelines in a way that safeguards pipeline integrity.

15.3 Pipeline repair and rehabilitation

Dick Williamson
TD Williamson, Tulsa, OK, USA

AN IMPORTANT ELEMENT IN ANY pipeline integrity management program is repair and rehabilitation, which dominates the corrective and reactive response agenda. Pipeline operators take painstaking steps to eliminate and mitigate the various threats that pose harm to their systems with predictive and preventive responses. However, as they fight the never-ending battle with corrosion; as pipeline medium changes occur that stretch material durability; as damage is inflicted from extreme weather events; as material or manufacturing faults are found; as third-party damage is discovered; and as systems are stressed due to production demands or harsh environments, pipeline operators will forever need robust repair and rehabilitation solutions within their corrective and reactive response armory.

As the whole global pipeline infrastructure ages, the need to repair and rehabilitate increases. Given many of these pipeline systems are beyond their design life, it is imperative that safe, efficient and economically viable systems can be used to minimize the disruption and maintain flow. For the majority of these assets the material is steel — so welding and hot tapping are proven methods of both intervention and repair from outside the pipe. Developments and welding practices for mainline construction have carried through to repair processes such as automated welding of shells and fittings, the use of advanced non-destructive evaluation for eliminating critical weld defects, and the use of numerical methods to ensure that post-weld heat treatment is minimized. In addition, improved methods such as engineering critical assessments and risk-based design are now commonplace. The natural evolution of these techniques will move to robotics and automated welding as well as advanced methods for removing pipe material other than saw and cutter. The use of composites will continue to be used externally to the pipe. Composite repairs, long thought of as a temporary fix, are now accepted as mainstream, long-term, solutions. And there are many types of composite repair systems that can be used for all configurations in the pipeline.

Another area of constant development is the use of internal isolation, connectors and clamps. Internal isolation, long used in the offshore environment, has had limited adoption onshore. High stresses and costs have been prohibitive. But that will change with new elastomeric materials and improved design principles, and in-line isolation will become an alternative to cut-out or hot tapping onshore. Connectors and leak repair clamps will also continue to be used offshore and their long-term viability for repair is well proven. As water depths increase, remote operated vehicle (ROV) deployed systems will continue to be the norm. ROV deployed mechanical clamps and connectors will be part of an important arsenal for emergency response for such deepwater applications.

Moving to the inside of the pipe, robotic applications are being developed for water, wastewater and lower-pressure gas distribution pipe, and the future will see this technology cross over to larger-diameter, longer-length transmission pipelines. There have also been developments in the use of internal liners made from materials such as high-density polyethylene, thermoplastic fluoropolymers, polyurethanes, and Kevlar. These are now being used where the products within the pipeline are aggressive from sour, bacterial — even for rock-based slurries. The use of nanotechnology-based solutions injected into the product flow would seem too futuristic. But platelet technology that can, in some applications, coagulate the product stream to reduce or seal a leak is already commercially available.

Looking to the future of pipelines, new materials are becoming available to handle increasingly aggressive pipeline products and extreme conditions. Non-metallic and metal matrix composites offer corrosion resistance and other physical benefits for harsh conditions. Furthermore, steel grades beyond X120 are becoming feasible. And as the products inside pipelines change to include hydrogen, anthropogenic carbon dioxide, etc., new ways of performing old approaches will be needed. This will present challenges to existing ways to repair and rehabilitate — and possibly require the development of new methods.

One imperative for any oil and gas operation is the concept of zero harm — pipelines are not exempt. Although pipelines remain by far the safest way to transport hydrocarbon energy, leaks and failures still occur. As is evident by reading any news article, greenhouse gases, fugitive emissions and global warming are on everyone's agenda. Hence, in the repair and rehabilitation arena, it is all about zero emissions. Several established methods are already geared toward primary containment — hot-tapping, in-line isolation and other hermetic techniques. But further developments will be required if loss of primary containment is to be fully combated. Huge gains have been seen recently in the development of process safety programs by looking at leading indicators rather than ones that are based upon lagging indicators. A cultural shift is happening from simply reporting of near misses to the use of risk tools and methods, and to the use of digital twins utilizing a digital replica of physical assets. Linking this to the industrial Internet provides access to all the information required to make quick decisions on a repair or emergency intervention, hence moving toward the goal of zero harm. As Benjamin Franklin wrote, "an ounce of prevention is worth a pound of cure".

One aspect that will loom large on the horizon for all pipeline operators — regardless of application — is that of the great crew change. As knowledgeable and experienced crews retire, operators will struggle to recruit, train and retain replacements. The degree to which operators cope with the challenges — in managing a workforce shifting from Generation X to Millennials — will distinguish the great from the good. Increasingly, this will lead to operators making tough choices regarding their operational focus — seeking to develop their teams to become experts in repeatable, operational tasks, while sourcing external experts to perform non-routine tasks. This will require industry-wide collaboration, and partnership with vendors, to reach new heights.

Repair and rehabilitation has come a long way from the use of wooden plugs (hammered into a leaking pipeline), wood-wrapped lathes (planks of wood fastened to the pipeline with wire) and "carrots" (a screw plug used to stop small leaks). Repairs are now heavily engineered, rigorously tested and have long track records in the field. The question is no longer where we go from here; the challenge is more how can the industry adapt and extend current, proven methods to the new materials, new products, new environments, new threats and new ways to deploy, all with the aim of zero harm.

15.4 Subsea engineering

Eric Jas
Atteris, Perth, WA, Australia

UNLIKE ONSHORE PIPELINES, WHICH ALSO transport water, sewage, slurry, and many other products, the vast majority of submarine pipelines convey hydrocarbon products. Hence, the future of the submarine pipeline, and associated subsea, industry will depend on the future of the (offshore) oil and gas industry.

Many people talk about how renewables will soon displace fossil fuels. Whilst electricity markets around the world are experiencing a fundamental change with the rapid rise of renewable energy technology, it is noted that electric energy consumption is only a part of all energy consumption. For instance, only one third of the natural gas extracted from the earth's crust and a small percentage of crude oil are currently used to generate electricity. However, a large proportion of coal is used to generate electricity globally. Thus, the rise of renewable electrical power generation technology is likely to impact the coal industry much more than the oil and gas industry. It is noted that whilst battery technology is still in its early stages of development, it has the potential to boost the sustainability of renewable electrical power generation technology as well as electric cars. The latter will have a greater impact on the consumption of crude oil derived products. However, it will take time to develop cost-effective and reliable batteries to enable large communities to solely rely on renewable electrical power generation for living and electric vehicles for transportation.

Natural gas has major benefits in the total energy mix. Its carbon footprint is relatively low, particularly when mixed with hydrogen. The storage of electricity and transportation of electrical power over very long distances, which is inevitable in renewable electrical power generation and consumption, has proven difficult and costly. Natural gas can be stored easily and transported very efficiently. It is a reliable and cost-effective source of energy.

Consequently, oil and gas pipelines, both onshore and offshore, will continue to play a role well into the future.

Pipelines in submarine environments have experienced significant change over the past decades. The industry continues to see the progressive increase in water depth. Several engineering thresholds have already been reached in this regard, in relation to the vertical distance to the seabed and the associated increasing hydrostatic pressure acting upon the pipeline. New pipeline installation techniques have been developed for deep water installation. Dynamically positioned pipelay vessels, introduced three decades ago, are now common. Further developments in submarine pipeline installation are expected, particularly in the field of automation.

Performing a hydrotest has been common for pipelines, both onshore and offshore. This necessary step in the pre-commissioning process has become a major challenge for deep water pipelines. The industry is now considering "dry commissioning" of submarine pipelines. This requires a shift in thinking and practices, in engineering, procurement, fabrication, installation, commissioning and operations.

Additionally, deeper water has an increased risk of buckling and / or collapse during installation, in particular for larger diameter gas trunklines, often driving the design towards thick-walled and, hence, costly pipelines. New engineering methods are being developed to ensure pipelines in these conditions can be made cost-effective.

Deep water oil and gas fields have also seen the development of flexible riser technology, including steel catenary risers. Longer lifetimes, higher pressures, higher temperatures and the need for high flow rates are some of the key drivers for improving flexible pipe materials and design. The introduction of composite materials offers an opportunity for improvements, including reduced weight that will enable larger diameters in deep water and better resistance to sour production fluids.

Corrosion continues to be a major threat to pipeline integrity. More superior coatings are likely to be developed in the quest to increase the lifetime and reliability of submarine pipeline systems. New pipeline inspection methods are being developed, aiming at improving the reliability of the collected data as well as avoiding pipeline system shutdowns and operational cost in general. Similarly, new pipeline repair methods are continuously being developed, aiming at quicker response and repair times for repair work at lower cost and higher safety standards.

Since computers were introduced in work in general, we have seen the development of computer hardware and software that assist in the planning, design, construction and operation of submarine pipeline systems and associated subsea equipment. The digitalization commenced with the introduction of the first computer. This process is ongoing with new developments occurring constantly.

Codes and standards are also continuously being challenged and improved as the industry moves further into (sometimes literally) unexplored waters. Much of the focus of the current submarine pipelines standards and recommended practices is on defining the appropriate methodologies, specifications and work practices that are to be applied. It is only recently, and only in some countries, that national competency standards for professional pipeline engineers have been defined and formally introduced into the industry. It is expected that the formalization of competency levels for professional pipeline (as well as subsea) engineers will be mandated further, thereby improving the performance, and consequently the reputation, of the industry.

Ultimately, submarine pipelines can only be justified if they are cost-competitive, reliable and safe to install and operate. In addition, the end-of-life situation must be accounted for sufficiently early, ideally during the early concept design phase of a new pipeline. Decommissioning may require complete or part removal of the subsea infrastructure. The impact onto the environment associated with this may exceed that of leaving the remnant materials in place. Hence, the latter should always be a consideration and allowed for in the legislation as a viable option. The ultimate aim is to ensure that the impact onto our natural environment is minimized over the life of a pipeline system.

There is one major aspect of submarine pipelines worth highlighting, which would become a threat if not addressed. Continuous improvement of the competency of the workforce, at all levels, is essential. Much focus should therefore be on the development and care of professionals using agreed competency standards. Only then will the industry continue to flourish.

15.5 Pipeline standards and regulations

Dr Alan Murray
Pipeline Engineering Centre, University of Calgary, Calgary, AB, Canada

A COMMONLY STATED OBJECTIVE OF pipeline integrity is to keep the product being transported inside the pipe, and if this is successful, then the public is kept safe and the environment is protected. Naturally, this must be achieved within the bounds set by operating profitability. However, pipelines are engineered products and sometimes fail, despite being designed, constructed and operated within the strictures imposed by the appropriate Standards and Regulations. This would seem to imply either an inadequacy in the requirements being set or, the ability of operators to comply with them; perhaps both. Hence there is a need for continual improvement. I have a marked reluctance, though, to attempt to predict the future content and direction of Standards and Regulations, not least because these documents, by their very nature, are generally retrospective and reflect the experience of their developers, by capturing what is known. I will however make recommendations in a few areas where I believe progress can and ought to be made. These include the need to streamline and merge standards by seeking out commonalities, making their content more accessible, and to consider emerging threats to pipeline integrity such as cyber-attacks. It is less challenging to offer recommendations than to make predictions!

It may be instructive to reflect on the pipeline standards-making process as this is key to structural change. In most jurisdictions the standards-making body forms a committee structure composed of subject matter experts that are broadly representative of the industry, often including Regulators. The composition and voting structure of the various committees are such that no single special interest grouping can prevail, and the content of the resulting standard evolves by consensus.

It follows that this consensus-seeking process produces a minimum standard which may, or may not, represent best practice. Any perceived shortfall, however, can be remedied by companies layering on top their own standard requirements (usually leading to increased cost), while a Regulator can assuage its concern by imposing a specific regulatory change. However, it is in everyone's interest to attempt to make a standard as robust and adaptable as possible. While the members of the various standards committees are described as volunteers, in most instances their participation is supported by their employers, whose interest they are expected to safeguard, if not actively promote. In the author's experience, a strong proprietorial interest develops among the standards committee membership, that can inhibit a future rationalization or merging with similar standards. In an era of increased globalization this attitude must surely change, particularly for components such as line pipe, valves and fittings, not least because vendors and manufacturers, and even entire economies such as China, have a vested interest in seeing a movement toward the increased use of ISO standards. The ISO structure is such that regional annexes can be appended to the parent standard, which permits consideration of specific issues such as harsh climatic conditions (low temperatures), unique environments (deep water), or different products (CO_2, slurries). In short, the ISO approach can provide the basis for omnibus standards for line pipe and components. Rationalizing standards, or the portions of them that address design, construction and operation for different types of pipeline, onshore/offshore, liquids/gas will be difficult, as evidenced by the current long, drawn out debate within ASME to merge the B31.4 and B31.8 codes. A task force drawn from both code committees has examined areas of overlap to capitalize on their technical strengths and to develop consistent approaches and definitions. Perhaps this will lead not only to a combined ASME code but an eventual merger with the ubiquitous Canadian Z662 pipeline standard to form a North American standard. This would open the ASME B31 standards to further advances in areas such as strain-based / limit states design and the use of probabilistic methods in pipeline maintenance and integrity. Implementation of a common framework would also be a first step towards developing a "Smart Code," an intelligent interactive tool that would guide the user to relevant sections of a standard. This has the potential to reduce errors and time spent in the using the code. The user would still take full responsibility for his decisions.

Almost all the national pipeline standards have their origin in the B31 codes and have diverged over time to account for unique local conditions and practices. However, enough commonality remains, firmly entrenched in sound engineering principles, to provide the basis for a truly global Smart Code that would transcend national borders. Embedded artificial intelligence algorithms would help the user account for local jurisdictional requirements while enabling all users to benefit from worldwide best practices, particularly those related to considering risk and land use. Historically standards have been developed and calibrated in accordance with "habit and repute"; that is, certain practices have been seen to work and to the extent that they can be codified, they have been. Such is the case with Class Location factors, the proxy for risk in the mid 1950s. Today we can do much better, with the availability of extensive aerial surveillance and sophisticated computer-modelling of atmospheric fluid dynamics, that accurate calculations can be made of exposure and consequence, so much so that is conceivable that pipeline standards should move toward providing the means of computing individual risk at a point location. This would be a big improvement on using an arbitrary setback distance of 220 yards – a furlong, a furrowed length, the distance that, in English Law, determined how far a horse was permitted to pull a plough before being allowed to rest! Nonetheless the class location factor and its twin the Potential Impact Radius, will remain if only to accommodate those with limited computational resources.

Turning now to Regulations. These have often lagged Standards and consequently have incorporated the relevant standard(s) in a process known as "adoption by reference." However, pipeline integrity is an area where Regulations have led the way. In

1999 the Canadian National Energy Board (NEB) promulgated its Onshore Pipeline Regulations, which contained a clear requirement for regulated companies to develop and implement an adequate and effective pipeline integrity management program. The Office of Pipeline Safety (OPS) in the United States, (now known as PHMSA, the agency of which it is a part), followed shortly thereafter with its Final Rule regarding Liquids (2001) and then Gas Pipelines (2003), with the requirement for an integrity management program (IMP).

After a suitable lapse to enable implementation, company audits were conducted by the NEB and OPS of these IMPs and deficiencies were found primarily in the areas of Change Management, Record Keeping, and Staff Training. Subsequent audits showed some improvements; however, they prompted a requirement for a broader view to be taken – a need for a comprehensive Safety and Loss Management System (SLMS). Central to the success of these programs is the premise that a company is in the best position to determine the current condition and hence fitness for service of its pipeline system. Twenty years after the introduction of IMPs many companies still struggle, and in attempting to ensure management commitment and support for programs, some jurisdictions require a senior Officer of the Company to be identified as the accountable party. Several pipeline standards now contain sufficiently detailed provisions for an SLMS to enable compliance where there is an overarching regulation. Directionally these programs should become the norm irrespective of a regulatory requirement.

The events of September 11, 2001, necessitated US and Canadian Pipeline Regulators take account of security threats, initially physical threats to infrastructure, more recently the threat of cyber-attacks on supervisory control systems. It is a difficult if not impossible task to provide physical protection to linear systems like pipelines other than at facilities. Nevertheless, aerial and satellite surveillance have been, and will continue to be, used for this purpose and regulations most likely in the form of performance /outcomes-based rather than prescription will be introduced in jurisdictions beyond Canada and the United States. Energy infrastructure has long been a target of cyber-attacks. Currently in North America pipelines conduct a self-examination of the vulnerability of their system, but this surely will change. The US National Institute of Standards and Technology has recently prepared an update to its Risk Management Framework (RMF), part of its cybersecurity framework, that can help pipeline companies more easily protect their critical assets from cybersecurity threats. Similarly, in Europe the United Nations Economic Commission for Europe (UNECE) will soon integrate its IEC 62443 series of standards into the Common Regulatory Framework on Cybersecurity (CRF). Doubtless pipeline Regulators throughout the world will peruse both these sources and reference them in total or in part.

Index

C

D

E

F

G

H

I

J

K

L

M

N

Q

R

T

U

Y

Z

All the supplemental documents linked by the QR codes and the embedded links can be accessed here: http://pipeline-threats.com/plthreats/

About the Editor

John Tiratsoo, 1948 - 2019

This book is the final contribution to the international pipeline engineering community from one of its great minds and personalities.

John Tiratsoo was one of the most well-known, well-liked, and recognized figures in the industry. His passion for knowledge transfer, education, and training is known by the whole pipeline engineering world. Over a span of 40 years he created, edited, and published many journals and magazines, including the **Journal of Pipeline Engineering** and **Pipelines International**.

He founded the renowned **PPIM** conference on **Pipeline Pigging & Integrity Management** and produced the first authoritative reference book on the subject, Pipeline Pigging and Integrity Technology, in 1985. The book is now in its fourth edition and remains an industry standard, while the conference celebrated its 30th year in 2018 and continues to draw almost 3,000 participants from all over the world each February.

He always believed the pipeline industry needed a single, professional voice and a structure for sharing experience and standardizing professional qualifications. To that end, John helped found the **Professional Institute of Pipeline Engineers** and, most recently, chaired the **Qualification Panel for the Pipeline Industry** in his ongoing effort to establish certification standards for professional competence. A devoted advocate for young people joining the industry, he was an ardent supporter of the **Young Pipeline Professionals** groups in the USA, Canada, Brazil and Australia.

John Tiratsoo's fertile mind and generous spirit had an enormous influence on the international industry and the people in it, and his life's work will continue to benefit everyone involved today, and tomorrow.

www.ingramcontent.com/pod-product-compliance
Lightning Source LLC
LaVergne TN
LVHW071340210926
848486LV00003B/195